Improving Organizational Interventions for Stress and Well-Being

This book brings together a number of experts in the field of organizational interventions for stress and well-being, and discusses the importance of process and context issues to the success or failure of such interventions. The book explores how context and process can be incorporated into program evaluation, providing examples of how this can be done, and offers insights that aim to improve working life.

Although there is a substantial body of research supporting a causal relationship between working conditions and employee stress and well-being, information on how to develop effective strategies to reduce or eliminate psychosocial risks in the workplace is much more scarce, ambiguous and inconclusive. Indeed, researchers in this field have so far attempted to evaluate the effectiveness of organizational interventions to improve workers' health and well-being, but little attention has been paid to the strategies and processes likely to enhance or undermine interventions. The focus of this volume will help to overcome this qualitative-quantitative divide.

This book discusses conceptual developments, practical applications and methodological issues in the field. As such it is suitable for students, practitioners and researchers in the fields of organizational psychology and clinical psychology, as well as human resources management, health & safety, medicine, occupational health, risk management, and public health.

Caroline Biron is a Chartered Psychologist and Associate Professor in Occupational Health and Safety Management in the Faculty of Administrative Sciences, and a member of the Chair in Occupational Health and Safety Management at Laval University, Québec, Canada. Her work on the intervention process won the Best Intervention Competition award at the Work, Stress & Health Conference 2011.

Maria Karanika-Murray is an Occupational Health Psychologist and Senior Lecturer in Psychology at Nottingham Trent University, UK. Her research focuses on the importance of the organizational context for employee health and well-being, and the assessment and management of work-related health and well-being.

Cary L. Cooper CBE, is Distinguished Professor of Organizational Psychology and Health at Lancaster University Management School, UK; Chair of the Academy of Social Sciences, and also Editor of the journal *Stress and Health*. He was honoured by the Queen with Commander of the British Empire for his contribution to occupational health.

Improving Organizational Interventions for Stress and Well-Being

Addressing process and context

Edited by Caroline Biron, Maria Karanika-Murray and Cary L. Cooper

LONDON AND NEW YORK

First published 2012
by Routledge
27 Church Road, Hove, East Sussex BN3 2FA

Simultaneously published in the USA and Canada
by Routledge
711 Third Avenue, New York NY 10017

First issued in paperback 2015

*Routledge is an imprint of the Taylor & Francis Group,
an informa business*

© 2012 Psychology Press
Chapter 14 © 2012 HMSO

The right of the editors to be identified as the authors of the editorial material, and of the authors for their individual chapters, has been asserted in accordance with sections 77 and 78 of the Copyright, Designs and Patents Act 1988.

All rights reserved. No part of this book may be reprinted or reproduced or utilised in any form or by any electronic, mechanical, or other means, now known or hereafter invented, including photocopying and recording, or in any information storage or retrieval system, without permission in writing from the publishers.

Trademark notice: Product or corporate names may be trademarks or registered trademarks, and are used only for identification and explanation without intent to infringe.

British Library Cataloguing in Publication Data
A catalogue record for this book is available from the British Library

Library of Congress Cataloging in Publication Data
  Improving organizational interventions for psychosocial stress and well-being: addressing process and context/edited by Caroline Biron, Maria Karanika-Murray, and Cary L. Cooper.
    p. cm.
  Includes bibliographical references and index.
  1. Work—Psychological aspects. 2. Job stress. 3. Stress (Psychology) 4. Industrial safety. 5. Organizational change. I. Biron, Caroline. II. Karanika-Murray, M. (Maria) III. Cooper, Cary L.
  BF481.I47 2012
  158.7—dc23
                                                    2011040913

ISBN13: 978-1-138-93318-7 (pbk)
ISBN13: 978-1-84872-056-5 (hbk)

Typeset in Times New Roman
by RefineCatch Limited Bungay, Suffolk

To my parents and my friends for their constant support, and to Chris, for his patience and for believing in me each time.

Caroline

To grandma Elena.

Maria

To all my wonderful PhD students, who made research in the field of occupational stress possible.

Cary

Contents

Foreword xi
NORBERT K. SEMMER

Preface xv
Contributors xvii

1 **Organizational interventions for stress and well-being – an overview** 1
CAROLINE BIRON, MARIA KARANIKA-MURRAY, AND CARY L. COOPER

PART 1
Challenges and methodological issues in organizational-level interventions 19

2 **Intervention development and implementation: understanding and addressing barriers to organizational-level interventions** 21
ANTHONY D. LAMONTAGNE, ANDREW J. NOBLET, AND PAUL A. LANDSBERGIS

3 **Taking a multi-faceted, multi-level, and integrated perspective for addressing psychosocial issues at the workplace** 39
NADINE MELLOR, MARIA KARANIKA-MURRAY, AND ELEANOR WAITE

4 **Research in organizational interventions to improve well-being: perspectives on organizational change and development** 59
LOIS E. TETRICK, JAMES CAMPBELL QUICK, AND PHILLIP L. GILMORE

5	**Psychosocial safety climate: a lead indicator of workplace psychological health and engagement and a precursor to intervention success** MAUREEN F. DOLLARD	77
6	**Perspectives on the intervention process as a special case of organizational change** STURLE D. TVEDT AND PER ØYSTEN SAKSVIK	102
7	**Does the intervention fit? An explanatory model of intervention success and failure in complex organizational environments** RAYMOND RANDALL AND KARINA M. NIELSEN	120
8	**How can qualitative studies help explain the role of context and process of interventions on occupational safety and health and on mental health at work?** GENEVIÈVE BARIL-GINGRAS, MARIE BELLEMARE, AND CHANTAL BRISSON	135
9	**What works, for whom, in which context? Researching organizational interventions on stress and well-being using realistic evaluation principles** CAROLINE BIRON	163

PART 2
Addressing process and context in practice 185

10	**Evaluation of an intervention to prevent mental health problems among correctional officers** RENÉE BOURBONNAIS, NATHALIE JAUVIN, JULIE DUSSAULT, AND MICHEL VÉZINA	187
11	**The vital role of line managers in managing psychosocial risks** RACHEL LEWIS, JOANNA YARKER, AND EMMA DONALDSON-FEILDER	216
12	**The impact of process issues on stress interventions in the emergency services** VIV BRUNSDEN, ROWENA HILL, AND KEVIN MAGUIRE	238

13 The development of smart and practical small group
    interventions for work stress   258
    JOHN KLEIN HESSELINK, NOORTJE WIEZER,
    HELEEN DEN BESTEN, AND ERNA DE KLEIJN

PART 3
Policy implications   283

14 **Implementation of the Management Standards for
    work-related stress in Great Britain**   285
    COLIN MACKAY, DAVID PALFERMAN, HANNAH SAUL,
    SIMON WEBSTER, AND CLAIRE PACKHAM

15 **Moving policy and practice forward: beyond
    prescriptions for job characteristics**   313
    KEVIN DANIELS, MARIA KARANIKA-MURRAY,
    NADINE MELLOR, AND MARC VAN VELDHOVEN

16 **Evidence-based practice – its contribution to
    learning in managing workplace health risks**   333
    ANDREW WEYMAN

PART 4
Conclusions   351

17 **Concluding comments: distilling the elements
    of successful organizational intervention implementation**   353
    MARIA KARANIKA-MURRAY, CAROLINE BIRON, AND CARY L. COOPER

    Index   362

Foreword

Norbert K. Semmer

As the editors state in their introductory chapter, results of intervention studies aimed at improving health and well-being by means of changing organizational characteristics have often been disappointing. Given the considerable evidence supporting the effects of working conditions on well-being and health, this state of affairs is a reason to be concerned. Are our models too simple, or even wrong? Have we underestimated the role of individual factors? Have we underestimated the tendency of social system to stay with, or revert to, established structures and behavior patterns (Katz & Kahn, 1978)? Have we underestimated the many difficulties associated with implementing even good ideas and concepts (Kristensen, 2005)?

For quite some time, we have seen a) rather few studies with strong designs, which have yielded mixed, and often rather disappointing, results, and b) rather many reports that were more optimistic but much less rigorous (cf. Semmer, 2006; 2011). Unfortunately, the more rigorous studies often contained only limited information about the factors that contributed to success or failure. By contrast, the less rigorous ones dealt with the issues, pointing out the enormous importance of factors related to process and context; however, this information frequently was provided in terms of post-hoc explanations of difficulties, rather than being a systematic focus of the project itself.

Fortunately, things are moving. It seems we are about to get beyond the point of pitting qualitative and quantitative approaches against each other, moving to the point of conceiving optimally adapted designs (Randall, Griffiths, & Cox, 2005), and to the point of assessing process and context issues in a way that is based on qualitative thinking and yet provides measures that can be used in a quantitative way as well. In my view, that is exactly what is needed.

This book, with its focus on process and context, therefore, is overdue. It moves issues that have often been talked about but rarely been investigated systematically, from a kind of informal backstage existence into the systematic focus. It discusses models of how context and process can be incorporated into program evaluation, it provides examples of how this can be done, and it yields insights that are promising to improve interventions.

The focus of this volume will help to overcome the qualitative-quantitative divide, which has been so unfortunate for the field. Furthermore, the volume broadens the focus. As several authors emphasize, health-related interventions are about organizational change, and aligning the field with that of organizational development is certainly necessary. And we can go even further: Modern research on psychotherapy has made tremendous progress in terms of process research; we could profit from some of these developments; these include a focus on the specific behaviors of change agents (cf. Orlinsky, Rønnestad, & Willutzki, 2004). Do we reflect enough on the competences needed for helping organizations to change? After all, we are endangered to persuade where we need to convince, to over-identify with specific suggestions and solutions, to become impatient when progress seems insufficient, to become defensive when setbacks occur, or to regard as resistance what might well be legitimate concerns of participants about unwanted side effects of the planned changes. I am confident that this volume will instigate efforts to tackle that kind of problem as well.

Finally, let me follow up on the remarks made by several authors about the potential danger of talking about stress interventions (or similar terminology). This is not simply a problem of terminology, but also a conceptual issue. It highlights the fact that interventions are unlikely to succeed unless they become an integral part of everyday operations, rather than a "health project". It also highlights the possibility that what we see as health, or stress, issues, might be seen by others as mainly representing other problems – as something that is a nuisance, as something that is demotivating, but possibly also as a harsh condition that they are proud to be able to deal with (cf. Meara, 1974). The health focus should not detract from acknowledging that the issues we are dealing with overlap to a considerable extent with issues discussed long ago in terms of fostering internal motivation and personal development (e.g., Hackman & Oldham, 1980; Hackman & Suttle, 1977), in terms of aligning technical and social systems to each other (Cherns, 1987; Clegg, 2000), or, more generally, in terms of quality of working life (Davis & Chernsa, 1971, 1972). What it boils down to in the end is, in my view, something like this: We are talking about work that people like, that motivates them, that gives them a sense of meaning and fulfillment, participation in social life, a sense of accomplishment, etc. It may not mainly be about health. It may "simply" be about: good work.

Achieving "good work" has been a concern of many over many years. We have made progress towards issues of how to conceptualize it, how to assess it, and what is does to people. We urgently need a better understanding of how to change existing structures and processes in a way that combines accumulated general knowledge with local structures and specific problems, and that supports the empowerment of people so that they can better align their work situation with their needs. Those who have contributed to this volume care about these issues, have carefully thought about them, and advance our understanding of the factors that often impede progress in organizational changes aimed at improving health and well-being: context and process. I hope it finds its readers, and I am confident it will.

References

Cherns, A. B. (1987). Principles of sociotechnical design revisited. *Human Relations 40*, 153–162.

Clegg, C. W. (2000). Sociotechnical principles for system design. *Applied Ergonomics, 31*, 463–477.

Davis, L. E., & Cherns, A. B. (Eds.) (1971, 1972). *The quality of working life* (Vols. 1 and 2). New York: Free Press.

Hackman, J. R., & Oldham, G. R. (1980). *Work redesign*. Reading, MA: Addison-Wesley.

Hackmanm, J. R., & Suttle, J. L. (1977). *Improving life at work*. Santa Monica, CA: Goodyear.

Katz, D., & Kahn, R. L. (1978). *The social psychology of organizations* (2nd ed.). New York: Wiley.

Kristensen, T. (2005). Intervention studies in occupational epidemiology. *Occupational and Environmental Health, 62*, 205–210.

Meara, H. (1974). Honor in dirty work: The case of American meat cutters and Turkish butchers. *Sociology of Work and Occupations, 1*, 259–283.

Orlinsky, D. E., Rønnestad, M. H., & Willutzki, U. (2004). Fifty years of psychotherapy process-outcome research: Continuity and change. In M. J. Lambert (Ed.), *Bergin and Garfield's Handbook of psychotherapy and behavior change* (5th ed., pp. 307–389). New York: Wiley.

Randall, R., Griffiths, A., & Cox, T. (2005). Evaluating organizational stress-management interventions using adapted study designs. *European Journal of Work and Organizational Psychology, 14*, 23–41.

Semmer, N. K. (2006). Job stress interventions and the organization of work. *Scandinavian Journal of Work, Environment and Health, 32*, 515–527.

Preface

The idea for this book was born at the 2008 conference of the British Psychological Society's Division of Occupational Psychology during which we realized how closely related our respective work was. The topic of process and context issues in organizational intervention research has been raised several times but no existing publication reflected the joint efforts of researchers and practitioners worldwide on this ambiguous and complex area. The broader context and need for this edited volume is reflected in a number of important developments in the area, including numerous calls for an increased focus on process issues (Cox, Karanika-Murray, Griffiths, & Houdmont, 2007; Goldenhar, LaMontagne, Heaney, & Landsbergis, 2001; Griffiths, 1999; Nytrø, Saksvik, Mikkelsen, Bohle, & Quinlan, 2000; Semmer, 2006) and a special issue in the journal *Work & Stress* that was published in 2010.

Organizations worldwide are facing major challenges in terms of changes in work and the economy, the introduction of new technology, and the increasing diversity of the workforce (Dewe & Kompier, 2008). In turn, these changes translate into pressures to manage the health and performance of the workforce and by extension the productivity of the organization, which in many cases are supported by legislation for organizations to look after their employees. In the UK, for example, the government's initiative to keep workers healthy and in an employment relationship by emphasizing prevention (Dame Carol Black's Report; Black, 2008) is directly in line with the purpose of this book. In order to understand why and how interventions on psychosocial risks and interventions to improve health and well-being produce certain outcomes, be they positive, null, or negative, more attention needs to be paid to the context and processes by which they are developed, implemented, and evaluated. A growing volume of scientific work from many of the contributors of this book over the last few years has helped to strengthen the agenda on organizational interventions and the management of psychosocial risks. With a view to furthering our understanding of what are the processes and contextual issues influencing organizational interventions on well-being, and with a view to guide future research in this area, we thought it would be useful to sum up the work of pioneer researchers in this area. With this volume, we hope to bring some answers to researchers, postgraduate students, and practitioners on how to improve the development, implementation, and evaluation of organizational initiatives aiming to reduce exposure to psychosocial risks, to promote healthy organizations, and healthy workers.

We would first like to thank all the contributors who kindly agreed to share their work and research experiences in this volume. We are very grateful for their hard work and the wonderful contributions they provided. The collection of work presented here reflects state-of-the-art knowledge in how to develop, implement, and evaluate organizational interventions to prevent stress and improve well-being in the workplace. We are thankful for the extremely useful comments provided by the three reviewers of the volume, namely Michael O'Driscoll, Philip Dewe, and Thomas R. Cunningham. We are also grateful to people who helped in preparing the final versions of the manuscripts at the Population Health Research Unit and the Chair in Occupational Health & Safety Management, both in Quebec city. More specifically, we would like to thank Chantal Brisson and Jonathan Mercier, as well as Sylvie Montreuil and Marie-Esther Paradis for their help with the last versions of the chapters.

Contributors

Geneviève Baril-Gingras, Laval University, Québec, Canada
Geneviève Baril-Gingras, PhD, is a Professor in the Industrial Relations Department at Laval University. She holds a master degree in ergonomics (UQAM, 1992) and a PhD in administrative sciences (Laval University, 2003). Her research interests concern conditions and processes leading to changes related to prevention following occupational health and safety or psychosocial work environment interventions. Her interests also encompass occupational health and safety management systems' implementation and public policy respecting workers' health. She explores the contribution of qualitative methods, in particular multiple case studies to these endeavours.

Marie Bellemare, Laval University, Québec, Canada
Marie Bellemare has a PhD in ergonomics from CNAM, France. She is a Professor in the Industrial Relations Department at Laval University where she teaches Ergonomics Intervention and Occupational Health and Safety. She has conducted many action-research projects in the industrial and service sectors focusing on primary prevention of work related health problems, particularly musculoskeletal disorders. She is also interested in the analysis of interventions conducted by the different actors in the field of work design and occupational Health and Safety prevention.

Heleen den Besten, TNO, Hoofddorp, the Netherlands
Heleen den Besten (MSc in psychology) works at TNO since 2006. She has done research and consultancy projects for governments and public and business organizations in the field of work and health (working conditions, mental health, employee satisfaction, workload, stress, health management and employability of older employees). She is involved in several European projects on workplace health promotion. The aim of these projects is to provide companies with good practice information on how to motivate employers and workers to become more involved in health promotion at the workplace.

Renée Bourbonnais, Rehabilitation Department, Laval University, Québec, Canada
Renée Bourbonnais has a PhD in epidemiology from University Paris V and a postdoctorate in occupational epidemiology. She is full professor in the Rehabilitation

department at Laval University. Since 1983, she has been a member of an interdisciplinary research group who have developed an original approach to the study of psychosocial work environment and health. She is also scientific coordinator of a Research group on Personal, Organizational and Social Interrelations at Work. She has studied the impact of restructuring in the health care system and is presently working on evaluative intervention research aimed at reducing adverse psychosocial work factors.

Chantal Brisson, Laval University, Québec, Canada

Chantal Brisson, PhD, is a Professor in the Social and Preventive Medicine Department at Laval University. She holds a Masters degree in industrial relations and a formation in occupation health. She also holds a PhD in epidemiology. She leads a multidisciplinary team (GIROST) regrouping nine researchers who conduct studies in aetiology and the prevention of health problems resulting from work organization. She has collaborated on many epidemiological studies conducted among various groups of workers. She has undertaken three prospective studies conducted among large populations (N=9000, 2200 and 1000) to examine the effect of specific work stressors on incidences and recurrences of health problems. Chantal Brisson held research investigator grants from the National Health Research and Development Program and the Canadian Institutes of Health Research and the Fonds de Recherche en Santé du Québec at the time of those studies.

Viv Brunsden, Nottingham Trent University, UK

Viv Brunsden is a Principal Lecturer in Psychology, a Chartered Psychologist and a Chartered Scientist. She is the Head of the Emergency Services Research Unit at Nottingham Trent University where she specialises in the psychology of the Fire & Rescue Service and in the psychology of disasters and emergencies. She is a member of the Research Expert Panel for the UK Government's Department of Communities & Local Government's Fire Research & Statistics Division and a member of the Emergency Planning Society's Educational Standards Group. She is also Chair of the British Psychological Society's (BPS) Standing Committee on Psychology Education for other groups and a member of the BPS Psychology Education Board.

Kevin Daniels, University of East Anglia, UK

Kevin Daniels is Professor of Organizational Behaviour, Norwich Business School, University of East Anglia. He has a PhD in Applied Psychology and is a Fellow of the British Psychological Society. His research interests revolve around job design, cognitive approaches to understanding affect and affect regulation, health, safety and innovation. He has served as Associate Editor of *Journal of Occupational and Organizational Psychology* and *Human Relations*.

Maureen F. Dollard, University of South Australia, Australia
Maureen F. Dollard is Professor of Work & Organizational Psychology, and Director of the Centre for Applied Psychological Research, and the Work and Stress Research Group at the University of South Australia. She is co-chair of the ICOH Scientific Committee on Work Organization and Psychosocial Factors. Her research on occupational stress, psychosocial safety climate, and ecological models of work stress is published in books and journals such as *Journal of Occupational and Organizational Psychology, Journal of Applied Psychology* and the *Journal of Occupational Health Psychology*. Books include Dollard, M.F., et al. (Eds), *Occupational Stress in the Service Professions* (2003), Taylor and Francis, London.

Emma Donaldson-Feilder, Affinity Health at Work, UK
Emma Donaldson-Feilder is a Registered Occupational Psychologist who specialises in helping organizations achieve sustainable business performance through improvements in the well-being and engagement of staff. Combining research and practitioner roles with writing and presenting on workplace well-being, Emma is: (i) part of a research team investigating the link between leadership/management and employee well-being and engagement, (ii) director of Affinity Health at Work, a specialist consultancy offering services to improve employee well-being and engagement and (iii) author of numerous publications, a regular presenter at professional and academic conferences and provider of expert comment on TV/radio on issues relating to health at work.

Julie Dussault, Centre de santé et de services sociaux de la Vieille-Capitale, Laval University, Canada
Julie Dussault (MS, Sociology, Université de Montréal) is a PhD candidate with a scholarship in Sociology at Laval University (Québec, Canada). She also holds a certificate in personal management from the École de Hautes Études Commerciales de Montréal (Québec, Canada). She has worked as a research professional for more than eight years. Her principal interests are psychosocial harassment at work, health and security at work concerns, personal management practices, work organization and workplace group norms.

Phillip L. Gilmore, Psychology Department, George Mason University, USA
Phillip Gilmore is a research assistant at George Mason University where he is pursuing his doctoral degree under the advisement of Lois Tetrick in the field of Psychology. His research focuses on how innovative performance can improve the health of individuals and organizations, and he is further interested in the conditions that foster cross-cultural learning between China and the USA. He studied Psychology, Chinese and Anthropology at Louisiana State University where he met his wife. They have one daughter and currently reside in northern Virginia.

John Klein Hesselink, TNO, Hoofddorp, the Netherlands
John Klein Hesselink (PhD) is a psychologist and has worked at the Dutch research institute TNO since 1990. Before this he was a researcher at Leiden

University and Rotterdam Erasmus University. As a TNO researcher he conducts many projects on work and health, occupational accidents, sickness absence, working circumstances, work stress, night and shift work, flexibility of work and employment, and labour market participation. As a consultant he assists Dutch companies in the implementation of new rosters and flexible employment and work stress solutions. His thesis in 2002 concerned the effect evaluation of a work stress intervention project for employees.

Rowena Hill, Nottingham Trent University, UK
Rowena Hill is a Senior Lecturer in Psychology, a Chartered Psychologist and a member of the Emergency Services Research Unit at Nottingham Trent University. She has been researching with the fire and rescue community for nine years now and has developed an extensive knowledge of both the practices of the Fire and Rescue Service (FRS) and the psychology of the FRS. Her particular focus is on the family-work interface and relatives' needs. She is currently leading a number of Knowledge Transfer Partnerships with fire related organizations. She serves on the editorial advisory boards for a number of fire related journals.

Nathalie Jauvin, Centre de santé et de services sociaux de la Vieille-Capitale, Laval University, Québec, Canada
Nathalie Jauvin (PhD, Sciences Humaines Appliquées, Université de Montréal) is a researcher at the Centre de services sociaux de la Vieille Capitale (affiliated to Laval University, Québec, Canada) and a member of RIPOST (Research on Personal, Organizational and Social Interrelations at Work). Her main interests revolve around workplace violence and bullying and mental health in the workplace. She has been involved in several studies, including a large quantitative and qualitative project on mental health and workplace violence among correctional officers. She is co-author of articles in *International Journal of Law and Psychiatry* and in *Revue d'épidémiologie et de santé publique*.

Erna de Kleijn, TNO, Hoofddorp, the Netherlands
Erna de Kleijn (PhD) is a sociologist and has worked at TNO, since 1991. She is a consultant and is involved in projects concerning work organization and working conditions. She has extensive experience in projects concerning issues such as improving efficiency and labour productivity and reducing work stress. In daily practice this means solving bottleneck situations in the conduction of work. For instance: inefficiencies in work processes, miscommunication and cooperation problems between departments, high levels of work stress and lack of output quality. In the recent past she worked mainly in the health care sector in close cooperation with management and workers.

Anthony D. LaMontagne, University of Melbourne, Australia
Anthony LaMontagne is an Associate Professor at the McCaughey Centre at the University of Melbourne (Australia). His interest is in developing scientific and public understanding of work as a social determinant of health, and

contributing to improvements in policy and practice aimed at protecting people from the harmful effects of work as well as optimising the health-promoting aspects. He has an international profile for cross-disciplinary applied intervention research in occupational health and health promotion, with interventions of interest ranging from psychosocial risk management to hazardous substance exposure controls, integrated occupational health and health promotion interventions, and national occupational health policies. www.sph.unimelb.edu.au/about/allstaff/lamontagne

Paul Landsbergis, Department of Environmental and Occupational Health Sciences, School of Public Health, State University of New York-Downstate Medical Center, USA
Paul Landsbergis' research focuses on work organization, socioeconomic position, hypertension, cardiovascular disease, psychological disorders and musculoskeletal disorders. He is a co-editor of *The Workplace and Cardiovascular Disease* (Hanley & Belfus, 2000) and *Unhealthy Work* (Baywood, 2009), and has published widely on job strain and cardiovascular disease, on new systems of work organization and worker health, and on interventions to reduce job stress and improve health. He has been a member of the National Research Council's *Committee on the Health and Safety Needs of Older Workers*, and the National Institute for Occupational Safety and Health's *Intervention Effectiveness Research Team*. www.downstate.edu/publichealth/departments/faculty/landsbergis.html

Rachel Lewis, Affinity Health at Work, and Kingston Business School, UK
Rachel Lewis PhD CPsychol is a Director of Affinity Health at Work, a Registered Occupational Psychologist and a senior lecturer and course director in Occupational Psychology at Kingston Business School. She combines her academic career with regular conference speaking, consultancy and training, focusing on the links between leadership, management and employee well-being and engagement.

Colin Mackay, Health & Safety Executive, UK
Dr Colin Mackay is Chief Psychologist in the Chief Scientists Advisory Group in the Health and Safety Executive's Science and Technology Group with a particular responsibility for technical policy aspects of, and research into, work-related stress, work-related upper limb disorders, human factors and behavioural aspects of health and safety. He is currently working on the implementation of HSE's Management Standards for work-related stress and occupational mental health policy more generally. He has worked extensively on evaluating interventions for stress and mental health problems in working populations. Colin has published extensively on the measurement of mood and aspects of psychological health. He is also a Visiting Professor at the Institute of Work, Health and Organizations (part of the Faculty of Medicine and Health Sciences) at the University of Nottingham.

Kevin Maguire, Nottingham Trent University, UK

Dr Kevin Maguire is a Chartered Psychologist and a Senior Lecturer at Nottingham Trent University where he is a member of the Work and Organizational Psychology Research Unit and the allied Work and Health Special Interest Group. Prior to entering academia he worked as an enforcer of health and safety law which has allowed him to bring grounded practitioner understandings into his academic work. His research focuses on understanding and improving working lives. He has studied a diverse range of occupational groups including: catering chefs, clergymen, construction workers, fire and rescue workers, police officers, and tunnellers.

Nadine Mellor, Health & Safety Laboratory, UK

Dr Nadine Mellor is a chartered occupational psychologist and UK government social research member. She works for the Health & Safety Laboratory as a principal psychologist and technical leader. Her research includes the design of preventive health and well-being strategies at the workplace, and the role of health and safety leadership in work environments. Nadine worked previously in France within the training department of a large international computer company as management trainer and consultant focusing on coaching and developing professionals and managers.

Karina M. Nielsen, National Research Centre for the Working Environment, Denmark

Karina M. Nielsen is Professor of Work and Organizational Psychology at the National Research Centre for the Working Environment, Denmark. Her research interests lie within the area of organizational interventions. She is particularly interested in ways to develop evaluation methods to understand how and why such interventions succeed or fail. Her work has been published in book chapters and in journals such as *Human Relations*, *Journal of Organizational Behavior* and *Journal of Occupational Health Psychology*. She has won several awards for her research including the early career achievement award (APA-NIOSH) and a top paper award in *Work & Stress*.

Andrew Noblet, Deakin Graduate School of Business, Deakin University, Australia

Andrew Noblet is an Associate Professor in Organizational Behaviour at Deakin University (Australia). Andrew's research interests are in the areas of occupational stress, organizational fairness, leader-member relationships and workplace health promotion. The results of his work have been published in numerous journals including the *International Journal of Human Resource Management*, *Journal of Public Administration Research & Theory*, and *Work & Stress*. He has also presented at leading conferences in the US, Europe and South-East Asia and has won several awards for his research. In addition to his scholarly research, Andrew provides advisory services to private and public-sector organizations and regularly undertakes employee needs assessments, leadership training and other organizational development initiatives.

Claire Packham, Health & Safety Executive, UK
Claire Packham has been working as statistician at the Health & Safety Executive for the past two years and is currently responsible for statistical support of the stress policy team. She has an undergraduate degree in Philosophy and Economics from the University of Edinburgh and a Masters degree in Human Rights from the University of Manchester, and has previously worked in the media industry as a research associate.

David Palferman, Health & Safety Executive, UK
David Palferman is a Senior Psychologist at the Health & Safety Executive (HSE), UK. He holds degrees in Psychology, Control Engineering and an MSc in Occupational Psychology. Since joining the HSE in 2004 he has provided technical support and training on the management of psychosocial risks in the workplace (stress, bullying and violence) to HSE regulatory inspectors, policy groups and to external stakeholders. Post the launch of the HSE Management Standards approach in 2004 this role has provided a rich opportunity to learn from the experience of literally thousands of organizations that have implemented the HSE approach to tackle the causes of stress in the workplace and the absence related to it.

James Campbell (Jim) Quick, University of Texas, Arlington, USA
James Campbell Quick, PhD, is John and Judy Goolsby Distinguished Professor, Goolsby Leadership Academy at The University of Texas at Arlington and Visiting Professor, Lancaster University Management School, UK. He is a Fellow of SIOP, APA, and American Institute of Stress. Jim was honoured with the *2002 Harry and Miriam Levinson Award* by the American Psychological Foundation and a 2009 University Award for Distinguished Record of Research by UT Arlington. Colonel (Ret) Quick was awarded The Legion of Merit by United States Air Force and the Maroon Citation by Colgate University. He is married to the former Sheri Grimes Schember.

Raymond Randall, Loughborough University, UK
Raymond Randall is Senior Lecturer, School of Business and Economics, Loughborough University, UK. His research interests focus on the enhancement of well-being, satisfaction and performance at work through the use of organizational-level interventions. His published research also tackles several of the methodological challenges that researchers face when attempting to measure the impact of these interventions. His work on these topics has been published in *Human Relations*, the *Journal of Organizational Behavior*, *Work & Stress* and the *European Journal of Work & Organizational Psychology*.

Per Øysten Saksvik, Norwegian University of Science & Technology, Norway
Per Øystein Saksvik is at present professor at the Department of Psychology, Norwegian University of Science and Technology where he also obtained his PhD in 1991 in Occupational Health Psychology. He has seven years

of experience as a researcher at the Institute of Social Research in Industry, Trondheim, Norway. He does research in occupational health and safety, organizational interventions, sickness absenteeism and presenteeism, and organizational change.

Hannah Saul, Health & Safety Executive, UK
Hannah Saul is a Research Officer in the Social Science Unit at the Health and Safety Executive. She joined the organization in 2009 with an MA in Sociological Research from the University of Sheffield. During her time at HSE Hannah has been responsible for managing a range of social research projects as well as conducting her own qualitative fieldwork. She has provided analytical advice to HSE's Health Policy Delivery Team on a number of areas, including options for appraising the Management Standards approach and exploring how medium sized organizations understand and manage occupational health risks.

Norbert K. Semmer, University of Bern, Switzerland
Norbert K. Semmer, PhD, is a professor of psychology of work and organizational psychology at the University of Benn, Switzerland. He studied psychology in Regensburg (Germany), Groningen (the Netherlands), and Berlin (Germany), and he received his PhD from the Technical University of Berlin in 1983. His major interests concern (1) Occupational Health Psychology: Stress and emotions at work and their relationship to health and well-being; currently pursuing the 'Stress as Offense to Self' approach; (2) Efficiency, quality, errors and safety; currently focusing on the influence of team coordination and communication on the performance of medical teams.

Lois E. Tetrick, George Mason University, USA
Lois E. Tetrick, PhD, is University Professor, George Mason University and Director of the Industrial Organizational Psychology Program. She is a fellow of the APA, SIOP, APS, and EAOHP. She has served as President of the Society for Industrial and Organizational Psychology and as Chair of the Human Resources Division of the Academy of Management. She has represented the Society for Industrial and Organizational Psychology (Division 14) on the American Psychological Association Council of Representatives and the Board of Scientific Affairs of the American Psychological Association. She gratefully acknowledges the support of her husband Bill Tetrick.

Sturle Tvedt, Norwegian University of Science & Technology, Norway
Sturle Danielsen Tvedt received his MSc in Psychology from the Norwegian University of Science and Technology (NTNU) in 2004, and has several years experience as a lecturer there, as well as experience as a part-time researcher at the SINTEF Research and Technology. He does research in occupational health and safety, organizational interventions, organizational change management, public risk communication and awareness, safety management, and operative psychology/simulator training.

Marc van Veldhoven, Department of Human Resource Studies, School of Social and Behavioral Sciences, Tilburg University, the Netherlands
Marc van Veldhoven (Master in Psychology/Tilburg University, 1987; PhD/ University of Groningen, 1996) worked as a practitioner in occupational health psychology for 15 years. He returned to academia in 2002, and is currently employed as full Professor "work, health and well-being" in the Department of Human Resource Studies at Tilburg University. His main interest is in building bridges between research on occupational health psychology and research on HRM. He is an associate editor for the *Journal of Occupational and Organizational Psychology*.

Michel Vézina, Social and Preventive Medicine Department, Laval University, Québec, Canada
Michel Vézina specializes in community health and has been a tenured professor at Laval University in the Social and Preventive Medicine department since 1983, and a consultant in workplace health at the Institut national de santé publique du Québec since 2000. He holds a Master's degree in public health from Harvard University. He has released numerous scientific publications on the effects of the organization of work on mental and cardiovascular health, and on psychological harassment at work. His expertise mainly concerns the social and psychological impacts of work and strategies that can be implemented to prevent them.

Eleanor Waite, University of Houston, USA
Eleanor Waite is a second year doctoral student in the Industrial/Organizational Psychology program at the University of Houston. Her research interests include occupational health and human performance. Eleanor also pursues a career in athletics and represented Scotland in the 3000 meter steeplechase at the 2010 Commonwealth Games in Delhi. She completed her BA in Economics and Psychology at Rice University in Houston, Texas in 2009.

Simon Webster, Health & Safety Executive, UK
Simon Webster has worked as a statistician for the Health & Safety Executive for five years and previously as a research associate and statistical consultant at Liverpool and Newcastle Universities. He graduated from Newcastle University with a Mathematics and Statistics degree in 2001 and has been continuously employed in the field of health statistics since then.

Andrew Weyman, University of Bath, UK
Dr Andrew Weyman has over 25 years experience as a human factors specialist in occupational health, workplace safety and well-being. He is currently a Senior Lecturer in the Department of Psychology at the University of Bath (2006–date). Earlier roles include Head of the Social and Organizational Factors Unit at the Health and Safety Laboratory, Buxton and working as Principal Government Social Researcher with the Health and Safety Executive. His specialist area is the

psychology of risk, in particular risk management systems, workplace health and safety climate/culture, intervention design and evaluation.

Noortje Wiezer, TNO, Hoofddorp, the Netherlands
Noortje Wiezer is a social scientist and has worked at TNO since 1998. She holds a PhD in Social and Behavioral Sciences from the University of Amsterdam. As a senior researcher she has done research on behalf of national government, sector organizations, social partners, the European Commission and the European Foundation, on prevention of psychosocial risk and on the effects of changes in organizations on mental health and well-being. She is currently coordinating a European Research Project on the relationship between restructuring and psychological health and well-being, funded by NEW OSH ERA (PSYRES: www.psyres.pl).

Joanna Yarker, Affinity Health at Work, UK
Joanna Yarker PhD CPsychol is a Director of Affinity Health at Work and a Registered Occupational Psychologist. She has previously held posts at Goldsmiths, University of London and the University of Nottingham. She writes regularly for trade and academic journals, and offers guidance to public and private sector organizations focused on improving workplace health.

1 Organizational interventions for stress and well-being – an overview

Caroline Biron, Maria Karanika-Murray, and Cary L. Cooper

> *"Interventions are fragile creatures. Rarely, if ever, is the 'same' program equally effective in all circumstances."*
>
> (Pawson, 2006, p. 30)

The literature on stress at work has been dominated by studies attempting to demonstrate the causal relationship between exposure to stressors, such as psychosocial risks in the workplace, workers' health, and organizational performance. With a world market increasingly specialized and rapidly changing, a healthy workforce becomes a competitive advantage for organizations. The costs associated with reduced performance due to sickness absenteeism and presenteeism are substantial enough for employers to gain from investing in keeping their workers present, healthy, and well (Black, 2008; Foresight Mental Capital and Wellbeing Project, 2008). Psychosocial risks factors such as high psychological demands, low decision latitude, low social support and low recognition contribute to the development of mental and physical health problems (Kivimäki et al., 2006; Marmot et al., 1997; Stansfeld and Candy, 2006; Ylipaavalniemi et al., 2005).

Yet, attempts at preventing stress and promoting health and well-being effectively are still at an embryonic stage. Research has typically focused on describing the effects of individual-level interventions aiming to help individuals deal with the sources of pressure at work. The number of studies evaluating the interventions to reduce exposure to psychosocial risks and promote health and well-being is still disproportionately small compared to individual-level interventions, and the results often inconsistent and modest (Parkes and Sparkes, 1998; Richardson and Rothstein, 2008; Ruotsalainen et al., 2006; Van Der Klink et al., 2001). Without clear evidence on what could be done to successfully prevent work-related stress and promote well-being, it is difficult for employers to know how to implement effective interventions that will produce the intended results. For policy-makers, the lack of evidence regarding how and why interventions produce their results also constitute a barrier to progress. To date, more emphasis has been placed on appraising the effects of interventions rather than reporting on the intervention itself and how it is implemented (Egan et al., 2009). Research on organizational interventions to prevent stress and improve well-being has been focused on *what* works and *for whom*, but not to *why* and *under what circumstances*.

In their editorial of the special issue of *Work & Stress*, Cox, Taris, and Nielsen (2010) highlight the disparity between the growing number of organizational intervention studies being published, and the "societal need for practically useful and effective interventions" (p. 17). The complex and diverse nature of organizational interventions implies, on the one hand, difficulties for researchers who have to ensure their methodology is rigorous. On the other hand, practitioners who have an in-depth and pragmatic experience of the intervention process may often be seen as lacking credibility in the research community (Anderson, 2007). This divide between research and practice has important implications for knowledge in the field of organizational interventions for occupational health. In practice, research has shown that organizational-level interventions often fail to be implemented or to produce the desired results (Biron, Gatrell, and Cooper, 2010; Fullan, 2003; Nielsen et al., 2006). This discrepancy between the progress made in intervention research and the need for practical guidelines for effective interventions indicates the necessity to be innovative in the way we conduct research on this topic.

Considering process and context

The lack of conclusive evidence on intervention effectiveness is worrying, given the large amounts of resources invested in designing and implementing organizational-level interventions. It has been suggested that interventions often fail not due to their content or design, but because contextual and process factors that might determine the success or failure of their implementation are omitted in evaluation studies (Biron et al., 2010; Cox et al. 2007; Egan et al., 2009; Nytrø et al., 2000; Randall, Griffiths, and Cox, 2005). Incorporating process-related factors and contextual issues in intervention research could optimise the fit of the intervention to the specific organizational context and thus improve implementation effectiveness and sustainability. Factors such as the level of management support, employee participation and perceptions, the social climate, the cultural maturity, level of ownership and readiness for change are examples of process and contextual issues likely to influence the implementation and the success of the intervention (Biron et al., 2010; Nielsen et al., 2006; Nielsen, Randall, and Albertsen, 2007; Nytrø et al., 2000; Saksvik et al., 2002; Tvedt, Saksvik, and Nytrø, 2009). Such considerations could improve current intervention evaluation research, and help to integrate the public health (prevention) and applied psychology (individual- and organization-focused) agendas (Lamontagne et al., 2007). Given the difficulties associated with their successful implementation and evaluation (Cooper, Dewe, and O'Driscoll, 2001; Kompier and Kristensen, 2001; Semmer, 2006), there is a striking paucity of studies considering why and how organizational-level stress interventions succeed or fail to be implemented, and how they produce their effects on workers and organizations.

Opening the door to more eclectic approaches to cumulate evidence

There are several reasons explaining this scarcity of research on organizational intervention implementation. The first and probably most influential reason is the fact that intervention studies are notoriously difficult to conduct. Organizations are constantly changing, and the stability required to evaluate the effects of an intervention is more an exception than the rule. Griffiths (1999) underlines how the requirements of organizations are often incompatible with the constraints of traditional research methods as used in the natural science paradigm. The type of evaluative research and the methodological requirements to claim causal explanation (i.e. interventions caused changes in measured outcomes) are too often incompatible with organizational realities (Cox et al., 2007). As argued by many, there is no need to abandon our efforts to accumulate evidence on the effectiveness of stress programmes, but a need to adapt our methods so that they are fit for the purpose of enquiring into this subject area (Cox et al., 2007; Nielsen, Taris, and Cox, 2010; Randall et al., 2005; Semmer, 2006). It is important to bear in mind that research has shown that the more rigorous the design, the more modest the documented results of the interventions. Indeed, Heaney and Goetzel (1997) critically reviewed evaluation studies of the health-related effects (i.e. health risk modification and reduction in worker absenteeism) of multi-component worksite health promotion programmes. Their results suggest that randomized-control trials (RCTs), considered as the "golden standard" in intervention research, have less probability of finding positive effects of a treatment compared to non-randomized studies that use comparison groups. Both RCT and non-randomized comparison group designs yield a lower probability of positive effects compared to studies with no comparison group at all. This suggests that studies using a less rigorous design are positively biased. Although RCTs and quasi-experimental research designs might allow stronger claims regarding the effectiveness of the program and are likely to yield results that are less optimistic than other types of research designs, there are many methodological and ethical constraints making them difficult to conduct in organizational settings. Moreover, the experimental and quasi-experimental research designs are emphasising summative evaluation at the expense of formative evaluation (Nielsen et al., 2010). In other words, the emphasis is placed on answering the question "does the intervention work" instead of "how and why does the intervention work".

Another reason explaining this scarcity of rigorous studies in this field is that traditionally, the scope of our inquiry in stress research has been limited by some of the methodological debates where quantitative methods are explicitly associated with positivism whereas qualitative methods are linked with more interpretive theoretical perspectives (Bergman, 2008b; Crotty, 1998). This polarization or dichotomization of quantitative and qualitative methods with claims of superiority of one method over another has not been helpful in furthering research on organizational stress interventions. Instead, it has hampered a more systematic and theoretically grounded application of a plurality of methods chosen to suit the problem under investigation. Originating from these "paradigm wars" specific unambiguous attributes have been assigned to qualitative and quantitative methods, making their distinctions clear cut. This clear division between paradigms is not always useful in intervention research on stress and well-being. Work psychology has been strongly rooted within the positivist theoretical perspective and the quantitative methods traditionally associated with it (Sparrow, 1999; Weiss and Rupp, 2011). As such, it has developed credibility as a discipline. As suggested by Sparrow (1999), many social science disciplines have adopted a reflexive approach and have been critically analysing their paradigmatic assumptions, but organizational psychology has not been characterized by paradigmatic change. As pointed out by Hollway (1991): "... virtually no debate about the status of the knowledge which makes up work psychology and this state of affairs is the result of the uncritical identification of work psychology with behavioural science, which in turn identifies with natural science" (p. 7). Johnson and Cassell (2002) indicate that work psychology has remained completely indifferent to these debates. However, as Johnson and Cassell (2002) point out, this over-reliance on positivism has not only brought some gains to the discipline, it also has dangers associated with it:

> We are not suggesting here that an approach such as postmodernism is either the right or wrong way of conducting research, but rather that work psychologists need to be aware of the current debates and the impact they have on the discipline. In criticizing the overemphasis on positivism within work psychology research we have highlighted the attendant problems of epistemological conformity and lack of reflexivity.
>
> (p. 138)

Indeed, work psychology has been characterized by a strong positivist perspective, without ever any questioning taking place as to how to expand the limits of the paradigm, use more eclectic methods to compensate for the weaknesses of a particular approach, or consider different methods as "mutually enriching partners in a common enterprise" (McKinlay, 1993, p. 113). As Bergman (2008a, p. 14) argues:

> Theorists and researchers engaging in mixed methods research design have to maintain a strangely schizophrenic position toward the division of labour between quantitative and qualitative methods: on the one hand, they must accept and emphasize the divergent qualities attributed to each paradigm; on the other hand, they put forward the proposal that the strength of each paradigm can be combined fruitfully within one single research design.

The collection of works presented in this volume originates from both qualitative and quantitative backgrounds. We contend that in addition to judging the methodological quality of a study by the methods used, we should also judge the methodological appropriateness of a method for answering particular research questions (McKinlay, 1993). This is in line with what Cox et al. (2007) conceptualize as choosing methods "fit for purpose". Fitness for purpose can be defined as the correct approach to obtaining data of appropriate quality (Thompson and Fearn, 1996) judged against the purpose of obtaining those data. The important point made by Cox et al. (2007) is that the choice of method should be determined by its being fit for purpose instead of by a paradigmatic orthodoxy.

Given the difficulties faced by researchers trying to evaluate the effectiveness of stress interventions, and the inherent changes generated by stress interventions, a question arises as to how to successfully develop, implement and evaluate a complex, generally multi-component stress intervention program within organizations where change is the norm as opposed to a one-off exercise (Peters, 1987). The knowledge about the role of process and contextual variables on organizational interventions is rather embryonic because researchers have mainly focused their attention on individual-level interventions instead of the organizational-level, and have attempted to determine *if* interventions are effective in reducing the negative consequences of stress at work. Because little attention is paid to the influences of process and contextual issues, they cannot explain *why* and *how* the intervention had certain effect(s) on outcome(s). The main thrust of this volume is to define, and examine the influence of process and contextual issues on the outcomes of work-oriented interventions.

Defining and refining process- and context-related factors

Despite the acknowledged need to evaluate process and contextual issues in occupational health interventions (Cox et al., 2007; Griffiths, 1999), little attention has been paid to this topic so far. However, it is necessary to be clear on what is meant by "process" and "context", and how these issues can be considered to enhance our understanding of interventions that are likely to be effective. The present volume offers a critical and pragmatic account of the work on process and contextual issues in an attempt to illustrate how they could be considered more extensively in the evaluation of organizational interventions to prevent stress and improve well-being. This section briefly reviews some of the definitions of the terms, which are further described by the contributors of this book.

The term "process evaluation" has been used in the past few decades to refer to various aspects and types of evaluation. As described in the historical review of process evaluation by Linnan and Steckler (2002), the term is not new and has been used at least since the 1960s by Schuman (1967) to refer to the study of the reasons why a program succeeds or fails. The components of this evaluation vary, but Linnan and Steckler (2002) propose a model comprising the context (broader physical, social, or political factors), reach (proportion of intended audience that participates), dose delivered (by the intervention's providers), dose received (extent of engagement of participants with the intervention), fidelity (quality and integrity of intervention delivered *vs* planned), and recruitment strategies. Program implementation refers to the composite score of reach, dose, and fidelity.

Broader conceptualizations of the term "process" have been used to refer to how an intervention has been carried out from its instigation. According to

Goldenhar et al. (2001), the term "process" encompasses all the phases of the intervention and not just aspects related to the implementation: "research studies that inform intervention development, and studies that evaluate whether the intervention was implemented as planned, complement effectiveness studies" (p. 617). This broader definition is in line with the one offered by Cox et al. (2007), for whom process refers to "the flow of activities; essentially who did what, when, why, and to what effect" (p. 353). Nytrø et al. (2000) conceptualize the term in a more specific way to refer to "individual, collective or management perceptions and actions in implementing any intervention and their influence on the overall result of the intervention" (p. 214). The components of a process evaluation and their influence on the outcomes may differ depending on which definition is used. Interventions are complex and the quality of their implementation is influenced by a wide range of contextual, individual, and collective factors, as well as by the actual content and design of the intervention. From its embryonic state, factors such as the characteristics of the intervention, the context, and the people who are involved in it (remotely or directly), play a role in determining whether the intervention will be developed and implemented, and how it will be evaluated (Biron, Cooper, and Bond, 2009). To understand the resulting effects on outcomes and to qualify the intervention as a success or failure, the components of the "black-box" have to be known at least to some extent. Black box is a metaphor which is used to describe evaluations with inadequate information about the contextual factors and processes which influence the relationship between the program and the effects (Nilsen, 2007).

Weiner, Lewis, and Linnan (2009) make an important distinction between the program theory, and the implementation theory. The program theory refers to the explanation of how the intervention is supposed to work, or in other words, how the intervention is likely to have its effects on the targeted problems (e.g. health, well-being). Weiner et al. (2009, p. 293) describe the implementation theory as follows: "It explains how or why implementation activities (e.g. planning, training and resource allocation) generate observed or desired program use (e.g. employee participation in program activities)". As shown in the review on implementation evaluation of complex social interventions by Egan et al. (2009), to date these factors have been left out of intervention studies and the existing tools allowing them to be taken into account are scarce and relatively underdeveloped. This volume aims to bridge some of these gaps in the field of organizational interventions for stress and well-being and to refine the definition, measurement, and roles of process and contextual issues. By considering the program theory, the implementation theory and the contextual influences, we can attempt to provide some answers regarding the reasons why intervention programs work, what the necessary or favourable conditions for them to be implemented successfully are, and interpret the effects on outcomes in a more meaningful way.

Stress and well-being terminology

Before describing the content of the volume in more detail, a few notes are due on the terminology used by the contributors of this book. There are controversies in the literature about the use of the term "stress", which is considered by some as vague and ambiguous, or even as a term being used to refer to just about any individual or organizational problem (Briner, 1997). The term stress and the risk reduction approach have also been criticized for being too negatively oriented, and not sufficiently emphasizing positive aspects such as what makes an individual flourish and achieve a state of well-being. This is particularly evident in the establishment of new scholarly movements within the past ten years that emphasise the study of more positively orientated phenomena (e.g. Caza and Cameron, 2009; Dutton and Glynn, 2008; Luthans and Avolio, 2009; Seligman and Csikszentmihalyi, 2000). As discussed by Norbert Semmer in the Foreword of this volume, focusing on reducing ill-health does not preclude an emphasis on promoting health, well-being, motivation, personal development and fulfilment at work. As argued by Biron, Cooper, and Gibbs (2011), although debates exist regarding the conceptual distinctiveness of the positive *vs* the traditional stress approach, both share the ultimate aim of helping organizations to manage and develop employees' psychological and physical health, to facilitate a positive organizational climate, and create healthy workplaces, which is the main thrust of this volume. Nevertheless, given the ambiguity around the stress terminology, this section aims to clarify a few points.

In work settings, extensive research has allowed for the identification of a wide range of stress theories, causes, consequences, moderating and mediating factors. As Cooper et al. (2001) point out, however, "this interest and popularity surrounding the term stress are not always useful" (p. 20). Indeed, there is still some confusion about the definition of concepts and their use as dependent, independent, moderating or mediating variables. Typically, the term "stress" is used to describe either the environmental factors that interfere with the well-being of individuals, the effects of these factors on the individual or, alternatively, the reactions individuals adopt to deal with these factors (Appley and Turnbull, 1986). Following the terminology proposed by Cooper et al. (2001) and Beehr and colleagues (Beehr, 1998; Beehr and Franz, 1987), the term "stress" is used in the present volume to describe a process, whereas the terms "stressors", "risk factors", and "psychosocial risks" are used to refer to events or situations in the workplace which can have a detrimental effect on employees. Strain is defined as the individual and collective responses of exposure to these stressors, and the reactions can be manifested either at a psychological, physical or behavioural level. Usually, "outcomes", "effects" and "impacts" are used to indicate the effects of strain on workers (psychological distress, exhaustion, psychosomatic complaints, health problems) and on organizations (increased absenteeism, presenteeism, productivity losses, etc.). This conceptualization implies that interventions aiming at reducing or eliminating negative effects of stress hence can either focus on the environment, on individuals and how they perceive their environment, or on alleviating the adverse consequences. At the conceptual level, there is some acceptance that stress is a transactional process between the individual and the environment, yet at the empirical level, the components are often studied separately from the process (Dewe, 2000).

Theoretically, interventions aiming to modify aspects of the work (work-oriented) or of the organization (organization-oriented) should reduce exposure or eliminate the source of stress (or enhance and foster a positive working climate or working conditions). These changes should, at least theoretically, be associated with decreased strain reactions, which in turn should be linked with decreased adverse consequences on individuals and organization (or positive states) (Semmer, 2006). However, the stress process and the more or less positive approach to this topic are not the main concerns of this volume: Organizational interventions, and the process and contextual issues that can impact their effectiveness are. Therefore, contributors draw on several approaches and stress models and on various types of outcomes. In this volume, the terms "outcomes" and "effects" are used in a broader way to refer to the changes brought by the interventions either on the sources of stress, strain, individual and organizational consequences of strain. To be more specific regarding the sources of stress, they are often defined as work organization, management, and organizational aspects which can have a negative impact on employees' psychological and physical health (Cox and Griffiths, 1995). Because of extensive studies on work-related stress, there is a certain consensus among the scientific community regarding the risk factors causing strain and leading to other negative organizational and individual consequences. These factors can be grouped on the basis of whether they concern the context within which work is produced, or the content of work itself (Cox, 1993). The term psychosocial risks is used interchangeably with the term "stressor" and concerns aspects of the design and management of work and its social and organizational contexts that have the potential for causing psychological or physical harm (Leka, Griffiths, and Cox, 2003).

The choice to use a broader and more eclectic framework to approach the topic of organizational interventions has two motives. First, we share the point of view expressed by Beehr and Franz (1987) who suggest that there are more important issues that need confronting other than the specific label we use. More clarity and specificity would surely be desirable. However, given the lack of studies in the intervention field and the disagreements still existing in the most thoroughly researched stress models (Beehr et al., 2001; de Lange et al., 2003), we side with Pawson (2006) in leaning toward a more "open-door policy on evidence" (p. 178). In this sense, the contributors in this volume draw from a broader repertoire of methods, studies, theories and approaches, as well as on conceptual and critical literature. The attention is thus focused on understanding interventions instead of on the aetiology of stress or well-being, their measurement issues, mechanisms, or on the extent of their consequences on individuals and organizations. There is substantial evidence on the causes of workplace illness, but little is known on how to prevent it. The second reason for not adopting only one particular stress or well-being model follows the reasoning provided by Semmer (2006):

> Ample evidence now shows that stress at work, especially when it is chronic, is a risk for psychological and physical health. This statement applies to a variety of stressors, such as high pressure, barriers to task accomplishments,

or social conflict, as well as to a lack of resources, such as control, social support, or recognition, and other rewards, and it applies to a variety of outcomes, such as depression, psychosomatic complaints, back pain, and cardiovascular disease. Given this state of affairs, it seems "natural" that the promotion of health and the prevention of health problems should predominantly focus on creating a work environment that does not induce an undue amount of stress and that compensates for unavoidable stresses by characteristics such as high control and high rewards.

(p. 515)

This reasoning summarizes the starting point of this volume. Given the scarcity of evidence on organizational stress intervention effects deplored by many (Briner, 1997; Cooper et al., 2001; Kristensen, 2008), the lack of attention paid to process and contextual issues (Cox et al., 2007; Egan et al., 2009; Johns, 2001; Nielsen et al., 2010), and the widely acknowledged costs of occupational stress (Cooper and Dewe, 2008), we offer a collection of chapters which are focused on the intervention process and context instead of on the specific terms used, relations between variables, and the stressor-strain aspect.

In conclusion, this volume emerged in the context where the consequences and adverse health effects of stress at work are widely acknowledged, but little is known regarding the ways of preventing it. Most research has focused on the effects of individual-level stress interventions, although there is clear and convincing evidence demonstrating the adverse health effects of work design and organizational risks such as workload, lack of control, of support, and of recognition. The scarce body of literature focusing on organizational-level interventions is somewhat promising, although inconsistent, and mainly takes a black box approach with little if any focus on the process and the reasons why outcomes were or were not produced. On the one hand, epistemological gatekeepers argue for more rigour in the stress and well-being intervention research, and on the other hand, all agree that there is still a scarcity of evidence on the effectiveness of organizational interventions to prevent stress and improve well-being. We therefore need to be scientifically rigorous, but also functional and pragmatic in how we approach our understanding of organizational interventions for stress and well-being. This volume emerged from this clear need to improve the manner in which initiatives focusing on reducing stressful work-related aspects are developed, implemented, and evaluated.

Content and organization of the volume

In Part 1, contributors highlight some of the difficulties and complexities encountered in research on organizational interventions for stress and well-being. They raise some of the methodological challenges encountered and suggest ways to overcome these challenges.

Anthony LaMontagne, Andrew Noblet, and Paul Landsbergis provide recent evidence on the development and implementation of interventions to address psychosocial risk in the workplace. They specifically outline current understandings of barriers to organizational-level interventions and ways to overcome those barriers.

Nadine Mellor, Maria Karanika-Murray, and Eleanor Waite suggest a more comprehensive approach to the management of psychological risks to workers' health. They review recent research into health and well-being and suggest consideration of job characteristics from a multi-faceted perspective to take into account processes that impinge on workers' health at multiple levels (individual, group, organization). They discuss positive job features that foster individuals' mental capital and empower individuals to protect and maintain their own health and well-being. The success of this comprehensive approach is dependent upon several conditions, of which the development of an occupational health climate is one.

The influence of climate on intervention implementation and on health and performance outcomes is discussed further by Maureen Dollard in the fourth chapter of this book. Maureen describes the Psychosocial Safety Climate (PSC), an emerging construct that reflects the organizational climate for employee psychological health and safety that is largely driven by management, and the broader organizational policies, practices and procedures (to which the safety literature has paid a lot of attention). Using interventions in two public sectors, she shows the influence of PSC on the subsequent quality (whether participants are listened to; whether trust develops) and progress (whether change occurred, whether actions were implemented) of a participatory action risk management stress intervention. PSC gave rise to better intervention implementation and best predicted improvements in individual outcomes such as reduced psychological distress and emotional exhaustion, and organizational ones, such as reduced intention to leave and sickness absence.

Lois Tetrick, James Quick, and Phillip Gilmore review the change management and organizational development literature to provide frameworks for guiding organizational interventions to enhance the health and well-being of organizations and their employees. Using two case studies they illustrate how organizational interventions can lead to psychologically and physically healthy work environments that support and enhance employees' well-being as well as the organization's well-being. As previously mentioned, the literature on change management has rarely been considered in the field of organizational interventions (Heaney, 2003), the two originating from different paradigms but nevertheless converging on a number of important issues.

The following chapter by Sturle Tvedt and Per Øysten Saksvik makes a case for studying interventions as cases of organizational change. They define process to refer to the implementation and adoption of change; how the change is planned, launched, and carried out. Their approach of healthy organizational change processes originates in the context of the Nordic participative tradition. These two chapters provide good examples of how we can draw and learn from other disciplines such as organizational change.

Raymond Randall and Karina Nielsen argue that the search for simple, linear, and universal intervention effects does not fit with the complexity of intervention processes and outcomes. For interventions to be effective, they need to fit the problem as it is perceived by employees and the context within which it occurs. The authors introduce the notion of fit between the individual and the intervention and that of fit between the environment and the intervention. Their model of intervention fit is intended to offer researchers and practitioners a framework for better appreciating the factors that influence the success (or otherwise) of interventions. This requires a more complex model of intervention design, implementation and outcomes variables than is typically used in intervention research. They also remind us that only customized solutions will do when it comes to organizational interventions for psychosocial health.

In line with the idea of complexity in evaluating "how" and "why" interventions produce their results, Geneviève Baril-Gingras, Marie Bellemare, and Chantal Brisson question what is relevant information in terms of process and contextual issues. Their chapter aims to illustrate the contribution of qualitative studies for describing the context and process of interventions, as well as the changes proposed and implemented and their influences on intermediate and final outcomes. Using qualitative multiple case studies of occupational health and safety (OHS) interventions, they expose factors related to context, process, and nature of proposed/implemented changes which were identified as relevant for OHS interventions. Their chapter is a step towards multidisciplinary reflection as they examine recent intervention studies on mental health at work and report on how they related to OHS interventions. The utilization of qualitative research methods and multidisciplinary approaches could support comparisons between interventions, and thus promise a better understanding of how and why interventions produce their outcomes.

As an illustration of the need for more eclectic methodologies to study organizational interventions, Caroline Biron suggests potential avenues for future research by using realistic evaluation theory (Pawson and Tilley, 1997) as a framework to analyze a case where the intervention failed to be implemented. According to realistic evaluation, there are no universal, "magic recipe", one size fits all interventions. Indeed, researchers should attempt to understand "what works, for whom and under what circumstances" (Pawson and Tilley, 1997, p. 342). Intervention programmes work in certain contexts, when certain mechanisms are triggered. Using the case of an implementation failure to apply realistic evaluation principles, Caroline illustrates how process and context issues can be used to further our understanding of the results yielded.

Part 2 of this volume aims to be applied and practical, and to raise some key implications of the role of process issues and contextual issues. The chapters in this section illustrate the role of process issues using empirical studies.

Renée Bourbonnais, Nathalie Jauvin, Julie Dussault, and Michel Vézina provide empirical evidence on the development, implementation, and effectiveness phases of a participative intervention research aimed at reducing adverse psychosocial factors and improve mental health of correctional officers, and complement the chapter by Anthony LaMontagne and colleagues by pinpointing factors facilitating or hindering this intervention process.

Rachel Lewis, Joanna Yarker, and Emma Donaldson-Feilder provide a discussion of the vital role of line managers in managing psychosocial risks. They look at how line managers, one of the key players in employees' work experience, can influence the success of interventions for psychosocial issues through their impact on process and contextual issues, as well as the direct role of the line manager in causing or reducing psychosocial risks.

Similarly, Viv Brunsden, Rowena Hill, and Kevin Maguire discuss issues around stress interventions in an "extreme" or "unique" occupational context of the emergency services, where stress is an integral part of the job, and describe ways of delivering stress interventions, alongside an evaluation of their likely success in this occupational context. They also show that for some occupations preventative approaches are not always feasible. Their chapter provides an example of the need to understand both the occupational environment and specific cultural context of an organization in order to design and apply stress interventions successfully.

Using a practitioner approach, John Klein Hesselink, Noortje Wiezer, Heleen den Besten, and Øma de Kleijn illustrate the development of small organizational work stress interventions by presenting five small-scale intervention studies conducted in the Netherlands. Their chapter highlights that when working with small groups, it is not feasible to meet the methodological requirements to claim causality, yet it is possible to adapt the methods in order to evaluate the outcomes of an organizational intervention.

In Part 3, contributors discuss the development, implementation, and evaluation of organizational interventions to prevent stress and improve well-being, taking a macro perspective and considering implications for policy.

Colin Mackay, David Palferman, Hannah Saul, Simon Webster, and Claire Packham describe the development, implementation, and evaluation of a 10-year priority programme devised by the Health & Safety Executive (HSE) to reduce the burden of occupational stress in Great Britain. Along with the Nordic countries (see chapters by Sturle Tvedt and Per øysten Saksvik, and the chapter by Kevin Daniels, Maria Karanika-Murray, Nadine Mellor, and Marc van Veldhoven) this is one of the few governments to elevate stress as an issue of national concern and to attempt to tackle work-related health at the national level. Several countries have either started to adopt this approach or are considering it. Colin and his colleagues describe the logic behind the adoption of this approach as well as the results obtained so far, and the lessons learnt.

Kevin Daniels, Maria Karanika-Murray, Nadine Mellor, and Marc van Veldhoven bring us a step further in the reflection on policy. Their chapter first reviews critically yet constructively policies and guidance for organizational practice based on prescriptions for job redesign. The job or workplace as the units of analysis typify many national and supra-national monitoring systems for work-related stress, and, in guidance and policy, is probably at its most sophisticated in systems such as the HSE's Management Standards for Work-Related Stress. They illustrate their arguments on what the Standards are expected to achieve by presenting a historical account of a similar approach adopted in the Netherlands that preceded the HSE approach. Their chapter then suggests areas that policy makers and practitioners may explore to ensure an evolution in guidance, policies and practices that better reflects current knowledge on job design, stress, and well-being.

In the last chapter of this book, Andrew Weyman reflects on the evidence-based prevention-orientated approach. Although intuitively appealing for employers, professionals, and policy makers, there is also a need for social researchers to provide tools and techniques allowing robust evidence on the intervention delivery process and the demonstration of its impacts.

Conclusion

So far, although there has been a clear call for more attention to be paid to contextual differences and the process by which interventions are developed, implemented, and evaluated, there is little guidance on how to take these factors into account both in research and in practice. Several reviews (Briner and Reynolds, 1999; Caulfield et al., 2004; Graveling et al., 2008; Hill et al., 2007; Parkes and Sparkes, 1998; Richardson and Rothstein, 2008; Van Der Klink et al., 2001) have concluded that:

1 there is insufficient evidence for firm conclusions on the effectiveness of organizational-level interventions to prevent stress and improve well-being,
2 research designs used are too varied or not considered as sufficiently strong, and
3 more research is needed to advocate the relevance of organizational-level stress interventions.

Given the inconsistencies of studies evaluating the effects of organizational-level stress interventions, this volume attempts to bridge some of these gaps by providing insights about the more neglected aspects of organizational interventions: *how* interventions are developed and implemented, *why* and *when* they succeed or fail to produce the intended effects on outcomes.

References

Anderson, N. (2007). The practitioner–researcher divide revisited: Strategic-level bridges and the roles of IWO psychologists. *Journal of Occupational & Organizational Psychology*, *80*(2), 175–183.

Appley, M. H., & Turnbull, R. (1986). *Dynamics of stress: Physiological, psychological and social perspectives*. New York, NY: Plenum Press.

Beehr, T. (1998). An organizational psychology meta-model of occupational stress. In C. Cooper (Ed.), *Theories of organizational stress* (pp. 6–27). New York: Oxford University Press.

Beehr, T., & Franz, T. (1987). The current debate about the meaning of job stress. *Journal of Organizational Behavior Management*, *8*(2), 5–18.

Beehr, T. A., Glaser, K. M., Canali, K. G., & Wallwey, D. A. (2001). Back to basics: Re-examination of Demand-Control Theory of occupational stress. *Work & Stress*, *15*(2), 115–130.

Bergman, M. M. (2008a). *Advances in mixed methods research*. London: Sage.

Bergman, M. M. (2008b). The straw men of the qualitative-quantitative divide and their influence on mixed methods research. In M. M. Bergman (Ed.), *Advances in mixed methods research*. London: Sage.

Biron, C., Cooper, C. L., & Bond, F. W. (2009). Mediators and moderators of organizational interventions to prevent occupational stress. In S. Cartwright & C. L. Cooper (Eds.), *Oxford Handbook of Organizational Well-being* (pp. 441–465). Oxford: Oxford University Press.

Biron, C., Cooper, C. L., & Gibbs, P. (2011). Stress interventions vs positive interventions: Apples and oranges? In K. S. Cameron & G. M. Spreitzer (Eds.), *Oxford Handbook of Positive Organizational Scholarship* (pp. 938–952). New York: Oxford University Press.

Biron, C., Gatrell, C., & Cooper, C. L. (2010). Autopsy of a failure: Evaluating process and contextual issues in an organizational-level work stress intervention. *International Journal of Stress Management, 17*(2), 135–158.

Black, C. (2008). Working for a Healthier Tomorrow London: TSO.

Briner, R. (1997). Improving stress assessment: Toward an evidence-based approach to organizational stress interventions. *Journal of Psychosomatic Research, 43*(1), 61–71.

Briner, R. B., & Reynolds, S. (1999). The costs, benefits, and limitations of organizational level stress interventions. *Journal of Organizational Behavior, 20*, 647–664.

Caulfield, N., Chang, D., Dollard, M. F., & Elshaug, C. (2004). A review of occupational stress interventions in Australia. *International Journal of Stress Management, 11*(2), 149–166.

Caza, A., & Cameron, K. S. (2009). Positive organizational scholarship: What does it achieve? In S. R. Clegg & C. L. Cooper (Eds.), *The Sage Handbook of Organizational Behavior* (Vol. 2: Macro approaches, pp. 99–116). London: Sage Publications.

Cooper, C., & Dewe, P. (2008). Well-being – absenteeism, presenteeism, costs and challenges. *Occup Med (Lond), 58*(8), 522–524. doi: 10.1093/occmed/kqn124

Cooper, C. L., Dewe, P. J., & O'Driscoll, M. P. (2001). *Organizational Stress: A Review and Critique of Theory, Research, and Applications*. Thousand Oaks, CA: Sage Publications.

Cox, T. (1993). *Stress Research and Stress Management: Putting Theory to Work*. Nottingham: Centre for Organizational Health and Development, University of Nottingham.

Cox, T., & Griffiths, A. J. (1995). The assessment of psychosocial hazards at work. In M. J. Shabracq, Winnubst J.A.M. & Cooper C. L. (Eds.), *Handbook of Work and Health Psychology*. Chichester: Wiley & Sons.

Cox, T., Karanika-Murray, M., Griffiths, A., & Houdmont, J. (2007). Evaluating organizational-level work stress interventions: Beyond traditional methods *Work & Stress, 21*(4), 348–362.

Cox, T., Taris, T. W., & Nielsen, K. (2010). Organizational interventions: Issues and challenges. *Work & Stress, 24*(3), 217–218.

Crotty, M. (1998). *The foundations of social research: Meaning and perspective in the research process*. London: Sage Publications.

de Lange, A. H., Taris, T. W., Kompier, M., Houtman, I. L. D., & Bongers, P. M. (2003). The Very Best of the Millennium: Longitudinal research and the Demand-Control-(Support) Model. *Journal of Occupational Health Psychology, 8*(4), 282–305.

Dewe, P. (2000). Measures of coping with stress at work: a review and critique. In P. Dewe, M. Leiter & T. Cox (Eds.), *Coping, Health and Organizations*. London and New York: Taylor & Francis.

Dutton, J. E., & Glynn, M. A. (2008). Positive organizational scholarship. In J. Barling & C. C. L. (Eds.), *The Sage Handbook of organizational behavior* (Vol. 1: Micro approaches, pp. 693–712). Thousand Oaks, CA: Sage Publications.

Egan, M., Bambra, C., Petticrew, M., & Whitehead, M. (2009). Reviewing evidence on complex social interventions: appraising implementation in systematic reviews of the health effects of organisational-level workplace interventions. *Journal of Epidemiology and Community Health, 63*(1), 4–11. doi: 10.1136/jech.2007.071233

Foresight Mental Capital and Wellbeing Project. (2008). Mid-adulthood – work and skills: interventions (pp. 171–225). London: The Government Office for Science.

Fullan, M. (2003). *Change Forces With a Vengeance.* London, New York: Routledge.

Goldenhar, L. M., LaMontagne, A. D., Katz, T., Heaney, C., & Landsbergis, P. (2001). The intervention research process in occupational safety and health: an overview from the National Occupational Research Agenda Intervention Effectiveness Research team. *Journal of Occupational and Environmental Medicine, 43,* 616–622.

Graveling, R., Crawford, J. O., Cowie, H., Amati, C., & Vohra, S. (2008). Workplace interventions that are effective for promoting mental well-being: Synopsis of the evidence of effectiveness and cost-effectiveness. Edinburgh: National Institute for Health and Clinical Excellence (NICE).

Griffiths, A. (1999). Organizational interventions: Facing the limits of the natural science paradigm. *Scandinavian Journal of Work and Environment Health, 25*(6), 589–596.

Heaney, C. A. (2003). Worksite health interventions: targets for change and strategies for attaining them. In J. C. Quick & L. E. Tetris (Eds.), *Handbook of Occupational Health Psychology.* Washington, D.C.: American Psychological Association.

Heaney, C. A., & Goetzel, R. Z. (1997). A review of health-related outcomes of multi-component worksite health promotion programs. *American Journal of Health Promotion, 11*(4), 290–307.

Hill, D., Lucy, D., Tyers, C., & James, L. (2007). What Works at Work? Review of evidence assessing the effectiveness of workplace interventions to prevent and manage common health problems. London, UK: Department for Work and Pensions.

Hollway, W. (1991). *Work psychology and organisational behaviour: Managing the individual at work.* London: Sage.

Johns, G. (2001). In praise of context. *Journal of Organizational Behavior, 22,* 31–42.

Jonhson, P., & Cassell, C. (2002). Epistemology and work psychology. *Journal of Occupational and Organizational Psychology, 74,* 125–143.

Kivimäki, M., Virtanen, M., Elovainio, M., Kouvonen, A., Väänänen, A., & Vahtera, J. (2006). Work stress in the etiology of coronary heart disease – a meta-analysis. *Scandinavian Journal of Work and Environmental Health, 32*(6 (special issue)), 431.

Kompier, M., & Kristensen, T. (2001). Organizational work stress intervention in a theoretical, methodological and practical context *Stress in the workplace: Past, present and future.* London: Whurr.

Kristensen, R. T. (2008). *Psychosocial intervention studies – opportunities and challenges.* Paper presented at the Third ICOH International Conference on Psychosocial Factors at Work., Quebec, Canada.

Lamontagne, A. D., Keegel, T., Louie, A. M., Ostry, A., & Landbergis, P. A. (2007). A Systematic Review of the Job-stress Intervention Evaluation Literature, 1990–2005. *International Journal of Occupational and Environmental Health, 13,* 268–280.

Leka, S., Griffiths, A. J., & Cox, T. (2003). *Guidelines on Work Organization and Stress.* Geneva: World Health Organization.

Linnan, L., & Steckler, A. (2002). Process evaluation for public health interventions and research: An overview. In A. Steckler & L. Linnan (Eds.), *Process evaluation for public health interventions and research* (pp. 1–21). San Fransisco: Jossey-Bass Publishers.

Luthans, F., & Avolio, B. J. (2009). The "point" of positive organizational behavior. *Journal of Organizational Behavior, 30*, 291–307.

Marmot, M., Bosma, H., Hemingway, H., Brunner, E., & Stansfeld, S. (1997). Contribution of job control and other risk factors to social variations in coronary heart disease incidence. *Lancet, 350*, 235–239.

McKinlay, J. B. (1993). The promotion of health through planned sociopolitical change: Challenges for research and policy. *Social Science & Medicine, 36*(109), 109–117.

Nielsen, K., Fredslund, H., Christensen, K. B., & Albertsen, K. (2006). Success or failure? Interpreting and understanding impact of interventions in four similar worksites. *Work & Stress, 20*(3), 272–287.

Nielsen, K., Randall, R., & Albertsen, K. (2007). Participants, appraisals of process issues and the effects of stress management interventions. *Journal of Organizational Behavior, 28*, 793–810.

Nielsen, K., Taris, T. W., & Cox, T. (2010). The future of organizational interventions: Addressing the challenges of today's organizations. *Work & Stress, 24*(3), 219–233.

Nilsen, P. (2007). The how and why of community-based injury prevention: A conceptual and evaluation model. *Safety Science, 45*, 501–521.

Nytrø, K., Saksvik, P. Ø., Mikkelsen, A., Bohle, P., & Quinlan, M. (2000). An appraisal of key factors in the implementation of occupational stress interventions. *Work & Stress, 14*(3), 213–225.

Parkes, K. L., & Sparkes, T. J. (1998). Organizational interventions to reduce work stress: Are they effective? A review of the literature. *HSE books* (p. 52). Sudbury: Health and Safety Executive.

Pawson, R. (2006). *Evidence-based policy – A realist perspective*. London: Sage Publications Ltd.

Pawson, R., & Tilley, N. (1997). *Realistic evaluation*. London: Sage Publications.

Peters, T. (1987). *Thriving on chaos*. Basingstoke: Macmillan.

Randall, R., Griffiths, A., & Cox, T. (2005). Evaluating organizational stress-management interventions using adapted study designs. *European Journal of Work and Organization Psychology, 14*(1), 23–41.

Richardson, K. M., & Rothstein, H. R. (2008). Effects of occupational stress management intervention programs: a meta-analysis. *Journal of Occupational Health Psychology, 13*(1), 69–93.

Ruotsalainen, J. H., Verbeek, J. H., Salmi, J. A., M., J., Persternack, I., & Husman, K. (2006). Evidence of the effectiveness of occupational health interventions. *American journal of Industrial Medicine, 49*, 865–872.

Saksvik, P. Ø., Nytrø, K., Dahl-Jorgensen, C., & Mikkelsen, A. (2002). A process evaluation of individual and organizational occupational stress and health interventions. *Work & Stress, 16*(1), 37–57.

Schuman, E. A. (1967). *Evaluative research*. New York: Russeil Sage Foundation.

Seligman, M. E. P., & Csikszentmihalyi, M. (2000). Positive psychology: An introduction. *American Psychologist, 55*, 5–14.

Semmer, N. K. (2006). Job stress interventions and the organization of work. *Scandinavian Journal of Work and Environmental Health, 32*(6, special issue), 515–527.

Sparrow, P. R. (1999). Editorial. *Journal of Occupational and Organizational Psychology, 72*(3), 261–264.

Stansfeld, S., & Candy, B. (2006). Psychosocial work environment and mental health – a meta-analytic review. *Scandinavian Journal of Work, Environment & Health, 32*(6), 443–462.

Thompson, M., & Fearn, T. (1996). What exactly is fitness for purpose in analytical measurement? *Analyst, 121*, 275–278.

Tvedt, S. D., Saksvik, P. Ø., & Nytrø, K. (2009). Does change process healthiness reduce the negative effects of organizational change on the psychosocial work environment? *Work & Stress, 23*, 80–98.

Van Der Klink, J. J. L., Blonk, W. W. B., Schene, A. H., & Van Dijk, F. J. H. (2001). The benefits of interventions for work-related stress. *American Journal of Public Health, 91*(2), 270–276.

Weiner, B. J., Lewis, M. A., & Linnan, L. A. (2009). Using organization theory to understand the determinants of effective implementation of worksite health promotion programs. *Health Education Research, 24*(2), 292–305. doi: 10.1093/her/cyn019

Weiss, H. M., & Rupp, D. E. (2011). Experiencing Work: An Essay on a Person-Centric Work Psychology. *Industrial and Organizational Psychology, 4*(1), 83–97. doi: 10.1111/j.1754-9434.2010.01302.x

Ylipaavalniemi, J., Kivimaki, M., Elovainio, M., Virtanen, M., Keltikangas-Jarvinen, L., & Vahtera, J. (2005). Psychosocial work characteristics and incidence of newly diagnosed depression: a prospective cohort study of three different models. *Social Science & Medicine, 61*(1), 111.

Part 1
Challenges and methodological issues in organizational-level interventions

2 Intervention development and implementation

Understanding and addressing barriers to organizational-level interventions

Anthony D. LaMontagne, Andrew J. Noblet, and Paul A. Landsbergis

This chapter summarizes recent evidence on the development and implementation of interventions to address psychosocial risk in the workplace, in particular outlining current understandings of barriers to organizational-level interventions and ways to overcome those barriers. This chapter and this book are premised on the need for organizational-level intervention to effectively manage workplace psychosocial risk. This premise is based on principles from relevant disciplines (e.g. public health, occupational health, health promotion, and organizational psychology (LaMontagne, Keegel, and Vallance, 2007) combined with empirical evidence from intervention research.

Introduction

A brief restatement of the basis of this premise is warranted as a prelude to the chapter. The international job stress intervention research literature has been the subject of a number of recent systematic reviews. The most comprehensive of these reviews (summarising 90 intervention studies published between 1990–2005) focused on interventions in which organizations set out to address job stress proactively (LaMontagne et al., 2007). This review is the most directly relevant to this chapter; it concluded that individually-focused approaches (e.g. coping and time management skill development) are effective at the individual level, favourably affecting individual-level outcomes such as health and health behaviours. Individual level interventions, however, tended not to have favourable impacts at the organizational level (e.g. reducing exposures or sickness absence). High systems approaches, defined as combining individual- and organization-directed interventions (e.g. addressing working conditions and organization of work), however, conferred both individual- and organizational-level benefits. Over the 1990–2005 period, there was a hopeful trend observed: intervention studies using high and moderate (organization-directed only) systems approaches represented a growing proportion of the job-stress intervention evaluation literature, possibly reflecting growing application of organizationally-directed approaches in practice internationally. Low systems approaches, however, were still the most common.

Two systematic reviews on psychosocial work environment interventions were published by the Cochrane Collaboration Public Health review group (www.cochrane.org) soon after the above-described review (Bambras et al., 2007; Egan et al., 2007). While these had more strict inclusion criteria to optimize causal inference, they also included natural experiments, or unintended changes in psychosocial stressors, such as from downsizing and restructuring. The systematic review described above included only interventions purposefully addressing job stress, hence these two subsequent Cochrane reviews provide important complementary findings. The first review of organizational-level interventions that increased job control found some evidence of health benefits (e.g. reductions in anxiety and depression) when employee control increased or (less consistently) when demands decreased or support increased (Egan et al., 2007). They also found evidence of worsening employee health from downsizing and restructuring (Egan et al., 2007). The second review of task restructuring interventions (Bambra et al., 2007) found that interventions that increased control resulted in improved health.

The Cochrane Public Health review group subsequently published an overarching "umbrella" summary of systematic reviews of the effects on health and health inequalities of organization changes to the psychosocial work environment (Bambra et al., 2009). In addition to including the two reviews described above, shift work, work scheduling, privatization, and restructuring were considered. Findings suggested that organizational-level changes to improve psychosocial working conditions can have important and beneficial effects on health.

This set of recent systematic reviews demonstrates that feasible and effective strategies for the prevention and control of workplace psychosocial risk at the organizational level are available. Despite the growing evidence in support of systems or comprehensive approaches as the most effective, prevalent practice in most OECD countries remains disproportionately focused on individual-level intervention with inadequate attention to organizational-level intervention (Giga et al., 2003; Hurrell and Murphy, 1996; LaMontagne et al., 2006; Leka et al., 2008). This chapter aims to explore why this is the case, and to propose strategies for retaining a focus on organizational change in the management of workplace psychosocial risk. While the main focus of this chapter is on challenges at the organizational, or internal, level (termed "micro-level" in dedicated section below), some challenges to organizations are due to external influences (termed "macro-level"). Accordingly, we conclude the chapter with an acknowledgement of the importance of external context and a brief outline of macro-level influences on organizational practices in the management of psychosocial risk (third section).

Micro-level challenges to organizational level interventions: focusing on the organization

The following section will focus on the internal, micro-level challenges that confront program organizers when planning and implementing organizational-level interventions. These challenges emanate from several processes that are considered crucial to the effectiveness of organizational health interventions. These processes include:

1 gaining management support
2 articulating the need for comprehensive worker- and work-directed interventions
3 establishing participatory processes, and
4 the early detection of opportunities and threats.

Gaining management support

The success of organizational change programs have long been known to rest heavily on the extent to which high-level managers support the need for change and are willing to commit considerable time and energy to developing and implementing appropriate initiatives (e.g. Kotter and Schlesinger, 1979; Lewin, 1947). More recent studies from the behavioural and occupational health sciences indicate that the support of organizational leaders and front-line managers is particularly important in the development of organizational level interventions aimed at addressing psychosocial risk factors (DeJoy et al., 2010; Nielsen et al., 2010; Polanyi et al., 2005). A key challenge early in an intervention's life-cycle is therefore to gain the support of high-level management.

There are both functional and symbolic reasons why the support of senior personnel is critical to the success of organization-level interventions (Noblet and LaMontagne, 2009). From a functional perspective, modifying operating systems, developing new policies, re-designing work practices and other large-scale reforms require the sustained commitment of personnel from all levels of the organizational hierarchy. Support from organizational leaders is therefore required to authorize the involvement of relevant personnel, to give people the time required to take part in the development of organizational health initiatives, to ensure all activities are adequately resourced and generally to under-write the goals of the interventions (DeJoy et al., 2010).

Securing the support of the CEO, company directors, and other executive-level staff can also have more symbolic benefits. Employee cynicism regarding the motives behind proposed interventions can be a major source of resistance to change, particularly in cases where there is a lack of trust between management and employees and where workers feel that management are not genuinely committed to achieving the stated goals (Oreg, 2006; Polanyi et al., 2005). The tangible support of senior management – through the allocation of time, funding, and the active involvement of top-level personnel – can send out the message that organizational leaders genuinely value employee well-being and are prepared to devote the resources needed to identify and address priority health issues.

A genuine commitment to change needs to extend to all levels of management especially those middle managers with direct line-management responsibilities. Like many change initiatives, the task of actually implementing organizational-level interventions is often delegated to front-line managers and hence the extent to which activities are implemented in the way they were intended rests heavily with this group (Kompier et al., 2000). Middle managers have been found to impede the planning and implementation of organizational health interventions, for example by altering the processes used to develop initiatives (Saksvik et al., 2002) or by preventing employees from spending time on the interventions themselves (Dahl-Jorgensen and Saksvik, 2005). Other research indicates that employees' perceptions of intervention outcomes were more likely to be positive when middle managers had taken responsibility for implementing the changes and had actively sought the involvement of employees in these activities (Nielsen and Randall, 2009). Overall, the organizational health research suggests that once top management has approved the proposed interventions, the next major task is to gain the day-to-day support of front-line managers.

Given the importance of management support in shaping the effectiveness of organizational-level interventions, program coordinators need to develop a set of strategies that can be used to capture the commitment of organizational leaders and to maintain this management "buy-in". According to the organizational development literature, three general strategies can help achieve this goal (adapted from Waddell, Cummings, and Worley, 2004). The first strategy is to sensitize the organization to the pressures of change. Fundamentally, this strategy involves identifying the external (e.g. increased market competition or changing societal expectations) and internal (e.g. high labour turnover and declining productivity) pressures that support the need for change. The second strategy is to reveal the discrepancies between current and desired states. That is, to highlight the deficiencies associated with the current situation and to acknowledge the benefits of the new approach. The final strategy that can be used to boost management support is to convey credible, positive expectations regarding what the change program can achieve and to outline the processes for developing and implementing these changes.

Making a business case for reducing psychosocial risks is one approach to generating positive expectations. A recent Australian report estimated the annual economic benefits of eliminating job strain-attributable depression through comprehensive worker- and work-directed intervention as at least equal to if not greater than the annual national workers compensation costs for "mental stress" claims (LaMontagne, Sanderson, and Cocker, 2010). More importantly, findings showed that the vast majority of economic benefits would accrue to employers (98 per cent), through reduced productivity and employee replacement costs, providing a clear business incentive for employers to invest in initiatives that reduce job stress and promote mental health. Arguments presenting a business case for organization-directed intervention provide a valuable complement to ethical and legal/regulatory arguments for persuading business leaders.

Positive expectations might also be generated by linking psychosocial risk management to something more broadly supported by upper and middle management. In many organizations, there is a growing interest in mental health and illness in a generic sense. In Australia, for example, there has been a substantial and sustained national initiative, called *beyondblue*, to promote mental health literacy, including a workplace program (www.beyondblue.org.au). There is strong demand from employers in this regard, driven by various influences including good will and the need to meet anti-discrimination obligations (Australian Human Rights Commission, 2010). One of the authors (ADL) recently partnered with *beyondblue* to develop and implement an integrated job stress and mental health promotion intervention approach. Though the project is on-going, it has already achieved a valuable outcome in the form of a significant increase in job stress content in the on-going and expanding *beyondblue* National Workplace Program. In short, management buy-in may sometimes be heightened by integrating psychosocial risk management efforts with complementary employee wellness, employment law, human resources, or other initiatives.

Although some of the more detailed information regarding current and desired states in a given organization may not be available until an extensive needs assessment has been undertaken, program coordinators would – in the first instance – need to gather sufficient background information to convince management that organizational health issues represent a key threat to the functioning of the firm and that effectively addressing these issues could result in substantial improvements for both the organization and its members. Psychosocial needs or risk assessment can be undertaken by itself or in combination with broader organizational concerns (LaMontagne and Keegel, 2011).

Once the intervention is underway, an important means for maintaining managerial support is to use the results of the program – in relation to processes and/or impact measures – as a way of demonstrating the continued relevance of the program. Operational and commercial imperatives often mean that managers are juggling a variety of competing interests at any one time. An ongoing challenge for program organizers is to keep senior personnel up-to-date with how the program is progressing and to ensure they are reminded of why the initiatives are important.

Articulating the need for comprehensive worker- and work-directed interventions

Despite the importance of developing organizational health interventions that seek to address the worker and the workplace (Noblet and LaMontagne, 2006; Semmer, 2006), evidence suggests that there is a general reluctance to tackle work- or organizationally-based risk factors. Organizations are much more likely to involve strategies directed at workers' lifestyle-related behaviours and generally fail to consider the direct impact that working conditions have on employee wellbeing and/or their indirect influence on people's ability to adopt more stress-resistant behaviours (Caulfield et al., 2004; Giga et al., 2003; LaMontagne et al., 2006; Landsbergis, 2009; Murta, Sanderson, and Oldenburg, 2007). Attempting to modify behavioural risk factors without taking into account the influence of organizational conditions may actually exacerbate health outcomes and hence there is a strong need to develop comprehensive multi-level interventions that address work and worker-oriented attributes (LaMontagne et al., 2007; LaMontagne, Keegel, and Vallance, 2007). Another challenging task confronting the organizers of work-based interventions is to convince managerial personnel to shift the focus from individual employees (their knowledge, attitudes and behaviours) to characteristics of both the worker and the workplace.

There are at least four reasons why it may be difficult persuading organizations to address organizational sources of ill-health. Proximal risk factors relating to employee behaviours (e.g. smoking, inactivity, stress management techniques) are generally more apparent to workplace representatives than distal risk conditions and thus getting managers to recognize the influence of these underlying conditions may take considerable effort (including undertaking a detailed assessment of how risk conditions contribute to risk behaviours) (Nielsen, Taris, and Cox, 2010; Polanyi et al., 2005). The second reason why it may be difficult generating support for organizational-based interventions is that work-based sources of stress and ill-health are often embedded in systems, practices, and cultures that have been maintained over a long period of time. Modifying these systems, challenging the status quo and overcoming organizational inertia may therefore be seen as a complex, time-consuming exercise that is "too hard" or not worth the expense, especially if senior personnel are not convinced that the interventions will lead to more positive outcomes or if they doubt the organization's capacity to implement the necessary reforms (Armenakis and Bedeian, 1999; Holt et al., 2007).

Mobilizing support for organizational interventions may be even more problematic when the external contexts in which organizations operate are not conducive to the desired changes. Economic downturn, intense competitive pressures, the demand for more flexible employment practices, the inherent nature of the industry, and other social and economic conditions outside the organization may result in organizations feeling that they need to maintain current work systems and practices simply as a way of surviving (Armenakis and Bedeian, 1999). The final reason why there may be resistance to organizational-level interventions is that psychosocial working conditions are often influenced by leader attributes including decision-making styles, communication skills, and timing and quality of support (Karasek and Theorell, 1990; Nielsen and Randall, 2008). Effectively addressing working conditions will inevitably involve managers acknowledging that their own leadership styles may be contributing to the problem and then developing the skills, knowledge, and willingness to overcome these weaknesses. Again, considerable resources may be required to help managers appreciate their own contribution to the organizational conditions in question and then to ensure they have the opportunities to develop more appropriate leadership styles.

Generating commitment to the comprehensive work- and worker-directed approach will depend heavily on the level of support gained from top management. In addition to the general strategies for gaining managerial support discussed in the previous section, an important means for generating commitment to addressing organizational sources of ill-health is through leadership development initiatives (Kelloway and Barling, 2010; Nielsen and Randall, 2008). These initiatives could help managers better understand the dynamic, bi-directional relationship between employee-level outcomes, risk behaviours and risk conditions and could refer to case studies that illustrate how the organization and its members can benefit from this approach (e.g. DeJoy et al., 2010; Maes et al., 1998; Sorensen et al., 2002). The management development initiatives would also need to focus heavily on helping front-line managers develop the skills required to drive the change program and to adopt more supportive management styles.

Establishing participatory processes

Another factor that can make important contributions to the effectiveness of organizational health interventions is the degree to which participatory processes are used to plan, implement, and evaluate the interventions. Participatory processes refer to those workplace systems, mechanisms, or other communicative activities where decision-making influence is shared between superiors, subordinates and other relevant stakeholders (adapted from Sagie and Koslowsky, 1996). The forms of participation as well as the degree to which decision-making influence is shared can vary along several dimensions including whether participatory exchanges are forced or voluntary, formal or informal, direct (individual participation) or indirect (representation on committees), and full authority or minimal consultation (Bordia et al., 2004). In an organizational change context, the target of the decision-making may also involve strategic considerations (such as *whether* the organization should change and *what* aspects of the organization need to change) and/or tactical issues (including *when* and *how* the changes should be developed and implemented) (Sagie and Koslowsky, 1996).

There are several compelling reasons why intervention coordinators should seek the involvement of relevant stakeholders. First, participatory approaches are a hallmark of effective comprehensive or systems approaches to psychosocial risk management (LaMontagne et al., 2007; Landsbergis, 2009). Participation is a concrete enactment of job control, demonstrates organizational fairness and justice, and builds mutual support among workers and between workers and supervisors. Participatory processes that engage the individuals and groups most affected by the issue in question can result in a range of positive outcomes for both employees and employers (e.g. LaMontagne, Keegel, Louie et al., 2007; LaMontagne and Keegel, 2011; Mikkelsen, Saksvik, and Landsbergis, 2000; Nielsen et al., 2010; O'Brien, 2002), including:

- more accurate problem identification and issue analysis, which can result in a better fit between the interventions and the organizational context. This is crucial because organizations usually require unique psychosocial risk management solutions, even if the process of intervention may be based on generic principles and frameworks;
- improved communication regarding the form and function of the interventions, leading to reduced change-related uncertainty, enhanced opportunities for support between participants and an overall reduction in resistance to change;

- heightened responsibility for the problems identified, contributing to improved motivation and a stronger commitment to the change strategies, and;
- enhanced capacity building and organizational learning, by equipping people with the skills and willingness to identify and address their own problems.

Despite the benefits that can be gained by employing participatory processes to develop organizational-level interventions, there are indications that the active involvement of employees tends to be the exception rather the norm. Nielsen, Taris, and Cox (2010, p. 226) note that the predominant approach to developing and implementing organizational-level interventions is to assume that employees are passive recipients of the changes and to adopt a top-down approach. This centralized approach is consistent with other areas of occupational health. In the case of workplace health promotion, for example, a review revealed that only 25 per cent of programs were implemented in response to employees' explicit needs and 14 per cent included employees as partners in their planning and implementation (Harden et al., 1999).

There are also concerns regarding the extent to which attempts to gain employees' insights are genuine and, along similar lines, whether participatory processes address issues that really matter to employees. The National Institute for Occupational Safety and Health, the peak OHS authority in the US, states that "... various worker participation or involvement strategies may often be more ceremonial than substantive, having little meaningful influence on worker empowerment ..." (NIOSH, 2002, pp. 15–16). An inability to tackle substantive organizational issues was also identified as a barrier in a participatory-based intervention study involving over 2200 retail employees (DeJoy et al., 2010). In this case, poor decision-making access to core business operations, coupled with the company's reluctance to provide sufficient time and resources to support participatory mechanisms, threatened the sustainability and institutionalization of the interventions.

Another factor influencing the effectiveness of participatory processes is the extent to which they capture the views and ideas of all relevant stakeholders. Studies examining the effectiveness of participatory-based interventions indicate that the groups who are particularly vulnerable to experiencing high levels of work-related ill-health – blue-collar workers, low status workers, women who are precariously employed (LaMontagne and Keegel, 2011; LaMontagne et al., 2012; Landsbergis, 2010) – are also less likely to have the skills, experience or opportunities to effectively engage with participatory processes (Elo et al., 2008). In the case of blue-collar employees, a study aimed at investigating the effects of employee participation in an organizational stress management program found that some blue-collar employees felt uncomfortable expressing their opinions in participatory meetings (Elo et al., 2008). The authors of this study also noted that the more positive effects of participatory interventions have been gained among white-collar employees – a group that is generally more accustomed to higher levels of communication and social interaction processes – and suggested that training and development would be required to facilitate the increased involvement of blue-collar workers.

The level of employee involvement in participatory activities is not only influenced by employees' capacity to take part in these discussions, but also by the extent to which managers' are willing and able to share decision-making responsibilities. Empowerment-based initiatives may represent a major threat to managers, particularly those accustomed to the more traditional "command and control" management roles (Parker and Williams, 2001). In these cases, managers may be reluctant to hand over decision-making responsibility to followers for fear that their own role, and therefore their position, may become redundant. However sharing decision-making responsibilities with followers and developing a team approach to problem-solving can give managers more time to focus on macro-level matters (such as external opportunities and threats, and the strategic direction of the group) and can actually enhance their leadership capacities (Ivancevich, Konopaske, and Matteson, 2011). Research indicates that training and development can be beneficial in helping managers move from micro-level transactional leadership styles to the "bigger picture" transformational approach (Nielsen et al., 2010).

If organizations genuinely want to achieve participation in psychosocial risk management and other organizational processes, steps must be taken to ensure that the participatory process is both accessible to, and safe for, lower-level employees in particular. Openly-communicated pledges from upper management in this regard are helpful, such as recommended by Kristensen et al. (2005) in the use of their Copenhagen Psychosocial Questionnaire (COPSOQ) needs assessment instruments, in which they articulate "soft guidelines" for use including protecting survey respondent confidentiality and the need for the organization to pledge to act on the findings of the needs assessment before conducting surveys (see www.ami.dk). The involvement of trade unions, where present, can afford some protection for participating employees as well. Other methods include the use of "Health Circles," as developed in Germany (Aust and Ducki, 2004). Organizations can also involve external organizational change or other consultants to conduct participatory needs assessment and intervention development, such as through the use of Future Inquiry methods (Blewett and Shaw, 2008). Future Inquiry includes structured assessments with groups of employees from various levels within an organization (in which lower-level employees may be unlikely to speak critically) and separate assessments with groups of employees from the same or similar levels within the organization (in which lower-level employees feel more comfortable to speak openly). Views are synthesized and made anonymous by the consultants before reporting back to the organization-wide group.

Notwithstanding recent developments described in the previous paragraph, the organizational health literature, in all, suggests that there are a number of notable shortcomings in the uptake and reach of participatory processes. Organizational-level interventions are less likely to be based on the active involvement employees and, when they are utilized, these processes tend to overlook core operational matters and lack the sustained support from high-level management. There are also concerns regarding the extent to which more vulnerable groups have the capacity to become involved in decisions that impact on their health and the willingness of managers to seek the involvement of their followers. Collectively these shortcomings represent significant barriers to maximizing the benefits of participatory decision-making, and a major challenge confronting practitioners and researchers involved in the development of organizational interventions is therefore to develop further strategies that can overcome these barriers.

Early detection of opportunities and threats

Developing organizational-level interventions that are based on the active involvement of all relevant stakeholders and have the backing of both employers and employees is a time consuming, resource-intensive exercise. However many of the benefits that can be gained through detailed planning can be lost if the interventions are poorly managed during the implementation phase (Noblet and LaMontagne, 2009). One aspect of program implementation that needs to be monitored very closely involves the emergence of new policies, practices, and other conditions that can impact on how the interventions are implemented and/or the effectiveness of those interventions. Internal and external forces that arise during the implementation stage have the potential to undermine or enhance the effectiveness of organizational interventions, depending on when and how they are addressed, and an important challenge for program coordinators is to develop mechanisms that can identify these emerging opportunities and threats early, while there is still time to form an appropriate response (i.e. to prevent or reduce the impact of the threat or to take advantage of the opportunity).

Research has identified a range of relatively unforeseen events and situations that can impact on important intervention processes or outcomes. In terms of potential threats, these include concurrent activities that divert people's attention away from the organizational-level initiatives and/or prevent them from taking part in the required processes (Elo et al., 2008), declining support from important stakeholders such as middle managers and "shop floor" employees (DeJoy et al., 2010), key personnel being relocated or made redundant (Mikkelsen et al., 2000), organizational reforms that undermine the goals of the intervention, including possible lay-offs, mergers and other significant organizational restructuring (Dahl-Jorgensen and Saksvik, 2005; Egan et al., 2009; Nielsen et al., 2010), resistance to proposed interventions due to a lack of trust and the potential threats posed by the interventions (Nielsen et al., 2010), and adverse economic conditions, such as national recessions or severe industry downturn (DeJoy et al., 2010; Egan et al., 2009).

There is far less research on the potential opportunities that may arise during the implementation of organizational-level interventions, however the general organizational change literature (Armenakis and Bedeian, 1999; Fernandez and Rainey, 2006; Kotter, 1995) suggests that these include using "short-term wins" to strengthen support for the program while reducing individual and organizational resistance; leveraging off this broader support-base to expand the range of sites, departments and work groups involved in the program; using the feedback from program participants to assess the quality of the intervention methods and, where appropriate, to make necessary modifications; identifying areas where there can be improved integration between the interventions and new or previously unknown practices, policies, or other work-based initiatives; and, identifying changes in the external environment that could be used to reinforce the relevance of the program and/or extend the reach of the initiatives (e.g. the introduction of new legislation that mandates conditions that promote health; case studies from industry groups demonstrating the processes used to achieve higher levels of employee engagement and retention).

Stable and predictable organizational environments are rarely found in today's dynamic, fast-paced business world (Nielsen et al., 2006) and the coordinating team needs to have systems in place to identify changes that may undermine the effectiveness of the program or, conversely, enhance its effectiveness. An important strategy for identifying emerging threats/opportunities is to ensure there is frequent and ongoing communication with key stakeholders. Managers, employees, HR personnel, OHS staff, and other members of the organization need to receive regular updates on the progress being made and the success or otherwise of the different methods (Demmer, 1995; Fernandez and Rainey, 2006). The perceptions of key stakeholders can have a dramatic impact on the longevity of the interventions and the coordinating team needs to be proactive (through formal and informal communication mechanisms) in seeking feedback from external groups. This two-way dialogue can not only help coordinators identify potential threats and opportunities, but can also be instrumental in raising the profile of the interventions, reducing misunderstanding and apprehension regarding the goals of the interventions and attracting broader participation and support (Jick, 2003; Kotter, 1995).

While two-way communications with relevant individuals and groups is important for the early detection of threats/opportunities, program coordinators need to have the capacity to respond quickly to these new situations and conditions. Contingency planning (i.e. planning to deal with possible outcomes before they occur) can help organizers develop this capacity (Noblet and LaMontagne, 2009). The process of developing appropriate contingencies begins with organizers first undertaking a detailed assessment of the possible situations and events that may arise during the implementation phase and that have the potential to undermine or enhance the effectiveness of those initiatives. Some threats and opportunities may be difficult (if not impossible) to identify in advance, however many others can be predicted, particularly if organizers have already developed a detailed understanding of the context in which the interventions are operating. Once coordinators have identified possible opportunities/threats, they then need to draw up a set of responses that can be implemented if the threat or opportunity materializes.

In closing this section, it is worth noting that organizational-level interventions will always be vulnerable to threats when they are regarded as "add-on" activities that are not central to the organization's core business operations (DeJoy et al., 2010; Fernandez and Rainey, 2006). An important ongoing challenge – which represents both an opportunity and a threat depending on how it is tackled – is to develop organizational health initiatives that become an integral part of the organization's operations and culture (i.e. institutionalized) (Fernandez and Rainey, 2006; Lewin, 1947). In many cases, mainstreaming processes and strategies that promote improved quality of working life will begin by starting small and focusing on discrete work areas or specific organizational issues (French, Bell, and Zawacki, 2005). This is an accepted way of piloting new approaches and stimulating widespread organizational development. However in order to bring about lasting change, coordinators can't afford to focus solely on the intervention at hand. Instead, they need to adopt a cyclical program planning, implementation and evaluation framework (e.g. Noblet and LaMontagne, 2009) that is geared towards developing a culture that values employee well-being and where health-related outcomes are high on the list of priorities for key decision-makers.

Macro-level challenges to organizational level intervention: the need for higher-level intervention to complement organizational focus

While factors internal to organizations are the most immediate influences on organizational approaches to psychosocial risk management, some of the challenges to retaining an organizational focus sit outside the organization in the broader labour market, the local or international economy, national cultures, political conditions, and regulatory and other policy influences. In this section, we briefly describe some of these, referring readers to other sources for more detailed discussion (see, for example, Keegel, Ostry, and LaMontagne, 2009; LaMontagne, 2010; Landsbergis, 2009; Landsbergis, Cahill, and Schnall, 1999). The purpose of this section is to acknowledge that organizations operate in a broader context that continually shapes their practice, leading to a need to account for secular trends that may hinder the desired organizational focus, as well as the need for higher-level intervention to facilitate a population shift towards a stronger organizational focus in psychosocial risk management.

Secular and other trends that may hinder the desired organizational-level focus include the growth of some widely used management practices aiming to maximize productivity, quality, and profitability without adequate consideration of employee and organizational health. These include new systems of work organization, such as "lean production" or "total quality management" (Landsbergis, Cahill and Schnall, 1999), as well as downsizing and restructuring practices (Bambra et al., 2009) which can lead to a deterioration of psychosocial working conditions and associated mental and physical illness. This has taken a particular form in the public sector. New Public Management (NPM) refers to a wave of public sector reforms during the 1980s and 1990s aimed at building more efficient and market-oriented civil services in various OECD countries (Noblet and Rodwell, 2009; Noblet, Rodwell, and McWilliams, 2006). Concerns have been raised regarding the health effects of NPM with studies showing a close relationship between managerialist reforms and increased levels of employee stress and dissatisfaction, as well as declining levels of organizational commitment (Korunka et al., 2003; Mikkelsen, Osgard, and Lovrich, 2000; Young, Worchel, and Woehr,

1998). These outcomes raise further doubts about the long-term sustainability of NPM-style reforms and increase the risk that the inefficiencies associated with the "old" bureaucratic public sector management systems will be replaced by strain-induced inefficiencies in the contemporary paradigm. These and other trends must be acknowledged and their full costs to employees and society revealed in order to support those organizations that resist them.

In some instances, occupational health and safety policy that in theory should be protecting and promoting worker health & safety can paradoxically undermine the desired organizational focus in psychosocial risk management. In Australia, and many other OECD countries, so-called "mental stress" claims continue to rise. Regulatory mandates to prevent and control psychosocial risk can conflict with workers' compensation insurance pressures to contain claims, which to some extent manifests in an over-emphasis on individual explanations of job stress problems, and consequently an over-emphasis on worker-directed intervention in policy and practice (Keegel, Ostry and LaMontagne, 2009; LaMontagne et al., 2008). Over time, this problem may be overcome through the growing legitimacy and acceptance of work-related mental illness and the need for comprehensive psychosocial risk management to prevent and control the problem, as well as by better coordination of occupational health and safety regulatory and workers' compensation policy.

Regulatory or policy intervention can also mandate or facilitate a shift towards a stronger organizational focus in psychosocial risk management. Examples include the UK Health & Safety Executive's Management Standards to address work-related stress (www.hse.gov.uk/stress/standards), and a 2004 framework agreement on work-related stress by major European employers and union federations (www.etuc.org/a/529), nicely articulating the need for both worker- and organization-directed intervention and offering concrete examples of the interdependence of the two (on p. 1):

> Preventing, eliminating or reducing problems of work-related stress can include various measures. These measures can be collective, individual, or both ...
> Such measures could include, for example:
>
> - Management and communication measures such as clarifying the company's objectives and the role of individual workers, ensuring adequate management support for individuals and teams, matching responsibility and control over work, improving work organization and processes, working conditions, and the environment.

More recently, a large European government and researcher collaborative project to develop a Psychosocial RIsk MAnagement – Excellence Framework (PRIMA-EF) has articulated a comprehensive best practice framework for psychosocial risk management in the workplace to the full range of policy and practice stakeholders (see http://primaef.org). PRIMA-EF provides the most comprehensive best practice guidance currently available internationally (Leka and Cox, 2008), and emphasizes an organizational focus in psychosocial risk management (Leka et al., 2008). These and similar policy developments provide valuable external support for developing more comprehensive, evidence-based psychosocial risk management at the organizational level.

Conclusions and future directions

This chapter has provided an overview of various challenges to achieving and retaining an organizational focus in psychosocial risk management. Recognizing and emphasizing the workplace or organization as the most immediate setting for these efforts, we have also described some of the external influences that shape organizational practice, and acknowledged the need to consider the role of external influences in the continuing effort to shift prevalent practice towards a stronger focus on the organizational level. Continuing innovation – optimally through collaborative efforts across the full range of workplace stakeholders – will be required to meet this challenge.

References

Armenakis, A. A., & Bedeian, A. G. (1999). Organizational change: A review of theory and research in the 1990s. *Journal of Management, 25*(3), 293–315.

Aust, B., & Ducki, A. (2004). Comprehensive health promotion interventions at the workplace: experiences with health circles in Germany. *Journal of Occupational Health Psychology, 9*(3), 258–270.

Australian Human Rights Commission. (2010). Workers with Mental Illness: a Practical Guide for Managers (p. 46). Sydney: Australian Human Rights Commission.

Bambra, C., Egan, M., Thomas, S., Petticrew, M., & Whitehead, M. (2007). The psychosocial and health effects of workplace reorganization. 2. A systematic review of task restructuring interventions. *Journal of Epidemiology and Community Health, 61*(12), 1028–1037.

Bambra, C., Gibson, M., Sowden, A. J., Wright, K., Whitehead, M., & Petticrew, M. (2009). Working for health? Evidence from systematic reviews on the effects on health and health inequalities of organisational changes to the psychosocial work environment. *Preventive Medicine, 48*(5), 454–461.

Blewett, V., & Shaw, A. (2008). Future Inquiry: participatory ergonomics at work. *Human Factors in Organizational Design and Management – IX*.

Bordia, P., Hunt, E., Paulsen, N., Tourish, D., & DiFonzo. (2004). Uncertainty during organizational change: Is it all about control? *European Journal of Work and Organizational Psychology, 13*(3), 345–365.

Caulfield, N., Chang, D., Dollard, M. F., & Elshaug, C. (2004). A review of occupational stress interventions in Australia. *International Journal of Stress Management, 11*, 149–166.

Dahl-Jorgensen, C., & Saksvik, P. O. (2005). The impact of two organizational interventions on the health of service sector workers. *International Journal Of Health Services: Planning, Administration, Evaluation, 35*(3), 529–549.

DeJoy, D. M., Wilson, M. G., Vandenberg, R. J., McGrath-Higgins, A. L., & Griffin-Blake, C. S. (2010). Assessing the impact of healthy work organization intervention. *Journal of Occupational & Organizational Psychology, 83*(1), 139–165.

Demmer, H. (1995). Worksite Health Promotion: How to go about it. *European Health Promotion Series No.4*. Copenhagen: World Health Organization (Europe).

Egan, M., Bambra, C., Petticrew, M., & Whitehead, M. (2009). Reviewing evidence on complex social interventions: appraising implementation in systematic reviews of the health effects of organisational-level workplace interventions. *Journal of Epidemiology and Community Health*, *63*(1), 4–11.

Egan, M., Bambra, C., Thomas, S., Petticrew, M., Whitehead, M., & Thomson, H. (2007). The psychosocial and health effects of workplace reorganization. 1. A systematic review of organisational-level interventions that aim to increase employee control. *Journal of Epidemiology and Community Health*, *61*(11), 945–954.

Elo, A. L., Ervasti, J., Kuosma, E., & Mattila, P. (2008). Evaluation of an organizational stress management program in a municipal public works organization. *Journal of Occupational Health Psychology*, *13*(1), 10–23.

Fernandez, S., & Rainey, H. G. (2006). Managing successful organizational change in the public sector. *Public Administration Review*, *66*(2), 168–176.

French, W., Bell, C., & Zawacki, R. (2005). *Organization Development and Transformation: Managing Effective Change*. New York: McGraw-Hill Irwin.

Giga, S. I., Noblet, A., Faragher, B., & Cooper, C. (2003). Organisational stress management interventions: A review of UK-based research. *The Australian Psychologist*, *38*(2), 158–164.

Harden, A., Peersman, G., Oliver, S., Mauthner, M., & Oakley, A. (1999). A systematic review of the effectiveness of health promotion interventions in the workplace. *Occupational Medicine*, *49*(8), 540–548.

Holt, D., Armenakis, A., Feild, H., & Harris, S. (2007). Readiness for organizational change. *The Journal of Applied Behavioral Science*, *43*(2), 232–255.

Hurrell, J. J. J., & Murphy, L. R. (1996). Occupational stress intervention. *American Journal of Industrial Medecine*, *29*(4), 338–341.

Ivancevich, J., Konopaske, R., & Matteson, M. (2011). *Organizational behavior and management* (9th Ed ed.). Boston: McGraw-Hill.

Jick, T. (2003). Implementing change. In T. Dick & M. Peiperl (Eds.), *Managing Change: Cases and Concepts* (pp. 174–219). New York: McGraw-Hill.

Karasek, R., & Theorell, T. (1990). *Healthy Work: Stress, Productivity, and the Reconstruction of Working Life*. New York: Basic Books, Inc., Publishers.

Keegel, T., Ostry, A., & LaMontagne, A. D. (2009). Job strain exposures versus stress-related Workers Compensation claims in Victoria (Australia): Developing a public health response to job stress. *Journal of Public Health Policy*, *30*, 17–39.

Kelloway, E. K., & Barling, J. (2010). Leadership development as an intervention in occupational health psychology. *Work & Stress*, *24*(3), 260–279.

Kompier, M. A., Aust, B., van den Berg, A. M., & Siegrist, J. (2000). Stress prevention in bus drivers: evaluation of 13 natural experiments. *Journal of Occupational Health Psychology*, *5*(1), 11–31.

Korunka, C., Scharitzer, D., Carayons, P., & Sainfort, F. (2003). Employee strain and job satisfaction related to an implementation of quality in a public service organization: A longitudinal study. *Work & Stress*, *17*(1), 52–72.

Kotter, J. P. & Schlesinger L. A. (1979). Choosing strategies for change. *Harvard Business Review*, *59*(2), 106–114.

Kotter, J. (1995). Leading Change: Why Transformation Efforts Fail. *Harvard Business Review, March–April*, 59–67.

Kristensen, T. S., Hannerz, H., Høgh, A., & Borg, V. (2005). The Copenhagen Psychosocial Questionnaire (COPSOQ). A tool for the assessment and improvement of the psychosocial work environment. *Scandinavian Journal of Work, Environment & Health, 31*, 438–449.

LaMontagne, A. D. (2010). Precarious employment: Adding a health inequalities perspective. [Commentary]. *Journal of Public Health Policy, 31*, 312–317.

LaMontagne, A. D., Keegel, T., Louie, A. M., Ostry, A., & Landsbergis, P. A. (2007). A systematic review of the job stress intervention evaluation literature: 1990–2005. *International Journal of Occupational & Environmental Health, 13*(3), 268–280.

LaMontagne, A. D., Keegel, T., & Vallance, D. A. (2007). Protecting & promoting mental health in the workplace: Developing a systems approach to job stress. *Health Promotion Journal of Australia, 18*(3), 221–228.

LaMontagne, A. D., Keegel, T., Vallance, D. A., Ostry, A., & Wolfe, R. (2008). Job strain – Attributable depression in a sample of working Australians: Assessing the contribution to health inequalities. *BMC Public Health, 8*(181), 9.

LaMontagne, A. D., & Keegel, T. G. (2011). Reducing Stress in the workplace (An Evidence Review: Full Report). Melbourne: Victorian Heath Promotion Foundation, 52 pages.www.vichealth.vic.gov.au/workplace.

LaMontagne, A. D., Louie, A., Keegel, T., Ostry, A., & Shaw, A. (2006). Workplace Stress in Victoria: Developing a Systems Approach. Melbourne: Victorian Health Promotion Foundation, 43 pages. www.vichealth.vic.gov.au/workplacestress (accessed January, 2012).

LaMontagne, A. D., Sanderson, K., & Cocker, F. (2010). Estimating the Economic Benefits of Eliminating Job Strain as a Risk Factor for Depression. Melbourne: Victorian Heath Promotion Foundation (VicHealth). See www.vichealth.vic.gov.au/jobstrain (accessed December 22, 2011).

LaMontagne, A. D., Smith, P. M., Louie, A. M., Quinlan, M., Ostry, A. S., & Shoveller, J. (2012). Psychosocial and other working conditions: Variation by employment arrangement in a sample of working Australians. *American Journal of Industrial Medicine, 55*, 93–106.

Landsbergis, P. A. (2009). Interventions to reduce job stress and improve work organization and worker health. In P. Schnall, M. Dobson, E. Rosskam, D. Gordon, P. A. Landsbergis & D. Baker (Eds.), *Unhealthy Work: Causes, Consequences, Cures* (pp. 193–209). Amityville, NY: Baywood Publishing.

Landsbergis, P. A. (2010). Assessing the contribution of working conditions to socioeconomic disparities in health: A commentary. *American Journal of Industrial Medecine, 53*(2), 95–103.

Landsbergis, P. A., Cahill, J., & Schnall, P. (1999). The impact of lean production and related new systems of work organization on worker health. *Journal of Occupational Health Psychology., 4*(2), 108–130.

Leka, S., & Cox, T. (Eds.). (2008). *PRIMA-EF: Guidance on the European Framework for Psychosocial Risk Management*. Geneva: WHO.

Leka, S., Vartia, M., Hassard, J., Pahkin, K., Sutela, S., Cox, T., and Lindstrom, K (2008). Best practice in interventions for the prevention and management of work-related stress and workplace violence and bullying. In S. Leka, Cox, T. (Ed.), *The European Framework for Psychosocial Risk Management: PRIMA-EF* (pp. 136–173). Nottingham, UK: Institute of Work, Health and Organizations.

Lewin, K. (1947). Frontiers in group dynamics. *Human Relations, 1*(1), 5–41.

Maes, S., Verhoeven, C., Kittel, F., & Scholten, H. (1998). Effects of a Dutch wellness-health program: The Brabantia Project. *American Journal of Public Health, 88*(7), 1037–1041.

Mikkelsen, A., Osgard, T., & Lovrich, N. (2000). Modeling the effects of organizational setting and individual coping style on employees subjective health, job satisfaction and commitment. *Public Administration Quarterly, 24*(3), 371–397.

Mikkelsen, A., Saksvik, P. ø., & Landsbergis, P. (2000). The impact of a participatory organizational intervention on job stress in community health care institutions. *Work & Stress, 14*(2), 156–170.

Murta, S., Sanderson, K., & Oldenburg, B. (2007). Process evaluation in occupational stress management programs. *American Journal of Health Promotion, 21*(4), 248–254.

Nielsen, K., Fredslund, H., Christensen, K., & Albertsen, K. (2006). Success or failure? Interpreting and understanding the impact of interventions in four similar worksites. *Work & Stress, 20*(3), 272–287.

Nielsen, K., & Randall, R. (2008). The effects of transformational leadership on followers' perceived work characteristics and psychological well-being: A longitudinal study. *Work & Stress, 22*(1), 16–32.

Nielsen, K., & Randall, R. (2009). Managers' active support when implementing teams: The impact of employee well-being. *Applied Psychology: Health and Well-being, 1*, 374–390.

Nielsen, K., Randall, R., Holten, A., & Gonzalez, E. R. (2010). Conducting organizational-level occupational health interventions: What works? *Work & Stress, 24*(3), 234–259.

Nielsen, K., Taris, T. W., & Cox, T. (2010). The future of organizational interventions: Addressing the challenges of today's organizations. *Work & Stress, 24*(3), 219–233.

NIOSH. (2002). *The Changing Organization of Work and the Safety and Health of Working People* (pp. 43). Cincinnati, Ohio: US Centers for Disease Control/National Institute for Occupational Safety & Health.

Noblet, A., & LaMontagne, A. D. (2006). The role of workplace health promotion in addressing job stress. *Health Promotion International, 21*(4), 346–353.

Noblet, A., & LaMontagne, A. D. (2009). The challenges of developing, implementing, and evaluating interventions. In S. Cartwright & C. L. Cooper (Eds.), *The Oxford Handbook Of Organizational Wellbeing* (pp. 466–496). Oxford: Oxford University Press.

Noblet, A. J., & Rodwell, J. J. (2009). Integrating Job Stress and Social Exchange Theories to Predict Employee Strain in Reformed Public Sector Organizations. *Journal of Public Administration Research and Theory 19*(3), 555–578.

Noblet, A. J., Rodwell, J. J., & McWilliams, J. (2006). Organisational Change in the Public Sector: Augmenting the Demand Control Model to Predict Employee Outcomes Under New Public Management. *Work & Stress, 20*(4), 335–352.

O'Brien, G. (2002). Participation as the key to successful change – a public sector case study. *Leadership and Organization Development Journal, 23*(8), 442–455.

Oreg, S. (2006). Personality, context, and resistance to organizational change. *European Journal of Work and Organizational Psychology, 15*(1), 73–101.

Parker, S., & Williams, H. (2001). *Effective Teamworking: Reducing the Psychosocial Risks.* Norwich: HSE Books.

Polanyi, M. F., Cole, D. C., Ferrier, S. E., & Facey, M. (2005). Paddling upstream: a contextual analysis of implementation of a workplace ergonomic policy at a large newspaper. *Applied Ergonomics, 36*(2), 231–239.

Sagie, A., & Koslowsky, M. (1996). Decision type, organisational control, and acceptance of change: An integrative approach to participative decision making. *Applied Psychology, 45*(1), 85–92.

Saksvik, P. O., Nytrø, K., Dahl-Jorgensen, C., & Mikkelsen, A. (2002). A process evaluation of individual and organizational occupational stress and health interventions. *Work & Stress, 16*(1), 37–57.

Semmer, N. K. (2006). Job stress interventions and the organization of work. *Scandinavian Journal of Work, Environment & Health, 32*(6), 515–527.

Sorensen, G., Stoddard, A., LaMontagne, A. D., Hunt, M. K., Emmons, K., Youngstrom, R., McLellan, D., and Christiani, D.C. (2002). A comprehensive worksite cancer prevention intervention: Behavior change results from a randomized controlled trial in manufacturing worksites. *Cancer Causes Control, 13*(6), 493–502.

Waddell, D., Cummings, T., & Worley, C. (2004). *Organization Development and Change.* Melbourne: Thomson Learning Australia.

Young, B. S., Worchel, S., & Woehr, D. J. (1998). Organizational commitment among public service employees. *Public Personnel Management, 27*(3), 339–348.

3 Taking a multi-faceted, multi-level, and integrated perspective for addressing psychosocial issues at the workplace

Nadine Mellor, Maria Karanika-Murray, and Eleanor Waite

Increasing levels of common mental health problems and musculoskeletal disorders are the major causes of employee sickness absence in Britain (Black, 2008). These trends are not dissimilar to those in other developed countries and have become a challenge in today's workplaces. Tackling these problems requires a more comprehensive approach to the management of psychological risks to workers' health than a traditional approach can offer. In this chapter, we discuss what a comprehensive perspective may encompass, and endorse a multi-faceted, multi-level, positive, and integrated approach. First, we examine recent research into health and well-being that suggests that job characteristics should be considered from a multi-faceted perspective (interplay between work factors and individual differences) and that processes that impinge on workers' health at multiple levels (i.e. individual, group, and organization) should be taken into account. Second, we discuss a facet of job characteristics which has been overlooked in the literature, that of the positive job features that foster individuals' mental capital and in turn empower individuals to protect and maintain their health and well-being. Third, given that many diseases (i.e. high blood pressure, diabetes, coronary heart disease) are directly linked to lifestyle factors (i.e. smoking, drinking and obesity) (Black, 2008), we advocate that the management of psychosocial risks needs to be integrated with the promotion of health-related behaviours. This integration has also been recommended by the World Health Organization (WHO), the International Labour Organization (ILO), and the European Agency for Safety & Health at Work (EU-OSHA). The success of this comprehensive approach is dependent upon several conditions, one of which is the development of an occupational health climate. We conclude this chapter by considering some of the necessary components of such a climate.

Multi-faceted perspective

Psychosocial factors at work, according to the ILO (1986), refer to interactions between the content of a job, the work organization and its management, and other environmental and organizational conditions, on the one hand, and employees' competencies and needs on the other. Psychosocial hazards are those interactions that may prove to have a hazardous influence on employees' health through their perceptions and experience (ILO, 1986). A more detailed definition given by Sauter et al. (1998) views psychosocial factors as being aspects of the job and work environment such as organizational climate or culture, work roles, interpersonal relationships at work, and the design and content of tasks (e.g. variety, meaning, scope, repetitiveness, etc.). The concept of psychosocial factors extends also to the extra-organizational environment (e.g. domestic demands) and aspects of the individual (e.g. personality and attitudes), which may influence the development of stress at work (Sauter et al., 1998).

Over the last four decades, a number of models have been proposed to explain how the organization and the person's characteristics and the interaction between the two can contribute to employee health and organizational outcomes such as productivity and job satisfaction (e.g. Cooper, 2000; Cox, 1993). Models vary in their breadth (e.g. number of work factors assessed) and the functional relationships (e.g. main effects, interactions, non-linear relationships) between constructs (Kelloway and Day, 2005). They also vary in terms of the primacy they give to sources of stress at the workplace. Stress models, such as Karasek and Theorell's (1990) Job Demand-Control model give primacy to work-related psychosocial factors in determining stress and ill-health outcomes (Harenstam, 2008). In contrast, other models (e.g. Hurrell and Murphy, 1992) have emphasized the role of individual differences and contextual factors (termed stress moderators) influencing the effects of job stressors on health and well-being (Sauter et al., 1998). Emphasis is placed on individuals' perceptions of their job and the coping strategies or resources they use to face the demands of their jobs.

Although both job and individual characteristics are important, in practice and policy stress tends to be operationalized in terms of the presence or absence of potentially adverse work characteristics such as job demands or control (e.g. National Institute for Occupational Safety & Health in the USA, Health & Safety Executive in Great Britain; Sauter et al., 1998). Organizational stress surveys do not include individual characteristics except socio-demographic factors (e.g. gender, grade, job tenure, etc.). The principles of primary prevention are applied to psychosocial stressors in the workplace and, in this regard, are consistent with a public health model of prevention where the first line of defence in this approach is to eradicate or reduce exposure to environmental pathogens (Sauter et al., 1998). There may be practical reasons for a sole focus on work factors such as length of survey, preference for simple analyses and non-intrusive questions. This can lead to overemphasizing primary prevention at the organizational level (i.e. work factors), and overlooking primary prevention at the individual level (i.e. personality). Overall, this approach gives a broad indication of the work issues but may lack specific information about what needs to be changed in the work environment to answer individuals' needs and enhance employee well-being.

The neglect of individual differences can have important implications. Quick et al. (1997) note that the stress response has a pivotal role in the organizational stressor-individual distress connection, adding that there are a variety of work stressors that will have no impact on some individuals, but can have major impact on others. Several modifiers of the stress response (e.g. cognitive appraisal, social support, anger, self-reliance) have been shown to influence individual vulnerability and the potential for the stress response to lead to healthy or unhealthy consequences (Campbell-Quick et al., 1997). In particular, recent research in job crafting (Wrzesniewski and Dutton, 2001) has moved beyond this approach, looking at how individuals appraise and respond to work events on a daily basis over a week or longer and the effect this has on well-being. With the use of experience-sampling methodologies (e.g. diaries), individuals can record work events and experiences, feelings and behaviours as they unfold in the work environment, making it possible to analyze fluctuations in momentary or daily states of well-being and the role of discrete situational influences as opposed to stable job features (Ilies, Schwind, and Heller, 2007).

There is also research to show that dynamic events provide richer explanations of an individual's work experience and that individuals shape their jobs by constant interpretations and decisions they make (Daniels, 2011). It is by playing an active role in shaping their job that individuals can maintain their well-being. Daniels and Harris (2005) demonstrated the importance of workers' control over their job in choosing their breaks, modifying their ways of working, solving their own problems, as well as having sufficient resources and support available to them. Thus, providing employees with job autonomy and support at the organizational or unit level may be important but not sufficient. For maximum benefit, workplace characteristics should be tailored to individual needs. Different individuals may require different levels and types of support. Some may need to be equipped with the necessary skills in order to solve problems, others may need training or help to cope with stressors, develop resilience, adapt to their work environment, and regulate their well-being effectively. Thus, combining organizational primary prevention with individual primary prevention is necessary.

Inevitably, other types of interventions will be needed to enhance employee well-being such as secondary or tertiary level interventions. The former attempts to modify the effects of existing causal factors by equipping employees with skills and resources, whilst the latter assists individuals in coping better with stressful demands or help in the rehabilitation of those suffering from ill-health. Despite evidence that some types of organizational interventions can be effective in reducing stressors, they may not always be the most effective interventions to enhance well-being. La Montagne et al.'s (2007) review of organizational and individual-level interventions suggests that superior results can be expected from combined types of interventions over a single type.

Other research shows the importance attributed to goals towards which individuals strive and their effects on well-being (Diener et al., 1999). Harris and Daniels (2007) found that individuals reported more unpleasant effects and lower job satisfaction when they believed that work demands impact adversely on achieving their goals. This suggests that if changing job characteristics is the sole consideration, with no attention to how people appraise their work in relation to their personal goals and well-being, this may lead to designing well-being interventions that suit specific groups who appraise job characteristics in the same way, but do not suit the needs of others who may pursue different goals or appraise job characteristics in a different way (Daniels, 2011). Altogether, these findings suggest that whilst annual stress audits or surveys may be useful to monitor broad health and stress trends over time (Quick, Campbell-Quick, and Nelson, 2000), they need to be supplemented by other forms of assessments capturing ongoing idiosyncratic needs at team and individual level. This implies that the role and competencies of line managers in closely monitoring employee health and wellbeing are essential (e.g. Donaldson-Feilder, Yarker, and Lewis, 2008; Lewis, Yarker, and Donaldson-Feilder, this volume).

Multi-level perspective

Whilst it is important to assess how individuals appraise their work environment and cope with it, it is also recommended to adopt a multi-level assessment of processes occurring not only at the individual level, but also at the team, divisional, corporate, and extra-organizational level. A multi-level perspective is grounded in the theoretical proposition that higher-level factors (e.g. group cohesion, group leadership, group collective efficacy, work demands) have a measurable and important impact on individual actions and perceptions (Bliese and Jex, 2002). Such perspective can help to identify the steps organizational actors need to take individually and collectively to yield organizational benefits (Klein, Tosi, and Cannella, 1999). Individual strains are partly a function of one's group membership. For example, individuals within groups with a supportive leadership climate tend to respond differently to stressors than those with an unsupportive leadership climate (Bliese and Jex, 2002). The wider organizational context is also salient because organizational policies, procedures, practices, culture, change, and structural characteristics have been shown to influence individuals' perceptions (Youssef and Luthans, 2010). For example, frequent changes may result in perceived risk and increased mobilization of adaptive mechanisms at the individual level (Youssef and Luthans, 2010). Whilst individuals differ in the way they interpret these factors, there is some consistency on how they perceive and react to their shared environments (Bliese and Jex, 1999).

Harenstam (2008) discusses how a multi-level approach applied to the job demand-control model (Karasek and Theorell, 1999) can identify whether the determinants of demands and control are rooted in organizational conditions or in factors related to the person. The author questions the accuracy of the influential job demand-control (JD-C) model in today's worklife. The JD-C model predicts that high demands combined with low decision authority and few opportunities for development constitute a work situation hazardous to employee health. However, many studies found that active work situations with high demands and high control also relate to ill-health (Harenstam, 2008). To better understand this phenomenon, research has continued to investigate the association of these work attributes with employee health. Compared to 30 years ago when the JD-C model was established, work conditions have dramatically changed. In modern work environments job demands may be more internalized in organizational conditions than related to the individual, and employees may no longer attempt to regulate these demands even if they have the influence to do so (Harenstam, 2008).

Using data from 51 organizations and 3,485 employees, Bolin, Marklund and Bliese (2008) found that 12 percent of the variance in job demands and 20 percent of variance in job control could be attributed to the organizational level. Employees were asked to assess aspects of their work (e.g. workload, time pressure, decision latitude). Data were aggregated at three levels: company level, worksite and department level. All three levels had an impact on job demands and control, but job demands were most affected by the worksite level, whereas control was most affected by the sub-unit (departmental level). One explanation for the findings regarding control may be that supervisors' management style has a crucial influence over employees' work. These findings show that the organization matters for both variables and the degree of organizational impact differs between the two. In another longitudinal multi-level study, Labriola et al. (2006), using data from 52 workplaces, found that long-term sickness absence was predicted by physical work factors (e.g. pushing/ pulling heavy loads, lifting more than 30 kg) at the individual level and by psychosocial work factors (management quality, work demands, control, and support) at the organizational level, concluding that concurrent interventions at both levels are needed to reduce absenteeism. In the same vein, Bond, Flaxman and Bunce (2008) suggest that training individuals on psychological flexibility, a type of individual intervention, may improve the benefits employees receive from an organizational intervention. Specifically, employees high in cognitive flexibility reported more job control as a consequence of organizational interventions aimed at improving employee health. Furthermore, this increase in job control was related to a reduction in employee absenteeism and improvements in mental health. Overall, the results from Bond, Flaxman and Bunce (2008) suggest that organizational interventions can have a positive impact on individual perceptions of their environment, which in turn may influence well-being outcomes.

Thus, a multi-level perspective can enable researchers and practitioners to build a comprehensive picture of work-related health issues. Although in many situations, simple group mean analyses may be appropriate and sufficient for implementing interventions (Bliese and Jex, 1999), a multi-level approach encompasses additional factors related to employee health and allows for a more complete understanding of the interplay of additional dynamics that affect well-being in the workplace.

Positive perspective: stress vs. well-being

It has been acknowledged (e.g. Carson and Barling, 2008; ILO, 1986) that psychosocial aspects of work are mainly investigated in the context of stress evaluations, and that they tend to focus on adverse effects such as emotional disturbances, behavioural problems, and ill-health. Hence, psychosocial factors at work are typically studied with a focus on negative outcomes. However, their potential positive influences on health and other aspects of life merit increased attention

(ILO, 1986). Positive psychology and research on subjective well-being have made significant advances in the way the world of work can bring positive benefits to individuals' health. Particular organizational features may enable positive outcomes in the workplace above and beyond removing known stressors (Turner, Barling, and Zacharatos, 2002). Some scholars (e.g. Kelloway, Teed, and Kelley, 2008) have called for the use of countervailing interventions to reduce stress at work, defined as interventions that increase the positive experience of work rather than solely reducing its negative aspects.

Although often used interchangeably, stress and well-being refer to different things. On one hand, "stress" is grounded in the stressor-strain theory or the *deficit* approach, which predicts that a range of unfavourable psychosocial work characteristics such as workload, a lack of autonomy or long hours will lead to lower levels of physical and psychosocial health and poor organizational outcomes such as higher absenteeism and lower productivity (e.g. Cox et al., 2000; Karasek and Theorell, 1990). On the other hand, "well-being" relates to the emotional and cognitive health of the individual. Two philosophical traditions, the hedonic tradition and the eudemonic tradition, have attempted to define and conceptualize well-being. The hedonic tradition focuses on *subjective affective* well-being, which is often equated with happiness, pleasure attainment, and pain avoidance (Ryan and Deci, 2001). Diener, Lucas, and Oishi (2002) define subjective well-being as individuals experiencing pleasant emotions, low levels of negative moods, and high life satisfaction. The eudemonic tradition purports that well-being is more than just happiness and needs to include meaning and the self-realization of human potential (Ryan and Deci, 2001). It emphasizes optimal functioning and personal expressiveness (Waterman, 1993). The term *psychological* well-being is used to qualify this more cognitive-orientated aspect. It includes making progress and striving towards long-term goals or states such as self-acceptance, personal growth, purpose in life, positive relations with others, environmental mastery, and autonomy (Ryff and Keyes, 1995).

Research contrasts the type of workplace characteristics and person characteristics that predict stress versus well-being. Well-being is not only related to the absence of negative factors at work, but also to the presence of positive aspects of the work experience (Cotton and Hart, 2003; Luthans et al., 2007). Improving organizational climate, for instance, rather than solely reducing the impact of adverse work conditions, and specifically improving employees' perceptions of leadership and management practices, clarity of roles, and goal alignment, has a stronger impact than merely reducing the presence of stressors. Several organizational behaviour theories, such as job satisfaction, organizational commitment, organizational citizenship behaviours, motivation, leadership, organizational justice, and social responsibility (Youssef and Luthans, 2010) emphasize the integral role that positive aspects of the work environment play in employee health. Incorporating these theories into the domain of positive psychology and employee well-being may provide organizations with more of the knowledge they need to successfully design and implement interventions aimed at improving employee health.

Positive work factors vs. psychosocial risks and the development of personal resources

Well-being is important in the management of psychosocial risks and process issues in organizational interventions. Developing individuals' well-being could potentially help enhance the organizational climate before, during and after a work-related intervention by reducing resistance to change. It is within a supportive organizational context with optimal work conditions and positive job features that individuals can develop personal resources and "psychological capital" (Luthans et al., 2007). Psychological resource theory suggests that constructs such as confidence, optimism, hope and resilience are personal resources or energies that can affect the actions and behaviours of individuals (Hobfoll, 2002). The concept of "psychological capital" encompasses a person's cognitive and emotional resources. It can comprise hope, self-efficacy, resilience, optimism, proactivity, and mindfulness (Foresight Mental Capital and Wellbeing project, 2008; Fullager and Holloway, 2009; Luthans et al., 2007). It includes individuals' cognitive ability, defined as how flexible and efficient they are at learning, and their "emotional intelligence," which encompasses social skills and resilience in the face of stress. These two characteristics provide estimates of an individual's ability to contribute effectively to society, and their appraisal of the personal quality of their life (Foresight Mental Capital and Wellbeing Project, 2008). Youssef and Luthans (2010) suggest that for individuals to develop self-efficacy, which has been shown to correlate with performance (Stajkovic and Luthans, 1998), the organization needs to provide its employees with training opportunities to enhance task mastery. Additionally, managers need to be role models, acknowledge employee performance, provide constructive feedback and emphasize the importance of physical and psychological well-being. Hope, another facet of psychological capital, can be developed through goal-setting, empowerment, contingency planning and motivational approaches (Youssef and Luthans, 2010). Therefore, addressing psychosocial risks in the workplace (i.e. making the work environment, and leadership more effective) may help employees to reach their potential, develop creativity, and find meaning in their work.

It is important to emphasize that personal resources are particularly necessary in situations where symptoms of stress have become chronic, and are unlikely to change even when the conditions that produced them have changed. In such instances, personalized interventions may be needed (Semmer, 2006). However, personal resources should not be viewed as needed only during times of adversity. As explained earlier, the organizational context with the provision of optimal work conditions is a means to nurture these resources on a continuous basis.

Moreover, research has highlighted the role of positive emotions in reducing stress, fostering creativity, and flexibility in thinking and problem solving (Folkman and Moskowitz, 2000). Under stressful conditions when negative emotions are predominant, positive emotions may provide a psychological respite, support continued coping efforts, and replenish resources that have been depleted by stress. Positive affect is also influenced and sustained by certain coping strategies including positive reappraisal, goal-directed problem-focused coping, proactive coping, and the infusion of ordinary events with positive meaning (Folkman and Moskowitz, 2000; Dewe and Kompier, 2008). The person who copes proactively is one who strives for improvement of work or life and builds up resources that assure progress and quality of functioning (Schwarzer, 2001). As noted by Jenkins et al. (2008), psychological factors that contribute to poor mental health are poorly developed coping skills and low self-esteem. Therefore, training employees on effective coping skills can be essential but should be offered only as a complement to comprehensive organizational interventions.

Integrating risk management with health promotion

The suggestion that psychosocial risks need to be tackled alongside health-related behaviour risks stems from the recognition that ill-health is not only caused by poor work conditions, but also by non-occupational factors such as smoking, unhealthy lifestyle, or poor diet (WHO, 2010). Addressing both individual and work-related risks simultaneously is increasingly recognized as an effective way of tackling psychosocial risks at the workplace. Several policy bodies (e.g. WHO, NIOSH, ILO, EU-OSHA) call for the consideration of health promotion at the workplace to reduce health-related risks. The influential review *The Health of Britain's Working Age Population: Working for a Healthier Tomorrow* (Black, 2008) encourages employers to go beyond compliance with health and safety and other relevant legislation and extend their agenda to promoting aspects of healthy lifestyles.

In a review of the evidence for addressing both psychosocial risks and health-related behaviour risks concurrently, Sorensen and Barbeau (2004) comment that workers' risk of disease is increased by exposures to both occupational hazards and risk-related behaviours; the workers at highest risks for exposure to hazardous working conditions are also those most likely to engage in risk-related health behaviours. Blue collar workers or workers in technical or routine occupations are more susceptible to injury and illness than individuals with professional jobs. Additionally, those exposed to job hazards are more likely to smoke and have unhealthy dietary habits than their unexposed counterparts.

Evidence indicating that improvements in physical health can improve mental health is starting to accumulate (Martin, Sanderson, and Cocker, 2009). For example, weight loss or physical activity can reduce depression and anxiety, although their effects are relatively small (Martin, Sanderson, and Cocker, 2009). The reverse is also true, such that mental health affects physical health (Goetzel and Ozminkowski, 2008). For example, work-related stress may have implications for immune functioning and physical stress symptoms (Denson, Spanovic, and Miller, 2009). Additionally, higher levels of well-being (defined as positive traits or positive emotions) have been associated with greater immune system response and pain tolerance (Howell, Kern, and Lyubomirsky, 2007) and with reduced hypothalamic-pituitary-adrenal reactivity (Chida and Hamer, 2008). Therefore, the management of psychosocial issues in the workplace is not an either-or decision regarding where to best direct resources and efforts. A holistic approach of employee health addressing multiple issues and levels is needed.

Thus, in practice, a comprehensive approach to employee health and well-being requires the integration of three broad inter-related components (Dejoy and Southern, 1993; Weiner, Lewis, and Linnan, 2009):

1   *The Occupational Safety and Health (OSH) system*, which falls under the traditional regulatory health and safety requirements. The system is preventive in nature and consists of carrying out regular risk assessments and managing physical and psychosocial work conditions.
2   *Worksite health promotion*, which focuses on health awareness, lifestyles changes and supportive environments.
3   *The management of ill-health*, which takes the form of health care services (e.g. employee assistance programmes) or disability or absence management.

Added to the complexity of integrating these three components is their interrelation with other internal organizational systems, such as human resources policies and procedures (e.g. recruitment, flexible working, learning and development, employee performance, career development, etc).

Integrating health promotion with a traditional health and safety system is a relatively new process, and a lack of cross-fertilization between the two has been observed (Dejoy and Southern, 1993). On the one hand, health promotion focuses on educational and individual health behaviour changes, whereas, on the other hand, an OSH system focuses primarily on changing organizational factors by identifying the physical, chemical and psychosocial hazards to protect workers' health (Dejoy and Southern, 1993). However, the acknowledgement in both fields that multi-level interventions (individual and organizationally focused) are needed has shifted the focus towards more inclusive approaches. Integration of health promotion and occupational health and safety can increase programme attendance for high-risk workers. Workers who reported that employers had made changes to reduce hazardous exposures on the job were more likely to have participated in smoking cessation and nutrition programmes than workers not reporting management changes (Sorensen and Barbeau, 2004). The authors further explain that employer efforts to create a safe and healthy work environment may foster a climate of trust, and, thereby, enhance workers' receptivity to messages on health behaviour change.

Recent EU guidance argues for an integrated management system of OSH including safety, work-related health and health promotion, and for mainstreaming such a management system into normal business practice. The ILO (www.ilo.org/safework/areasofwork/lang–en/WCMS_108331/index.htm) also suggests that health promotion at the workplace is effective when it is integrated into OSH management systems to prevent accidents and diseases and to protect workers' health. It is outlined that an integrated system "cannot be implemented by taking the different management systems already in place and including them in a single set of manuals; integration must be carefully planned and implemented in a balanced way" (Smith, 2001, p. 31).

Integration in practice: the challenges

The importance organizations place on work-related health and well-being vary, and as a consequence, so do the resources attributed to it. Some take a reactive approach, focusing on the management of risks, whereas others manage OSH more systematically and proactively (EU-OSHA, 2010). They range from a minimal commitment observing minimal responsibility and legal requirements, to considerable investment and a more proactive approach to employee well-being. The latter is the ideal scenario for sustainable management of work-related health, as it ensures that the conditions in the organizational context are in place to support relevant initiatives. However, not all organizations invest the necessary knowledge, skills, and resources into facilitating a proactive approach to work-related health and well-being. Such a proactive approach depends on the decision-makers' attitudes and senior management's commitment to work-related health and well-being programmes.

Mainstreaming the management of work-related health and well-being into normal business practice can help both the resource-poor and the resource-rich organizations by ensuring that the essential context is in place to support interventions targeting employee health and well-being. This concept is not new, as many legal frameworks that support systematic OSH management highlight the importance of prevention, integration, and participation (although sufficient guidance as to how integration can be achieved is lacking). Case study examples are available where OSH management has been integrated within quality and/or environmental management systems, as an element of total quality management, into overall management, or as an integral part of HR management (EU-OSHA, 2010, p. 166).

Organizational practice suggests that cooperation between departments and professionals, and senior management involvement are key for implementing corporate health programmes. A study by Jordan et al. (2003) of UK organizations who implemented a comprehensive stress prevention strategy shows that those organizations who adopted a coordinated approach making use of expertise in three key areas – Occupational Health, Human Resources, and Health and Safety – were far more likely to exhibit a coherent strategy than those wherein stress prevention was considered to be the responsibility of only one of these fields. Furthermore, with the presence of senior management commitment, strategies tended to be clearer and more effective when top management figures were involved in their formulation (Blaug, Kenyon, and Lekhi, 2007). OSH professionals are viewed as having a key role in leading the OSH changes but sufficient authority needs to be granted to them for them to enact these changes (Swuste and Arnoldy, 2003). They need to understand organizational vision/strategy (Birkner and Birkner, 1999), use managers' language (Deacon, 1998; Gregory, Lukes, and Gregory, 2002), cultivate strong relationships with managers, present strong business cases, and think strategically (Deacon, 1998; EU-OSHA, 2010, p. 40).

Developing an occupational health climate

Underlying any management system is the psychological fabric that makes individuals trust, adhere to, or reject any organizational arrangements available to them. With regard to a health management system, research on climate can help shed light on how an occupational health climate can be initiated and maintained. Managing workplace health and establishing a positive occupational health climate includes managing psychosocial risks, as described above, with attention given to the provision of optimal job features. It will also require a certain level of consistency between employee changes in behaviour and a change in organizational culture. In the field of workplace safety, researchers have suggested combining individual changes with culture changes, which are often viewed as antagonistic (Dejoy, 2005). For example, individuals can be trained to adopt safe behaviours, such as wearing protective devices, and yet imperatives of productivity (at the organizational level) may lead them to ignore basic safety rules and procedures, putting them at risk. Similarly, an individual attending a training course on stress coping strategies may return to a stressful work environment, where no effort is made to adapt the work to the individual, so the new skills may be of no use. In order to achieve a successful health climate, the organization and the employees must align their commitment to health.

Climate stands for "the way we do things around here" (Schneider, 1990) and generally represents employees' shared perceptions regarding the rules, regulations and policies that an organization has in place, as well as the behaviours that get rewarded and supported in the organization (Hofmann and Mark, 2006). For example, hospitals outline procedures for lifting patients. However, adherence to lifting policies is dependant upon the extent to which following the rules to the outlined procedures is encouraged and reinforced. Research has shown that organizations with stronger safety climates have fewer nurse back injuries (Hofmann and Mark, 2006), providing evidence that creating a strong climate allows organizations to successfully emphasize a priority of interest. Much of the research on climate is grounded in social exchange theory (Blau, 1964) and the norm of reciprocity (Gouldner, 1960). A balanced social exchange helps form a quality relationship because each member is motivated to help the other achieve his or her goals. Additionally, exchange relationships are generally characterized by more support, which is an important resource for employees, and has been found to act as a buffer against workplace stressors (Wallace et al., 2009). Together, the theories suggest that employees feel a responsibility to pay back organizations and individuals for the treatment they have received (Eisenberger et al., 1986). The positive outcomes of investing in a climate that promotes employee health are grounded in the idea that if an organization shows commitment to enhancing employee health, employees will feel obligated to repay the organization for the favourable investment the organization has made in them.

When applying the norm of reciprocity to occupational climate research, it is important to consider the complexities that arise as a result of conflicts from competing demands (Zohar, 2002). Developing a successful health climate involves the acknowledgement of these competing demands. For example, if an organization's commitment to employee health suddenly dissipates due to the demand of increased productivity, employees' perception of the priority of health will be affected, thereby decreasing the shared perception of the importance of health in the organization (Zohar, 2010).

With regard to health promotion, research has supported this proposition by showing a positive relationship between discretionary investments in health promotion and workplace safety climate and commitment (Mearns et al., 2010). In one specific study (Mearns et al., 2010), medical personnel working on offshore installations ranked their specific installation on five areas: information regarding the provision of healthy activities, provision of health checks, involvement in health promotion, availability of occupational health training, and the installation's participation in a health promotion rewards scheme. Overall, scores in these areas were used to reflect each installation's commitment to occupational health promotion activities. Installations responding favourably to the questions concerning workplace health promotion investment exhibited a stronger commitment to safety and health climate. Thus, workers value health promotion investment and, as suggested by the norm of reciprocity, feel an inclination to increase their commitment to the organization in return (Mearns et al., 2010).

Evidence of the benefits of investing in employee health promotion programmes (e.g. physical exercise, balanced diet, smoking cessation programmes) has been found in several meta-analyses (e.g. Kuoppala, Lamminpää, and Husman, 2008; Parks and Steelman, 2008). Participation in wellness programmes is associated with reduced sickness absence, increased job satisfaction and increased overall worker well-being and mental health. Overall, these results suggest that fostering a climate that promotes health is important because of the positive outcomes on employees' health and the organizational benefits.

Another tool for creating a strong climate is effective leadership. Leadership's status as an antecedent of strong climate has stood firm for a number of years (e.g. Dragoni, 2005; Kozlowski and Doherty, 1989; Lewin, Lippitt, and White, 1939). First, employees often look to supervisors to gain a firm understanding of organizational goals and priorities. Inconsistent behaviour on the part of the supervisor can signal conflicting priorities to employees, weakening the desired organizational climate (Zohar, 2002). A comprehensive look at leadership dimensions and safety climate by Zohar (2002) indicated that a supervisor's concern for employee well-being as a consequence of engaging in close, personal relationships with employees resulted in adherence to safety practices, creating a stronger safety climate and safer behaviour.

Literature has looked at specific types of leadership to gain a firmer understanding of the characteristics of leaders who have the ability to spread a strong climate throughout the organization. Seeing that social exchange relationships are a tenet of strong organizational climates, high quality leader-subordinate relationships are particularly important for implementing successful climates. Transformational and transactional leadership have received extensive attention in the climate literature (e.g. Skakon et al., 2010; Zohar, 2002). Transformational leadership is characterized by idealized influence, inspirational motivation, intellectual stimulation, and individualized consideration (Bass, 1999). Together, these dimensions create a leader who takes a concern in individual employees, resulting in high quality leader-subordinate relationships. Transactional leadership is conceptualized as contingent reward and management-by-exception (Bass and Riggio, 2006). In transactional leadership, individuals are rewarded for effortful behaviour and are monitored and corrected on their actions, providing them with guidance on the behaviour that the organization expects from them.

A review of 30 years of empirical research (Skakon et al., 2010) found strong consistent support for the significant effects of leadership behaviour and leader-employee relationships on reducing employee stress and enhancing affective health. Specifically, transformational leadership style resulted in less stress and burnout, and greater affective well-being for employees. The results were mixed on the effects of transactional leadership, with only a couple of studies showing a positive relationship with employee health (i.e. Medley and Larochelle, 1995; Sosik and Godshalk, 2000). Literature focusing on safety climate argues that both transformational and transactional leadership qualities are required to create a strong culture. Together, the leadership styles provide the motivation employees need to adhere to the climate, and the guidelines on expected performance within the climate (Zohar, 2002).

Several studies explain why a considering, empowering, and engaging style of leadership has the potential to positively influence employees' psychological well-being. Arnold et al. (2007) and Nielsen et al. (2008) found that the psychological mechanisms linking leadership to well-being were employees' perceptions of their work as meaningful, rewarding and fulfilling, and the ability to see achievements. This link was partly accounted for by the manager's leadership style. Organizations seeking a climate emphasizing health should consider coaching leaders to invest in closer relationships with employees, and to provide meaningful rewards and feedback on performance.

A final way of developing shared cognitions among employees regarding occupational health climate is through increased communication. Once organizational policies and priorities are vocalized, it is up to the employees to determine the espoused vs. enacted policies (Zohar, 2010). Although the organization may highlight specific priorities, it is only enacted behaviours that provide employees with information regarding the type of rewarded and supported behaviours. High levels of communication and interaction give workers the opportunity to identify the true values of the organization. For example, an organization may say they are committed to reducing job demands on an employee in an effort to maintain employee health and mitigate stress, but if tough economic times invade the organization and the response is to cut workers and increase tasks for employees, then this serves as a clear indicator that the vocalized policies of the organization are in conflict with the enacted policies of the organization. Communication of such events allows employees to determine how policies and practices play out in the work environment. Social consensus and sense-making of organizational values is strengthened in an environment with increased opportunity for communication. Without communication, it is impossible for employees to develop a shared view of organizational policies and practices, which is a key ingredient for a strong climate.

Developing an occupational health climate involves some complexities in regard to communication. Health issues are often personal, making disclosure about health concerns more difficult. For example, relaying information about practices related to avoiding needle stick injuries in patients and discussing general safety practices is much less personal than disseminating information about interventions to help employees deal with chronic pain and depression. Corbière and Shen (2006) investigated the effects of psychological interventions for employees returning to work after absence due to mental health problems or physical injuries. Overall, their study supported the positive impact that organizational interventions and increased communication regarding the issue can have on the mental and physical progression of employees re-entering the workforce after time off due to mental health problems or physical injures. Specifically, the review highlighted the importance of relationships with co-workers and supervisors, and suggested that improvements in communication among workers have employee benefits (Nuttman-Sharwtz and Ginsburg, 2002). Previous research on disclosure has suggested how revealing personal information can help develop close relationships, especially when the information is more intimate (Levesque, Steciuk, and Ledley, 2002). Disclosure and communication improvements may signal to employees that the communication is supportive of them and understands their health issues. However, individuals can have fear and distrust in the workplace, which may prevent the development of effective health communication, thereby hindering the development of a health climate. Therefore, an added complexity in creating an occupational health climate is the issue of disclosure of personal health information. Organizations should facilitate this by establishing strong ethical guidelines about reassurance that confidentiality of any medical personal information will be maintained.

There is much room for advancement in research and knowledge within the occupational health climate literature. First, there are several complexities that need to be addressed. For example, research should investigate the best way for organizations to address and communicate health initiatives and for employees to reveal personal health information. Second, a consensus needs to be reached on what policies and practices should be classified as promoting employee health. Investigating what organizations can do to promote employee mental and physical health further will help researchers and practitioners advance the occupational health climate literature.

Conclusions

New research developments have shown what a comprehensive approach to the management of psychosocial risks can encompass. There are multiple ways of reducing the potential adverse effects of psychosocial risks in the workplace. Focusing on changing the unfavourable work features, fostering the positive ones, and equipping the person with adequate skills to deal with adverse conditions is more likely to produce beneficial effects. This proactive approach is in contrast with traditional approaches, which focus on changing only the negative work factors. In order to achieve maximum benefits, organizations should combine interventions in terms of different levels (e.g. the individual and the organization) and foci (e.g. work conditions and health behaviours). An integrated management system that focuses on promoting individual health behaviours and a health climate through initiating changes in both the work environment and employee behaviours is also essential for successfully managing psychosocial issues in the workplace. However, its implementation is not without challenges. Such systems will require a strong occupational health climate with cooperation between departments and various professionals (e.g. line management, human resources, occupational health, health and safety, trade unions) and active senior and line management involvement. Mainstreaming health and safety into normal business practices would also give a better chance of success in promoting employee well-being.

Notes

Correspondence for this chapter should be addressed to: Dr Nadine Mellor, Health & Safety Laboratory, HSL, Harpur Hill, Buxton SK17 9JN, United Kingdom. Email: Nadine.Mellor@hsl.gov.uk

Acknowledgements

Maria Karanika-Murray's contribution to this chapter has been partially supported by the Economic and Social Research Council's First Grant Scheme grant number RES-061-25-0344.

References

Arnold, K. A., Turner, N., Barling, J., Kelloway, E. K., & Mckee, M. C. (2007). Transformational leadership and psychological well-being: The mediating role of meaningful work. *Journal of Occupational Health Psychology*, *12*, 193–203.

Bass, B. (1999). Two decades of research and development in transformational leadership. *European Journal of Work & Organizational Psychology*, *8*, 9–32.

Bass, B., & Riggio, R. (2006). *Transformational Leadership* (2nd ed.). Mahwah: Lawrence Erlbaum.

Birkner, L. R., & Birkner, R. K. (1999). Measuring the value of occupational hygiene and safety. *Occupational Hazards*, *61*(4), 23–24.

Black, C. (2008). C*ross-government Health, Work and Well-being Programme. Working for a Healthier Tomorrow: Dame Carole Black's Review of the Health of Britain's Working Age Population*. London: The Stationery Office.

Blau, P. M. (1964). *Exchange and power in social life*. New York: John Wiley & Sons.

Blaug, R., Kenyon, A., & Lekhi, R. (2007). *Stress at Work*. London: The Work Foundation.

Bliese, P. D., & Jex, M. J. (1999). Incorporating multiple levels of analysis into occupational stress research. *Work and Stress*, *13*(1), 1–6.

Bliese, P. D., & Jex, M. J. (2002). Integrating multilevel analyses and occupational stress theory. *Historical and Current Perspectives on Stress and Health*, *2*, 217–259.

Bolin, M., Marklund, S., & Bliese, P. (2008). Organizational impact on psychosocial working conditions. *Work, 30,* 451–459.

Bond, F. W., Flaxman, P. E., & Bunce, D. (2008). The influence of psychological flexibility on work redesign: Mediated moderation of a work reorganization intervention. *Journal of Applied Psychology, 93,* 645–654.

Campbell-Quick, J. C., Quick, J. D., Nelson, D., & Hurrell, J. (1998). *Preventive Stress Management in Organizations.* Washington DC: American Psychological Association.

Carson, J., & Barling, J. (2008). *Work and well-being.* In J. Barling & C. L. Cooper (Eds.), *The Sage Handbook of Organizational Behavior Volume 1: Micro approaches.* London, UK: Sage Publications.

Chida, Y., & Hamer, M. (2008). Chronic psychosocial factors and acute physiological responses to laboratory-induced stress in healthy populations: A quantitative review of 30 years of investigations. *Psychological Bulletin, 134,* 829–885.

Cooper, C. L. (2000). *Theories of Organizational Stress.* Oxford: University Press.

Corbière, M., & Shen, J. (2006). A systematic review of psychological return-to-work interventions for people with mental health problems and/or physical injuries. *Canadian Journal of Community Mental Health, 25,* 261–288.

Cotton, P., & Hart, P. M. (2003). Occupational well-being and performance: A review of organization health research. *The Australian Psychologist, 38*(2), 118–127

Cox, T. (1993). *Stress research and stress management: Putting theory to work.* Sudbury: HSE Books.

Cox, T., Griffiths, A., Barlow, C. A., Randall, R., Thomson, L., & Rial-González, E. (2000). *Organizational interventions for work stress: a risk management approach.* Sudbury: HSE Books.

Daniels, K. (2011). Stress and well-being are still issues and something still needs to be done: Or why agency and interpretation are important for policy and practice. In G. P. Hodgkinson & J. K. Ford (Eds.), *International review of industrial and organizational psychology* (Vol. 25). Chichester: Wiley.

Daniels, K., & Harris, C. (2005). A daily diary study of coping in the context of the job demands-control-support model. *Journal of Vocational Behavior, 66,* 219–237.

Deacon, S. (1998). The value of health and safety to a successful business. *Health and Safety Bulletin, 272,* 9–15.

DeJoy, D. M. (2005). Behavior change versus culture change: Divergent approaches to managing workplace safety. *Safety Science, 43,* 105–129.

DeJoy, D. M., & Southern, D. J. (1993). An integrative perspective on worksite health promotion. *Journal of Occupational Medicine, 35,* 1221–1230.

Denson, T. F., Spanovic, M., & Miller, N. (2009). Cognitive appraisals and emotions predict cortisol and immune responses: A meta-analysis of acute laboratory social stressors and emotion inductions. *Psychological Bulletin, 135,* 823–853.

Dewe, P., & Kompier, M. (2008). *Foresight Mental Capital and Wellbeing Project. Well-being and work: Future Challenges.* London: The Government Office for Science.

Diener, E., Lucas, R. E., & Oishi, S. (2002). Subjective well-being Part V. In C. R. Snyder & S. J. Lopez (Eds.), *The handbook of positive psychology.* New York: Oxford University Press.

Diener, E., Suh, E. M., Lucas, R. E., & Smith, H. L. (1999). Subjective well-being: Three decades of progress. *Psychological Bulletin, 125,* 276–302.

Donaldson-Feilder, E., Yarker, J., & Lewis, R. (2008). Line management competence: The key to preventing and reducing stress at work. *Strategic HR Review, 7*, 11–16.

Dragoni, L. (2005). Understanding the emergence of state goal-orientation in organizational work groups: The role of leadership and multilevel climate perceptions. *Journal of Applied Psychology, 90*, 1084–1095.

Eisenberger, R., Huntington, R., Hutchinson, S., & Sowa, D. (1986). Perceived organizational support. *Journal of Applied Psychology, 71*, 500–507.

European Agency for Safety & Health at Work. (2007). *Expert forecast on emerging psychosocial risks related to occupational safety and health*. Luxembourg: Office for Official Publications of the European Communities.

European Agency for Safety & Health at Work. (2010). *Mainstreaming OSH into business management*. Luxembourg: Office for Official Publications of the European Communities.

Folkman, S., & Moskowitz, J. (2000). Positive affect and the other side of coping. *American Psychologist, 55*, 647–654.

Foresight Mental Capital and Wellbeing Project. (2008). *Mental capital and well-being: Making the most of ourselves in the 21st century*. Final Project report. London: The Government Office for Science.

Fullager, C. J., & Holloway, E. K. (2009). Flow at work: An experience sampling approach. *Journal of Occupational and Organizational Psychology, 82*, 595–615.

Goetzel, R. Z., & Ozminkowski, R. J. (2008). The health and cost benefits of work site health promotion programs, *Annual Review of Public Health, 29*, 303–323.

Gouldner, A. W. (1960). The norm of reciprocity. *American Sociological Review, 25*, 161–178.

Gregory, J. W., Lukes, E., & Gregory, C. A. (2002). Using financial metrics to prove and communicate value to management: Occupational health nurses as key players on the management team. *AAOHN Journal, 50*(9), 400–405.

Harenstam, A. (2008). Organizational approach to studies of job demands, control and health. *Scandinavian Journal of Work Environment & Health, 6*, 144–149.

Harris, C., & Daniels, K. (2007). The role of stressor appraisals in psychological well-being and physical symptom reporting. *European Journal of Work and Organizational Psychology, 16*, 407–431.

Hobfoll, S. (2002). Social and psychological resources and adaptation. *Review of General Psychology, 6*, 307–324.

Hofmann, D., & Mark, B. (2006). An investigation of the relationship between safety climate and medication errors as well as other nurse and patient outcomes. *Personnel Psychology, 59*, 847–869.

Howell, R. T., Kern, M. L., & Lyubomirsky, S. (2007). Health benefits: Meta-analytically determining the impact of well-being on objective health outcomes. *Health Psychology Review, 1*, 83–136.

Hurrell, J. J., & Murphy, L. R. (1992). An overview of occupational stress and health. In W. M. Rom (Ed.), *Environmental and Occupational Medicine* (2nd ed.) (pp 675–684). Boston: Little Brown.

Ilies, R., Schwind, K. M., & Heller, D. (2007). Employee well-being: A multilevel model linking work and non-work domains. *European Journal of Work and Organizational Psychology, 16*, 326–341.

International Labour Organization. (1986). *Psychosocial Factors at Work: Recognition and Control*. Report to the Joint ILO/WHO Committee on Occupational Health. Ninth Session. Occupational Safety and Health Series No. 56. Geneva: International Labour Organization.

Jenkins, R., Meltzer, H., Jones, P. B., Brugha, T., Bebbington, P., Farrell, M., Crepaz-Keay, D., & Knapp, M. (2008). *Foresight Mental Capital and Wellbeing Project. Mental Health: Future Challenges.* London: The Government Office for Science.

Jordan, J., Gurr, E., Tinline, G., Giga, S., Faragher, B., & Cooper, C. (2003). *Beacons of Excellence in Stress Prevention.* Norwich: HSE Books.

Karasek, R., & Theorell, T. (1990). *Healthy work: Stress, Productivity, and the Reconstruction of Working Life.* New York: Basic Books.

Kelloway, E. K., & Day, A. L. (2005). Building healthy organizations: What we know so far, *Canadian Journal of Behavioural Science, 37,* 223–236.

Kelloway, E. K., Teed, M., & Kelley, E. (2008). The psychosocial environment: Towards an agenda for research. *International Journal of Workplace Health Management, 1,* 50–64.

Klein, K. J., Tosi, H., & Cannella, A. A. (1999). Multilevel theory building: Benefits, barriers, and new developments. *Academy of Management Review, 24,* 243–248.

Kozlowski, S. W., & Doherty, M. L. (1989). Integration of climate and leadership: Examination of a neglected issue. *Journal of Applied Psychology, 74,* 546–553.

Kuoppala, J., Lamminpää, A., & Husman, P. (2008). Work health promotion, job well-being, and sickness absences: A systematic review and meta-analysis. *Journal of Occupational and Environmental Medicine, 50,* 1216–1227.

Labriola, M., Christensen, K. B., Lund, T., Lindhardt, M. N., & Diderichsen, D. (2006). Multilevel analysis of workplace and individual risk factors for long term sickness absence. *Journal of Occupational and Environmental Medicine, 48*(9), 923–929.

LaMontagne, A. D., Keegel, T., Louie, A. M., Ostry, A., & Landsbergis, P. A. (2007). A systematic review of the job-stress intervention evaluation literature 1990–2005. *International Journal of Occupational and Environmental Medicine, 13,* 268–280.

Lazarus, R. S., & Folkman, S. (1984). *Stress, appraisal and coping.* New York: Springer.

Levesque, M. J., Steciuk, M., & Ledley, C. (2002). Self-disclosure patterns among well acquainted individuals: Disclosers, confidants and unique relationships. *Social Behavior and Personality, 30,* 579–592.

Lewin, K., Lippitt, R., & White, R. K. (1939). Patterns of aggressive behavior in experimentally created social climates. *Journal of Social Psychology, 10,* 271–299.

Luthans, F., Avolio, B., Avey, J., & Norman, S. (2007). Psychological capital: Measurement and relationship with performance and satisfaction. *Personnel Psychology, 60,* 541–572.

Martin, A., Sanderson, K., & Cocker, F. (2009). Meta-analysis of the effects of health promotion intervention in the workplace on depression and anxiety symptoms. *Scandinavian Journal of Work Environmental Health, 35*(1), 7–18.

Mearns, K., Hope, L., Ford, M. T., & Tetrick, L. E. (2010). Investment in workforce health: Exploring implications for workforce safety climate and commitment. *Accident Analysis and Prevention, 42,* 1445–1454.

Medley, F., & Larochelle, D. (1995). Transformational Leadership and Job Satisfaction. *Nursing Management, 26,* 64JJ–64NN.

National Institute for Occupational Safety & Health. (1986). *Proposed National Strategy for the Prevention of leading work-related Diseases and Injuries. Occupational Cardiovascular Diseases.* U.S. Department of health and human services: Public Health Service Centres for Disease Control.

Nielsen, K., Randall, R., Yarker, J., & Brenner, S. O. (2008). The effects of transformational leadership on followers' perceived work characteristics and psychological well-being: A longitudinal study. *Work & Stress, 22*(1), 16–32.

Nuttman-Shwartz, O., & Ginsburg, R. (2002). Early rehabilitation program after workplace injuries. *Employee Assistance Quarterly*, *17*, 17–32.

Parks, K. M. & Stilman, L. A. (2008). Organizational Wellness Programs: A meta-analysis. *Journal of Occupational Health Psychology*, *13*, 58–68.

Quick, J. D., Campbell-Quick, J., and Nelson, D. L. (2000). The theory of preventive stress management in organizations. In: C. L. Cooper (Ed.), *Theories of organizational stress*. Oxford: University Press.

Quick, J. C., Quick, J. C., Nelson, D. L., and Hurrell, J. J. Jr. (1997). *Preventive Stress Management in Organizations*. Washington, D.C.: American Psychological Association.

Ryan, R. M., & Deci, E. L (2001). On happiness and human potentials: A review of research on hedonic and eudaimonic well-being. *Annual Review of Psychology*, *52*, 141–166.

Ryff, C. D., & Keyes, C. L. M. (1995). The structure of psychological well-being revisited. *Journal of Personality and Social Psychology*, *69*(4), 719–727.

Sauter, S. L., Hurrell, J. J., Murphy, L. R., & Levi, L. (1998). Psychosocial and organizational factors. Part V. In J. M. Stellman (Ed.), *Encyclopaedia of Occupational Health and Safety vol 2* (4th Ed.). Geneva: International Labour Organization.

Schneider, B. (1990). The climate for service: An application of the climate construct. In B. Schneider (Ed.), *Organizational Climate and Culture* (pp. 383–412). San Francisco: Jossey-Bass.

Schwarzer, R. (2001). Stress, resources, and proactive coping. *Applied Psychology: An International Journal*, *50*, 400–407.

Semmer, N. K. (2006). Job stress interventions and the organization of work. *Scandinavian Journal of Work Environmental Health*, *32*(6), 515–527.

Skakon, J., Nielsen, K., Borg, V., & Guzman, J. (2010). Are leaders' well-being, behaviours and style associated with the affective well-being of their employees? A systematic review of three decades of research. *Work & Stress*, *24*, 107–139.

Smith, D. (2001). *IMS: The Framework*. British Standards Institution: IMS Risk Solutions.

Sorensen, G., & Barbeau, E. (2004). *Steps to a healthier US workforce: Integrating occupational health and safety and worksite health promotion: State of the Science*. Paper presented at the meeting of the National Institute of Occupational Safety & Health, Washington DC.

Sosik, J., & Godshalk, V. (2000). Leadership styles, mentoring functions received, and job-related stress: A conceptual model and preliminary study. *Journal of Organizational Behavior*, *21*, 365–390.

Stajkovic, A. D., & Luthans, F. (1998). Self-efficacy and work-related performance: A meta-analysis. *Psychological Bulletin*, *124*, 240–261.

Swuste, P., & Arnoldy, F. (2003). The safety adviser/manager as agent of organizational change: A new challenge to expert training. *Safety Science*, *41*(1), 15–27.

Turner, N., Barling, J., & Zacharatos, A. (2002). Positive psychology at work. In C. R. Snyder, & S. Lopez (Eds.), *The Handbook of Positive Psychology* (pp. 715–730). Oxford: Oxford University Press.

Wallace, J., Edwards, B., Arnold, T., Frazier, M., & Finch, D. (2009). Work stressors, role-based performance, and the moderating influence of organizational support. *Journal of Applied Psychology*, *94*, 254–262.

Waterman, A. S. (1993). Two conceptions of happiness: Contrasts of personal expressiveness (eudaimonia) and hedonic enjoyment. *Journal of Personality and Social Psychology*, *64*(4), 678–691. doi: 10.1037/0022-3514.64.4.678

Weiner, B. J., Lewis, M. A., & Linnan, L. A. (2009). Using organization theory to understand the determinants of effective implementation of worksite health promotion programs. *Health Education Research, 24*(2), 292–305.

World Health Organization. (2010). *WHO Healthy Workplace. Framework and Model: Background and Supporting Literature and Practices. Healthy workplaces: A model for action: For employers, workers, policymakers and practitioners*, Geneva.

Wrzesniewski, A., & Dutton, J. (2001). Crafting a job: Revisioning employees as active crafters of their work. *Academy of Management Review, 26*, 179–201.

Youssef, C. M., & Luthans, F. (2010). An integrated model of psychological capital in the workplace. In A. Linley, S. Harrington, & N. Garcea (Eds.), *Oxford Handbook of Positive Psychology and Work*. New York: Oxford University Press.

Zohar, D. (2002). The effects of leadership dimensions, safety climate, and assigned priorities on minor injuries in work groups. *Journal of Organizational Behaviour, 23*, 75–92.

Zohar, D. (2010). Thirty years of safety climate research: Reflections and future directions. *Accident Analysis & Prevention, 42*, 1517–1522.

4 Research in organizational interventions to improve well-being

Perspectives on organizational change and development

Lois E. Tetrick, James Campbell Quick, and Phillip L. Gilmore

Introduction

It is a long recognized "truth" that organizations must change to survive (e.g. Armenakis and Harris, 2009). Organizational change can be stressful for employees (Rafferty and Griffin, 2006), which sets up a double bind situation. For organizations to remain healthy, they must manage change in such a way that employees' well-being is not compromised. Further, some organizational changes may be necessitated to prevent employee ill-health and promote employee well-being. This chapter reviews the change management and organizational development literature to provide frameworks for guiding organizational interventions to enhance the health and well-being of organizations and their employees.

The chapter begins with a discussion of the concept of well-being, which has a long legacy in philosophy and more recently in psychology and public health (Quick and Quick, in press). This chapter is anchored in the psychological and prevention perspectives of public health. The next major section of the chapter examines change management and organizational development. The third major section focuses on organizational interventions for enhancing employee well-being. The chapter concludes with two systemic cases of successful organizational intervention, one in Lincoln Industries and one in the United States Air Force. Each organization is unique and senior leaders must craft ways to enhance individual and organizational well-being suited to their industrial context, employee population, unique strengths, and particular vulnerabilities and risks.

Well-being

Before discussing organizational interventions to improve well-being, it is important to define what we mean by well-being. There are several conceptualizations of well-being to be considered. While philosophy has a long legacy of attention to the concept of well-being, more recently the domains of psychology and public health, with their strong emphasis on prevention, have taken "well-being" into their lexicon and modified or adapted the concept (Quick and Quick, in press). Well-being is not the same as the concept of health and yet the two concepts overlap.

Therefore, perhaps one of the most common ways of viewing well-being has been through equating well-being with physical and psychological health (e.g. Griffin and Clarke, 2011; Hammer and Zimmerman, 2011). This aligns rather well with the psychological and public health perspectives on well-being. From this point, it is important to consider our definition of health, which we use as a proxy for well-being. The World Health Organization in 1946 defined health as not just the absence of illness but as a state of optimal functioning. Recent trends in the literature on individual and organizational health have followed suit recognizing the importance of considering both the positive and negative aspects of health and well-being for individuals and organizations (Hofmann and Tetrick, 2003; Macik-Frey, Quick, and Nelson, 2007).

The integration of positive health in the conceptualization of well-being has important implications for organizational interventions. At the individual level, research has accumulated on the effects of positive emotions on individuals' decision-making (Isen and Labroo, 2003), creativity and innovative behaviours (Eubanks, Murphy, and Mumford, 2010; Thrash et al., 2010), resiliency (Fredrickson et al., 2003) and adaptability to stressful situations (Tugade and Fredrickson, 2007). Similarly, evidence is beginning to accumulate to suggest the importance of positive organizational behaviour (Cameron, Dutton, and Quinn, 2003; Luthans and Avolio, 2009; Luthans, Youssef, and Avolio, 2007) to individual and organizational well-being.

It is our perspective in this chapter that it is important to attend to both positive and negative aspects of the work environment to develop and implement interventions to promote the well-being of employees and prevent harm to employees. As Cameron (2007) pointed out, positive and negative dynamics in organizations are often intertwined. Further, we take the perspective that organizational well-being is dependent on employee well-being. The traditional indicators of organizational health, such as financial viability, as well as more recent indicators, such as grievances and absenteeism rates (Hofmann and Tetrick, 2003), are inextricably related to the psychosocial and physical environment within the organization.

In addition to the positive-negative duality concerning health, there is an individual-organizational duality to consider. Just as positive and negative are intertwined in organizations, so too are the individual and organizational perspectives. Principle one in the theory of preventive stress management in organizations is: Individual and organizational health are interdependent (Quick et al., 1997, pp. 150–151). This interdependency is a key point of departure in considering change interventions for enhancing health and well-being in organizations. In addition, the interdependency leads to interaction effects between the individual and the organization as change processes are initiated. We turn now to change management and organizational development as process approaches to making work-life better.

Change management and organizational development

Organizational development concerns the implementation of planned change efforts, generally managed from the top down, designed to improve organizational health and the survival of the organization (Martins, 2011). Contemporary organizational change management scholarship has been dominated by research focused on changing the patterns of employees' and leaders' behavioural processes and collective perceptions in order to increase productivity. Given that peoples' behavioural patterns and perceptions significantly influence well-being, we believe that previous research in this area is relevant to organizational health interventions as well. For this reason, we review the important organizational change elements that deserve the attention of practitioners and researchers.

Lewin's work has greatly influenced change management scholars (1947, 1951). Most contemporary models of organizational change are recognizable in the general Lewinian form: unfreeze, change, and refreeze (Martins, 2011). In his process model, Lewin posited that first it is necessary to "break social habits" or "unfreeze" the situation. Then it is possible to actually take action to move the group or organization to a new state or level ("change"), which when reached, change managers should "refreeze" or institutionalize. Martins (2011) argues that most other models of organizational change have Lewin's three-phase model as an underlying framework. Martins (2011) provides a review of the many organizational change and development theories and we will not review all of them here. However, common among these intervention theories of organizational change and development are the identification of the problem or discrepancy, the diagnosis of the problem, the development of a plan for an intervention, the implementation of the plan, and finally the evaluation and monitoring of the new level of performance. It appears that these intervention theories (see Martins, 2011) may be useful in modifying the psychosocial work environment to improve the well-being of organizations and their employees.

Organizational change and organizational development scholars recognize that the need for change can occur as a result of change in the organization's external environment as well as changes in the internal environment (Armenakis and Harris, 2009; Holden et al., 2008). Much of the literature on organizational change and development takes an episodic view of organizational change; later in the chapter, we elaborate on models that further refine the definition of episodic change. In addition, change efforts can be continuous and incremental, and change efforts can differ by scale such as organization-wide versus departmental or division-wide (By, 2005). The nature of the pressure of change and the scope of change may influence the design and implementation of the change effort. Further, if organizations and groups are continuously changed as espoused by Lewin, then health optimizing organizational change efforts must be monitored and adapted. After considering the many elements of organizational change, it may be no surprise that Balogun and Hope Haily (2004) as cited in By's (2005) report, find that only about 30 percent of all change programs are successful.

62 *Challenges and methodological issues*

Designers and implementers of organizational change efforts may not always need to control, manipulate, or predict every element of an organizational change event. However, they should be aware of all of the elements. We believe there exist five major threats to the success of organizational change efforts:

1   lack of an adequate framework for implementing organizational change;
2   a failure to accurately identify the problem;
3   an inaccurate diagnosis of the problem and its root causes;
4   a lack of fidelity in the implementation of the planned intervention; and
5   an inability to adequately evaluate the effects of the change effort, or an insufficient timeframe in order to be able to observe the effects.

Perspectives on effective change

Much of the organizational change and development literature has focused on management as initiators and controllers of the change process. Armenakis and Harris (2009), for example, take a change recipient (e.g. employees) approach in order to understand the motivation of the change recipients to change and to support the change effort. In their review of the literature, Armenakis and Harris (2009) identified six themes that characterize change recipient receptivity to change efforts; understanding recipient receptivity is important in understanding the effectiveness of illness prevention and health promotion organizational interventions in which employees and organizational entities are the recipients. These six themes are:

1   readiness to change,
2   participation of the change recipients in the change effort,
3   accurate diagnosis of the need for change,
4   taking a positive approach for creating readiness for change,
5   strategically leading to support the five key beliefs underlying the motivation to change, and
6   continuous assessment of reactions to the change effort.

These themes are consistent with Kanter, Stein, and Jick's (1992) *The Challenge of Organizational Change*, Kotter's *Leading Change* (1996), and Luecke's *Managing Change and Transition* (2003) (see By, 2005, for a review and integration; also see Young, 2009, for a meta-model of change identifying additional commonalities amongst a broad range of change literatures).

Readiness to change

Armenakis and Harris (2009) define readiness to change in terms of five key beliefs underlying the motivation to change, which are:

1   discrepancy
2   appropriateness

3   efficacy
4   principal support, and
5   valence.

They define these key beliefs as:

> *Discrepancy* refers to the belief that a change is needed; that there is a significant gap between the current state of the organization and what it should be. *Appropriateness* reflects the belief that a specific change designed to address a discrepancy is the correct one for the situation. *Efficacy* refers to the belief that the change recipient and the organization can successfully implement a change. *Principal support* is the belief that the *formal* leaders (*vertical* change agents) in an organization are committed to the success of a change and that it is not going to be a passing fad. In specifically defining sources of principal support, we include the *opinion leaders* who can serve as *horizontal* change agents. Finally, *valence* reflects the belief that the change is beneficial to the change recipient; there is something of benefit in it for them.
>
> (Armenakis and Harris, 2009, p. 129)

Armenakis et al. (2007) in a study of a top management team's reactions to an organizational restructuring, found support for these five key beliefs and posited that these five beliefs actually reflect readiness for change, a multidimensional construct. Following this line of research, Holt et al. (2007) published their scale measuring readiness for change which supported four of the five key beliefs (e.g. appropriateness, efficacy, management support, and personal valence). Neves (2009), in an investigation of the implementation of a new performance appraisal system, examined the effects of only two of the five beliefs, appropriateness and self-efficacy, on affective commitment, level of individual change, and intention to leave the organization. Neves's (2009) results substantiated the positive relation between perceived appropriateness of the change and the recipients' commitment to change. In Neves's (2009) study, self-efficacy did not have a direct effect on commitment to change although it was directly related to intention to leave. One caveat should be noted, however. The measure of self-efficacy used by Neves may not have adequately captured efficacy as defined by Armenakis et al. (2007).

Participation

The second theme identified by Armenakis and Harris (2009) is that change recipients need to be involved in the change process. This notion is consistent with action research, and this means that change recipients need to be engaged in the diagnosis of the problem, the development of the change effort, and the implementation of the organizational intervention. Armenakis and Harris (2009) argue that change recipient participation can directly affect key beliefs underlying the motivation to change. Jimmieson, Peach, and White (2008) found support for the importance of participation in gaining employees' support for organizational change. In their study, Jimmieson, Peach, and White (2008) applied the Theory of Planned Behaviour (Ajzen, 1991; Ajzen and Fishbein, 1980) and incorporated the two aspects of common process models of change – participation in the change process and receipt of timely and accurate communication about the change process. Jimmieson, Peach, and White (2008) found that among a sample of employees whose organization was undergoing a change effort to relocate to a new office building, participation predicted employees' intentions to support the change effort, and this effect was partially mediated by coworkers' subjective norm that employees should support the relocation. Contrary to what was expected, participation was not related to employees' attitudes toward the relocation nor their perceived control over the change. Indeed, the employees did not truly possess any behavioural control over the relocation; that was not an option. Similarly, Jimmieson, Peach, and White (2008) found that the effect of communication about the change effort was partially mediated by both subjective norms and perceived behavioural control. Again, attitudes toward change were not significantly related to communication. Therefore, this study's results support the importance of change recipients' participation in organizational interventions, and we suggest that participation may have an independent, direct effect on the effectiveness of a change effort as well as potentially changing employees' beliefs about the change effort.

Diagnosis

The third theme involves accurate organizational diagnosis. Diagnosis is *dia*: through and *gnosis*: knowledge. Diagnosis is highly appropriate in organizations as a process for informed understanding. Just as Lewin (1947) indicates that it is important to understand the current situation before attempting to initiate a change effort, Armenakis and Harris (2009) propose that a thorough organizational diagnosis is critical for designing appropriate change interventions to resolve the discrepancy. In fact, Judge and Douglas (2009) suggest that one of the reasons that so many planned organizational change efforts fail is the lack of valid and reliable instruments to monitor an organization's capacity to change. Better instrumentation and attention to the process of both diagnosis and implementation is necessary for uncovering and mindfully examining the root causes of the discrepancy gap. Effective diagnostic techniques may be a mechanism for influencing change recipients' key beliefs that the change is needed, the change effort is appropriate, and there is a collective efficacy regarding the ability to successfully implement the change. The results of Jimmieson et al. (2008), reported above, align with this third theme. Further, this theme emphasizes the need to be able to conceptually and operationally define relevant phenomena within the organization as well as the external environment. By fully diagnosing the situation, organizational interventions can be optimally designed to move the organization and its employees toward a satisfactory solution.

Creating readiness

The fourth theme is creating readiness for change. Armenakis and Harris (2009) argue that framing Lewin's first stage of the change model, unfreezing, in terms of creating a readiness for change is more consistent with a positive approach to organizational change than focusing on resistance to change. Ford, Ford, and D'Amelio (2008) suggest that change agents can actually create resistance to change in their attempt to make sense of change recipients' reactions to the change or they can create a readiness to change. Also, change agents can create resistance to change by breaking agreements during the change process. Neves and Caetano (2006) applied Social Exchange Theory (Blau, 1964) to examine the role of trust in one's supervisor and control over the change process in predicting employees' affective commitment. As Neves and Caetano (2006) anticipated, employees whose organization had implemented a new Quality Management System had higher levels of commitment when their trust in their supervisor was higher. Perceived control over the change had a moderating effect such that the relationship between trust in the supervisor and commitment was stronger when they had low levels of perceived control over the change. Further, when the employees perceived higher levels of control over the change, they had higher levels of commitment. This study thus supports the notion that trust in one's supervisor can have a positive effect on employees' readiness for change as posited by Ford et al. (2008).

Influence the key beliefs toward change

The fifth theme covers influence strategies that change agents use to shape the five key beliefs. Armenakis and Harris (2009) suggest there are seven strategies:

1   active participation
2   persuasive communication
3   management of internal/external information
4   human resources management practices
5   formalization activities
6   diffusion practices, and
7   rites and ceremonies.

In addition to perceived control over the change and trust in one's supervisor as described above, it has been demonstrated that characteristics of the change process have an effect on employees' readiness for change and their reactions to the changes. For example, Rafferty and Griffin (2006), using Lazarus and Folkman's (1984) cognitive appraisal theory of stress, were able to identify three characteristics of organizational change that were related to employees' adjustment to organizational change (change frequency, the degree of planning for change, and the degree to which change transformed the workplace). Frequency of change increased psychological uncertainty and planning for change decreased psychological uncertainty. The resulting psychological uncertainty was positively related to the intention to leave the organization and negatively related to job satisfaction. More importantly, perhaps, supportive leaders actually were found to lower recipients' perceptions of frequency of change, lower the impact of the change for recipients, reduce psychological uncertainty while increasing perceptions that change is being planned. When one looks at the items included in Rafferty and Griffin's (2006) measure of the three characteristics of change, most of the seven strategies identified by Armenakis and Harris (2009) are included.

Organizational change agents influence humans' identities

It is apparent that change agents and change leadership are critical in organizational interventions. Organizational change creates uncertainty (Rafferty and Griffin, 2006) and this uncertainty may be perceived by employees as a threat to their work-based identities (Karp and Helgø, 2008; van Dijk and van Dick, 2009). As Karp and Helgø (2008) state, "[t]he real threat when a leader is designing and implementing a grand change programme is to more than economic well-being and a job; the threat is to people's identities" (pp. 88–89). Van Dijk and van Dick (2009) argue that it may not be "change" that employees are resisting but the "change management" process. Based on their study of organizational mergers, van Dijk and van Dick (2009) suggest that conceptualizing resistance to change within a framework of work-related identities is useful to understanding employees' and change agents' motivation to change. Karp and Helgø (2008) similarly suggest that identity formation is critical for successful organizational change efforts and that change leaders need to serve as role models and communicators of the core values and purpose of the organization. Further, change agents facilitate readiness for change by affecting the language, memes and symbols within the organization. Changing these cultural artifacts is important in leading change (Senge et al., 1999; Weick, 2001) as the organization's cultural artifacts enhance work-related identities and motivation.

The above studies have focused on leaders' (i.e. change agents') behaviours in specific change situations. Herold et al. (2008) took a somewhat broader perspective on the role of leadership in successful organizational change efforts. They examined the effects of trans-situational leadership behaviours (viz., transformational leadership behaviours) and change-specific leadership behaviours on employees' commitment to an organizational change. In a study of change efforts in 30 organizations, Herold et al. (2008) found that transformational leadership, not change-specific leadership, results in greater commitment to a specific change. In addition, the impact of change-specific leadership was a function of both transformational leadership and the extent to which the change impacted an individual's job. The results of Herold et al.'s (2008) study are consistent with previous research that demonstrates the positive effect transformational leadership has on individuals' motivation to change (i.e. to aspire). The fact that change-specific leadership did not have the expected effect on change criteria further suggests that a consistent pattern of leadership behaviours may be quite important as contrasted with leaders' more momentary behaviours.

Assessing change

Lastly, the sixth theme mentioned by Armenakis and Harris (2009) is the assessment of the change. Evaluation of the change process is essential for understanding the fidelity of the implementation, the underlying reasons for the effects of the intervention, and the sustainability of the change. Martins (2011) describes the variety of methodological questions that arise when assessing organizational change efforts. Martins (2011) discusses the use of qualitative and/or quantitative methods, what level of analysis is needed (e.g. organizational level, group level, and individual level), the timing of the assessment(s) and the validity and reliability of the measures of central constructs. Meta-analytic results (e.g. Guzzo, Jette, and Katzell, 1985; Halfhill et al., 2002) indicate that the largest effect sizes on productivity are associated with training, goal setting, and social technological systems. However, when the outcomes were attitudes, the effect sizes of organizational interventions are much more varied with no clear pattern of significant effects emerging (Armenakis and Bedeian, 1999). If one defines organizational well-being in terms of productivity, it appears that there is support for the effectiveness of organizational interventions on organizational well-being. If one takes into account employee behaviours and perceptions of well-being, we expect previous organizational change models to similarly impact well-being. When it comes to changing attitudes however, strong evidence is lacking for the effectiveness of organizational interventions. As defined previously, well-being is composed of behaviours, perceptions and attitudes. In the following sections, we discuss what examples currently exist of organizational interventions targeted at well-being.

Organizational interventions for enhancing employee well-being

As suggested in Halfhill et al. (2002), the effectiveness of organizational interventions on employees' well-being may have smaller effect sizes than do organizational interventions targeting employees' productivity. Interestingly, the literature on organizational change and development seems to take a more macro, organization-wide perspective and maintain a focus on organizational level outcomes and primarily outcomes associated with the economic well-being of the organization. Interventions implemented to enhance employee well-being tend to be more targeted toward the prevention of specific aspects of ill-health such as stress reduction. However, interventions concerned with organizational well-being and those concerned with employee well-being tend to focus more on fixing problems and preventing illness and injuries than promoting well-being.

Literature on the effectiveness of organizational interventions in enhancing employee well-being is still small, but empirical evidence is growing. These interventions are not guided by a single theoretical perspective, much as Martins (2011) indicated is the case for organizational development. It may be that the diversity in theoretical perspectives is driven by the focus of the intervention. For example, if the underlying root cause or the primary risk factor is deemed to be systemic and work-process oriented, a macroergonomic theory (Hendrick, 1991) may be employed. If the problem is viewed as one requiring individual employees to change their behaviour, the Transtheoretical Model of Change (Prochaska, 2000) may be adopted. Therefore, we will organize our discussion of organizational interventions to prevent ill-health and to promote employee well-being around risk factors or bases for the interventions, keeping in mind that it is not possible to review all organizational interventions that have been reported in the literature.

Macroergonomics

The organizational change and organizational development literature has focused on a top-down approach and generally takes a more macro perspective. There has been an interesting development in the ergonomics field beginning in the early 1990s which has taken a more micro-to-macro approach in the field of human factors and ergonomics. This field has been named macroergonomics (Hendrick, 1991). Holden et al. (2008) proposed 30 change management principles that they argue are critical for organizational interventions and field research. These principles parallel those found in the organizational change literature (e.g. Armenakis and Harris, 2009) although the principles range from meta-principles such as the need to embrace a systems approach to more specific considerations like scanning, benchmarking, and assessing readiness for change. Other principles are included as well such as the need for teams, leaders, champions, and end user involvement; the need to cope with the unexpected; gaining "buy-in"; the need for training, resources, and top management support; and the need to sustain the change over the long run. In taking this approach, it has become clear that an organizational intervention is a multilevel process (Karsh and Brown, 2010). Therefore, assessments of the impact of the organizational intervention need to take into account the fact that the data are nested within levels – supervisor, unit, department, organization – and effects need to be attributed to these levels.

There is an increasing number of empirical studies adopting the macroergonomics view in evaluating organizational interventions. For example, Gaël, René, and Christine (2009) in their investigation of the commercial sea fishing industry, an industry with the highest risk of injury, compared a micro-ergonomic approach to risk reduction and a macro-ergonomic approach. In the micro-ergonomic approach, technology was provided to the individual worker (i.e. the skipper of the fishing craft), and the researchers were interested in the worker's use of the equipment, essentially a component of the total system. In the macro-ergonomic approach to risk reduction, the researchers focused on the fishing company as a system which took a participatory approach for safety improvement and developed a climate for safety. Contrary to their expectations, neither system worked as anticipated. The micro-ergonomic approach resulted in a misuse of the technology to enhance the catch (i.e. financial performance), which Gaël et al. (2009) attributed to the skippers' resilience (defined as exceptional skills and "know-how").

The macro-ergonomic approach resulted in fewer shipwrecks but actually more injuries; Gaël et al. (2009) suggested that this approach resulted in shifting the decision from loss of men to loss of vessels. This study highlights the importance in considering the financial well-being of the organization and external factors such as legal regulations when developing and implementing organizational interventions to enhance employee well-being.

Stages of change

The transtheoretical model (TTM) of change (Prochaska, 2000; Prochaska, Prochaska, and Levesque, 2001) has generated considerable research examining individuals' behavioural changes. This theory grew out of health psychology but has recently been adapted to organizational interventions. One of the core aspects of the TTM is the stages of change, which refers to the concept that individuals go from pre-contemplation (resistance/denial of the need to change), to contemplation (recognition of the need to change), to preparation (intending to change), to action (engaging in the change), to maintenance (sticking with the behavioural changes that were enacted). TTM posits that it is important to design and implement interventions that are consistent with the stage of change in which the individual presently resides. Based on this model, Whysall, Haslam and Haslam (2007) applied the stages of change framework to assess management and worker readiness to change in order to reduce the risk of work-related musculoskeletal disorders (WRMD). Perhaps not surprisingly, Whysall et al. (2007) found distinctly different stage-of-change patterns between managers and their workers. Workers tended to be in the pre-contemplation stage, the preparation stage, and the relapse stage; managers tended to be in the pre-contemplation stage (but fewer than among the workers) and the maintenance stage. Since managers often initiate organizational interventions (Armenakis and Harris, 2009; Hendrick, 1991), it might be expected that compared to their workers, more managers may have moved into the contemplation or later stages of change as they lead change. What may be more perplexing is that more managers tended to be in the maintenance stage. The question may be whether the managers are ill-informed of the extent of implementation of WRMD reducing interventions (i.e. premature maintenance) or whether they are resistant to changing from prior policies and practices (i.e. maintenance of the pre-change state).

Another recent study using the TTM looks at the adoption of workplace coaching skills (Grant, 2010). Grant's study represents a more positive organizational behaviour perspective; in that, the change agents were not seeking to implement an intervention to reduce a "problem". Rather, the change agents wanted to implement an intervention to improve employee well-being. Coaching has typically been considered a means for individual change, but this study also suggested that coaching may be viewed as an organizational intervention when broadly applied to change the organization. Although the empirical evidence is very limited, coaching has been suggested to result in more positive, humanistic and motivating communication styles which can create the establishment of a coaching culture. Grant (2010) found support for the relation of the stages of change and the perceived pros and cons of adopting coaching across the stages. As might be expected, levels of self-efficacy were higher for managers who were in the maintenance stage. Somewhat surprisingly, there were no significant differences between the rated pros and cons of coaching depending on which stage of change the manager was in. Grant interpreted this as meaning that the benefits of coaching were "obvious to all" (Grant, 2010, p. 72). Counter to what was expected, the managers in this study who were actually using coaching did not report greater well-being for themselves, and the study did not include information as to whether coaching enhanced the followers' well-being.

The literature on TTM has only recently been extended to organizational interventions targeted at well-being. To date there is mixed support for this approach for designing and implementing organizational interventions. One current challenge appears to be the meaningful distinction of the stage of change in which people exist at any point in time they reside in, and then being able to identify the specific behavioural, perceptual and motivational features associated with each particular stage. Future research that incorporates the full model may be more successful (Prochaska, 2000).

Leadership

Leadership has been found to be a major factor in organizational interventions in the organizational change and organizational development literature. This appears also to be the case for organizational interventions to prevent ill-health and promote well-being. There are several leadership theories, but it seems that transformational leadership is one theory that has attracted the attention of occupational health scholars (Mullen and Kelloway, 2011). Transformational leadership posits that leaders are sources of inspirational and intellectual stimulation for their employees. Based on similar logic that Grant (2010) employed for predicting the positive effect of "manager as coach," it has been posited that leaders have a direct positive effect on their followers' well-being (Nielsen et al., 2008).

Kuoppala et al. (2008) in their systematic review and meta-analysis found moderately strong support for leadership being positively related to employees' well-being. Futher, Bono et al. (2007) suggest that transformational leaders can affect followers' well-being indirectly by evoking positive emotions in their interactions. (For a review of the literature on the positive effects of transformational leadership on employee well-being and occupational safety, see Mullen and Kelloway, 2011.) The indirect effect of transformational leadership was also supported by Nielsen et al. (2008). In Nielsen et al.'s longitudinal study, they found that the effect of transformational leadership on employee well-being was mediated by perceived work characteristics. Nielsen et al. also found that employees' well-being at one point in time predicted their report of their leaders' behaviour 18 months later, suggesting that employees' well-being and perceptions of their leaders' behaviours are reciprocally related. Skakon et al. (2010) conducted a systematic review of three decades of research on the relation between followers' well-being and leaders' well-being, behaviours, and leadership style. This review demonstrated some support for the relation between leader stress and well-being with employee stress and well-being. However, the authors note that most of the studies were cross-sectional and did not assess the effects of an actual organizational intervention.

A central part of the leader's role is encouraging employee well-being because of its interdependency and instrumental contribution to organizational health (Macik-Frey et al., 2009). A key here may be the next generation theory from transformational leadership, that is "authentic leadership" (Luthans and Avolio, in Mack-Frey et al., 2009). This process model of leadership is one that results in self-awareness and self-regulated positive behaviours that are demonstrated by leaders and followers alike. In line with this process model and under the rubric of leadership is the importance of organizational justice and its impact on health in organizations (Cropanzano and Wright, 2011). Of the three forms of justice, which are distributive justice, procedural justice, and interactional justice, the latter is centrally related to leader-follower dynamics. Specifically, the perceived fairness of the interpersonal treatment that a follower receives from his or her leader has an impact on the follower's health and well-being. Hence, there are performance and economic incentives for leaders to treat followers and all employees with justice and fairness.

Taken together, we must conclude that in the context of organizational interventions, empirical research on the effects of leadership on leader and employee well-being is sparse. Further, this is an area for growth. Given the central role of leaders in the organizational change and organizational development research, it is important that quality research be conducted specifically examining the effectiveness of leadership as one component of organizational change efforts to enhance employee well-being.

Two systemic cases in organizational and employee well-being

To conclude this chapter, we would like to describe two cases of organizations adopting a systemic approach to organizational and employee well-being. The first case is that of Lincoln Industries in Lincoln, Nebraska. The second case is that of the San Antonio Air Logistics Center, the United States Air Force's largest logistics and maintenance depot that passed from existence in July 2001. Each of these two cases gives a different systemic view of interventions for enhancing organizational health and employee well-being.

Lincoln Industries

Lincoln Industries is a high performance metal finishing company located in Lincoln, Nebraska (www.lincolnindustries.com). They have received multiple awards for their culture of wellness. For example, Lincoln Industries has been named as one of the top 25 Best Places to Work among medium-sized firms and were chosen as a national Innovation in Prevention Award winner by the Department of Health and Human Services (HHS) for its efforts in promoting healthy lifestyles in its community (www.lincolnindustries.com/wellness). Consistent with the literature on organizational change and organizational development, Lincoln Industries has built a culture of wellness (Vyhildal, 2010). Wellness is defined as not only healthy lifestyle issues such as nutrition, physical fitness, and health education but also as safety participation and work behaviour including work-life balance. The program is integrated into "Life at Lincoln Industries" and involves employees at all levels of the organization. For example, wellness is part of the performance management system including individual wellness goals for each employee. Employees receive quarterly physicals and have a variety of wellness programs available at work to choose from including positive benefits for positive behaviours (e.g. tobacco-free discounts). Further, and perhaps most importantly, there is top management commitment to the wellness program that drives their strategic planning model.

In designing, implementing, maintaining, and updating the wellness program, Lincoln Industries has engaged employees and attended to their motivations for wellness and incorporated this into the programs. In addition, wellness has been "hardwired" into the organizational culture and leadership development programs (Vyhildal, 2010), and it appears to be working. Lincoln Industries has seen a drop in the number of employees who smoke, a reduction in workers' compensation costs, and their average health care costs per person are substantially below the industry average (www.lincolnindustries.com/wellness/benefits).

San Antonio Air Logistics Center

In the early 1990s, the San Antonio Air Logistics Center, Air Force Material Command (AFMC) was the largest of the U.S. Air Force's five logistics and maintenance depots carrying 40 per cent of the service's workload. In July 1995, President Bill Clinton signed the BRAC order that would close this Center by 2001. What transpired over that six year period was a uniquely challenging case for a positive leadership intervention to ensure the health and well-being of thousands of federal civil servants while supporting the flying mission of the U.S. Air Force.

The commanding general of the Center worked in collaboration with the USAF's Office of the Surgeon General to put in place a full-time, active duty clinical organizational psychologist who would lead and implement preventive interventions anchored in public health. Klunder (2008) used a systemic approach through preventive stress management by first forming a Board at the Center that included chaplains, occupational physicians, security forces, public health officials, civilian personnel officers, and several other offices whose interest was the well-being of the force. The Board next worked to identify the most high-risk personnel through public health surveillance, ultimately identifying about 1–3 per cent of the workforce, or approximately 300 employees, who were in one way or another at high risk of psychological, physical, or financial difficulty. Klunder (2008) designed a range of preventive interventions to first help this high risk group of employees and, subsequently, others. Over the six year period, this systemic and comprehensive set of interventions initiated by the commanding general led to the prevention of fatalities through suicide or homicide (there were none!) and cost avoidance of over $33,000,000 in complaints that the Equal Employment Opportunity personnel said were "never filed" because of Klunder's early interventions.

These two systemic case examples provide perspectives on organizational interventions that may lead to health and well-being as well as avoidance of distress and death. However, based on the design of the Wellness Program at Lincoln Industries, it appears that it is possible to create psychologically and physically healthy work environments that support and enhance employees' well-being as well as the organization's well-being. Hopefully, after having considered the contemporary organizational change models and perspectives laid out in this chapter, the reader is able to see more clearly the important elements of organizational change efforts.

References

Ajzen, I. (1991). The theory of planned behavior. *Organizational Behavior and Human Decision Processes*, 50, 179–211.

Ajzen, I., & Fishbein, M. (1980). *Understanding attitudes and predicting social behavior.* Upper Saddle River, NJ: Prentice Hall.

Armenakis, A. A., & Bedeian, A. G. (1999). Organizational change: A review of theory and research in the 1990s. *Journal of Management*, 25(3), 293–315. doi:10.1177/014920639902500303

Armenakis, A. A., & Harris, S. G. (2009). Reflections: Our journey in organizational change research and practice. *Journal of Change Management*, 9, 127–142. doi:10.1080/14697010902879079

Armenakis, A. A., Harris, S. G., Cole, M. S., Fillmer, J. L., & Self, D. R. (2007). A top management team's reactions to organizational transformation: The diagnostic benefits of five key change sentiments. *Journal of Change Management*, 7(3–4), 273–290. doi:10.1080/14697010701771014

Balogun, J. & Hope Hailey, V. (2004). *Exploring Strategic Change.* (2nd ed.). Prentice Hall, London.

Blau, P. M. (1964). *Exchange and Power in Social Life*. New York: Wiley.

Bono, J. E., Foldes, H. J., Vinson, G., & Muros, J. P. (2007). Workplace emotions: The role of supervision and leadership. *Journal of Applied Psychology*, 92(5), 1357–1367. doi:10.1037/0021-9010.92.5.1357

By, R. T. (2005). Organizational change management: A critical review. *Journal of Change Management*, 5, 369–380. doi:10.1080/14697010500359250

Cameron, K. S., Dutton, J. E., & Quinn, R. (2003). *Positive Organizational Scholarship.* San Francisco, CA: Berrett-Koehler.

Cameron, K. S. (2007). Forgiveness in organizations. In D. L. Nelson & C. L. Cooper (Eds.) *Positive Organizational Behavior:* 129–142. Thousand Oaks, CA: Sage.

Cropanzano, R., & Wright, T. A. (2011). The impact of organizational justice on occupational health. In J.C. Quick & L.E. Tetrick (Eds.) *Handbook of Occupational Health Psychology, Second Edition*: pp. 205–219. Washington, DC: American Psychological Association.

Eubanks, D. L., Murphy, S. T., & Mumford, M. D. (2010). Intuition as an influence on creative problem-solving: The effects of intuition, positive affect, and training. *Creativity Research Journal, 22(2)*, 170–184. doi:10.1080/10400419.2010.481513

Ford, J. D., Ford, L. W., & D'Amelio, A. (2008). Resistance to change: The rest of the story. *Academy of Management Review, 33*, 362–377.

Fredrickson, B. L., Tugade, M. M., Waugh, C. E., & Larkin, G. (2003). What good are positive emotions in crises? A prospective study of resilience and emotions following the terrorist attacks on the United States on September 11th, 2001. *Journal of Personality and Social Psychology, 84*, 365–376.

Gaël, M., René, A., & Christine, C. (2009). How good micro/macro ergonomics may improve resilience, but not necessarily safety. *Safety Science, 47(2)*, 285–294. doi:10.1016/j.ssci.2008.03.002

Grant, A. M. (2010). It takes time: A stages of change perspective on the adoption of workplace coaching skills. *Journal of Change Management, 10(1)*, 61–77. doi:10.1080/14697010903549440

Griffin, M. A., & Clarke, S. (2011). Stress and well-being at work. In S. Zedeck (Ed.), *Handbook of Industrial and Organizational Psychology, Vol. 3*, pp. 359–398. Washington, DC: APA Books.

Guzzo, R. A., Jette, R. D., & Katzell, R. A. (1985). The effects of psychologically based intervention programs on worker productivity: A meta-analysis. *Personnel Psychology, 38(2)*, 275–291. doi:10.1111/j.1744-6570.1985.tb00547.x

Halfhill, T. R., Huff, J. W., Johnson, D. A., Ballentine, R. D., & Beyerlein, M. M. (2002). Interventions that work (and some that don't): An executive summary of the organizational change literature. In R. L. Lowman (Ed.), *The California School of Organizational Studies Handbook of Organizational Consulting Psychology: A comprehensive guide to theory, skills, and techniques* (pp. 619–644). San Francisco: Jossey-Bass.

Hammer, L. B., & Zimmerman, K. L. (2011). Quality of work life. In S. Zedeck (Ed.), *Handbook of Industrial and Organizational Psychology, Vol. 3*, pp. 399–431. Washington, DC: APA Books.

Hendrick, H. W. (1991). Ergonomics in organizational design and management. *Ergonomics, 34*, 743–756.

Herold, D. M., Fedor, D. B., Caldwell, S. & Liu, Y. (2008). The effects of transformational and change leadership on employees' commitment to a change: A multilevel study. *Journal of Applied Psychology, 93*, 346–357. doi: 10.1037/0021-9010.93.2.346

Hofmann, D. A. & Tetrick, L. E. (2003). The etiology of the concept of health: Implications for "organizing" individual and organizational health. In D. A. Hofmann & L. E. Tetrick (Eds.) *Health and safety in organizations* (pp. 1–28). San Francisco, DA: Jossey-Bass.

Holden, R. J., Or, C. K. L., Alper, S. J., Rivera, A. J., & Karsh, B. (2008). A change management framework for macroergonomic field research. *Applied Ergonomics, 39*, 459–474.

Holt, D. T., Armenakis, A. A., Feild, H. S., & Harris, S. G. (2007). Readiness for Organizational Change: The Systematic Development of a Scale. *Journal of Applied Behavioral Science, 43(2)*, 232–255. doi:10.1177/0021886306295295

Isen, A. M., & Labroo, A. A. (2003). Some ways in which positive affect facilitates decision making and judgment. In S. L. Schneider & J. Shanteau (Eds.), *Cambridge series on judgment and decision making. Emerging perspectives on judgment and decision research* (pp. 365–393). New York: Cambridge University Press.

Jimmieson, N. L., Peach, M., & White, K. M. (2008). Utilizing the Theory of Planned Behavior to Inform Change Management: An Investigation of Employee Intentions to

Support Organizational Change. *The Journal of Applied Behavioral Science, 44(2)*, 237–262.

Judge, W., & Douglas, T. (2009). Organizational change capacity: The systematic development of a scale. *Journal of Organizational Change Management, 22(6)*, 635–649. doi:10.1108/09534810910997041

Kanter, R. M., Stein, G. A. & Jick, T. D. (1992). *The Challenge of Organizational Change*. New York: The Free Press.

Karp, T. & Helgø, T. I. T. (2008). From Change Management to Change Leadership: Embracing Chaotic Change in Public Service Organizations. *Journal of Change Management, 8*(1), 85–96. doi:10.1080/14697010801937648

Karsh, B.-T., & Brown, R. (2010). Macroergonomics and patient safety: The impact of levels on theory, measurement, analysis and intervention in patient safety research. *Applied Ergonomics, 41(5)*, 674–681. doi:10.1016/j.apergo.2009.12.007

Klunder, C.S. (2008). Preventive stress management at work: The case of the San Antonio Air Logistics Center, Air Force Materiel Command (AFMC). Managing & Leading: SPIM Conference and Institutes, San Antonio, 29 February.

Kotter, J. P. (1996). *Leading Change*. Boston, MA: Harvard Business School Press.

Kuoppala, J., Lamminpää, A., Liira, J., & Vainio, H. (2008). Leadership, job well-being, and health effects – A systematic review and a meta-analysis. *Journal of Occupational and Environmental Medicine, 50(8)*, 904–915. doi:10.1097/JOM.0b013e31817e918d

Lazarus, R. S., & Folkman, S. (1984). *Stress, appraisal, and coping*. New York: Springer.

Lewin, K. (1947). Frontiers in group dynamics: concept, method and reality in social science; social equilibria and social change. *Human Relations, 1*, 5–41. doi:10.1177/001872674700100103

Lewin, K. (1951) *Field theory in social science; selected theoretical papers*. D. Cartwright (Ed.). New York: Harper & Row.

Lincoln Industries Wellness Program. (5 December 2010) www.lincolnindustries.com/wellness.

Luecke, R. (2003). *Managing Change and Transition*. Boston, MA: Harvard Business School Press.

Luthans, F., & Avolio, B. J. (2009). The "point" of positive organizational behavior. *Journal of Organizational Behavior, 30(2)*, 291–307. doi:10.1002/job.589

Luthans, F., Youssef, C. M., & Avolio, B. J. (2007). *Psychological capital*. Oxford: Oxford University Press.

Macik-Frey, M., Quick, J. C., & Nelson, D. L. (2007). Advances in occupational health: From a stressful beginning to a positive future. *Journal of Management, 33*, 809–840.

Macik-Frey, M., Quick, J. D., Quick, J. C., & Nelson, D. L. (2009). *Occupational health psychology: From preventive medicine to psychologically healthy workplaces*. In A.-S. Antoniou, C. L. Cooper, G. P. Chrousos, C. D. Spielberger, & M. W. Eysenck (Eds.) *Handbook of Managerial Behavior and Occupational Health*: 3–19. Cheltenham, UK: Edward Elgar (Greek and English).

Martins, L. L. (2011). Organizational change and development. In S. Zedeck (Ed.), *Handbook of Industrial and Organizational Psychology, Vol. 3*, pp. 691–728. Washington, DC: APA Books.

Mullen, J., & Kelloway, E. K. (2011). Occupational health and safety leadership. In J. C. Quick and L. E. Tetrick (Eds.), *Handbook of Occupational Health Psychology*, 2nd edition, pp. 357–372.

Neves, P. (2009). Readiness for change: Contributions for employee's level of individual change and turnover intentions. *Journal of Change Management, 9(2)*, 215–231.

Neves, P., & Caetano, A. (2006). Social exchange processes in organizational change: The roles of trust and control. *Journal of Change Management*, 6, 351–364.

Nielsen, K., Randall, R., Yarker, J., & Brenner, S.-O. (2008). The effects of transformational leadership on followers' perceived work characteristics and psychological well-being: A longitudinal study. *Work & Stress, 22(1)*, 16–32. doi:10.1080/02678370801979430

Prochaska, J. M. (2000). A transtheoretical model for assessing organizational change: A study of family service agencies' movement to time limited therapy. *Families in Society*, *81*, 76–84.

Prochaska, J. M., Prochaska, J. O., & Levesque, D. A. (2001). A transtheoretical approach to changing organizations. *Administration and Policy in Mental Health, 28(4)*, 247–261. doi:10.1023/A:1011155212811

Quick, J. C., Quick, J. D., Nelson, D. L., & Hurrell, J. J. Jr. (1997). *Preventive stress management in organizations*. Washington, DC: American Psychological Association.

Quick, J. C., & Quick, J. D. (In press). Executive well-being. In K. Cameron & A. Caza (Eds.) *Part VII – Happiness and Organizations, Handbook of Happiness*. Cambridge and New York: Oxford University Press.

Rafferty, A. E., & Griffin, M. A. (2006). Perceptions of organizational change: A stress and coping perspective. *Journal of Applied Psychology*, *91*, 1154–1162. doi: 10.1037/0021-9010.91.5.1154

Senge, P., Kleiner, A., Roberts, C., Ross, R., Roth, G., & Smith, B. (1999). *The dance of change – the challenges of sustaining momentum in learning organizations – a complexity approach to change*. London: Nicholas Brealey Publishing.

Skakon, J., Nielsen, K., Borg, V., & Guzman, J. (2010). Are leaders' well-being, behaviours and style associated with the affective well-being of their employees? A systematic review of three decades of research. *Work & Stress, 24(2)*, 107–139. doi:10.1080/02678 373.2010.495262

Thrash, T. M., Maruskin, L. A., Cassidy, S. E., Fryer, J. W., & Ryan, R. M. (2010). Mediating between the muse and the masses: Inspiration and the actualization of creative ideas. *Journal of Personality and Social Psychology, 98(3)*, 469–487. doi:10.1037/a0017907

Tugade, M. M., & Fredrickson, B. L. (2007). Regulation of positive emotions: Emotion regulation strategies that promote resilience. *Journal of Happiness Studies*, *8*, 311–333.

van Dijk, R. & van Dick, R. (2009). Navigating Organizational Change: Change Leaders, Employee Resistance and Work-based Identities. *Journal of Change Management*, 9(2), 143–163. doi:10.1080/14697010902879087

Vyhlidal, T. (2010). Lincoln Industries. In *Employer Experiences: Lessons from Award-Winning Organizations*. Psychologically Health Workplace Conference. Washington, DC.

Weick, K. E. (2001). *Making sense of the organization*. Malden: Blackwell Publishing.

Whysall, Z. J., Haslam, C., & Haslam, R. (2007). Developing the Stage of Change Approach for the Reduction of Work-related Musculoskeletal Disorders. *Journal of Health Psychology, 12(1)*, 184–197. doi:10.1177/1359105307071753

World Health Organization. (1946). [www.who.int/bulletin/archives/80(12)981.pdf *WHO definition of Health*], Preamble to the Constitution of the World Health Organization as adopted by the International Health Conference, New York, 19–22 June 1946; signed on 22 July 1946 by the representatives of 61 States (Official Records of the World Health Organization, no. 2, p. 100) and entered into force on 7 April 1948.

Young, M. (2009). A meta model of change. *Journal of Organizational Change Management*, *22*, 524–548. doi:10.1108/-9534810910983488

5 Psychosocial safety climate

A lead indicator of workplace psychological health and engagement and a precursor to intervention success

Maureen F. Dollard

Introduction

Work-related stress is increasingly recognized as a central occupational health and safety (OHS) issue in many organizations, with the potential to have a significant detrimental impact on employees' health, psychological well-being and work effectiveness. Interventions to address work-related stress are increasing but are mainly secondary, targeting the skills, perceptions and beliefs of individual workers, or tertiary, focusing on treating injured workers (Caulfield et al., 2004; Giga et al., 2003; Jordan et al., 2003). By comparison interventions at the organizational level have been largely ignored (Leka, Cox, and Zwetsloot, 2008). This is not too surprising and follows both theoretical and empirical trends in the area that focus on task level psychosocial factors and individual factors rather than organizational factors as causal stress agents (Kang et al., 2008). However there is growing evidence that causes of work stress may be more distal than is commonly assumed, emerging from the organizational context (e.g., climate; organizational structure) or extra-organizational factors (Van Den Bossche, Smulders, and Houtman, 2006). In accord with the logic of a hierarchy of causes, for greatest effect it is imperative to tackle causes at the highest levels. Therefore there is an urgent need to develop and apply multi-level theory to inform primary prevention approaches so that interventions may address the true source of the problem.

Psychosocial safety climate (PSC) is an emerging construct that reflects the management value position and philosophy about work stress, and management priority of regard for the psychological health of workers versus production imperatives of the organization. We contend that PSC is a lead indicator of psychosocial hazards – "a cause of the causes" – and work stress, and is a crucial organizational level target for primary stress intervention. In this chapter we introduce theory and evidence for this position. Therefore, PSC is what an intervention should aim to improve. Every component of an intervention aimed to improve psychological health – the participatory process, the communication process, the support from management, and the priority of psychological health – together builds PSC.

A particular complication in work stress intervention research is the lack of evidence about the successful elements of primary interventions. In particular

78 *Challenges and methodological issues*

hardly any research has explored intervention starting conditions and their impact on the process of implementation and the impact of the intervention. This chapter thus reports on the development and implementation of an organizational primary prevention strategy focused on risk management, with a particular focus on the starting role of the PSC.

Evolution of the psychosocial safety climate construct and theory

The effects of psychosocial safety climate (PSC) were first observed in the 2004–2005 *Stress Prevention Study* in the Victorian Budget Sector, Australia. In that study, parts of which are described here, we assessed elements of best practice in work stress prevention interventions and found that these elements showed strong and consistent predictive effects with many psychosocial work conditions and stress outcomes.

This led to the identification of a new construct, PSC, and a theory building process began to develop. Theory building involves a descriptive stage and a normative stage each with three steps:

1 observation,
2 classification, and
3 defining relationships.

The process cycles through these steps iteratively (Carlile and Christensen, 2005). These steps are discussed next.

Step 1: Observation

This involves description and development of the construct, and defines what it is and how it operates (Carlile and Christensen, 2005). PSC refers to policies, practices, and procedures (PPP) for the protection of worker psychological health and safety (Dollard, 2007) that are largely driven by management (Dollard and Bakker, 2010). Psychosocial safety relates to freedom from psychological and social risk or harm (Dollard and Bakker, 2010).

PSC is a property of the organization

PSC is a facet-specific component of organizational climate, a "climate for psychosocial health and safety" (Dollard and Bakker, 2010, p. 580). PSC is an attribute of the organization rather than the individual (Glick, 1985). As a climate construct PSC may be assessed in terms of shared perceptions. This is usually achieved in organizational climate research by aggregating individual data from a team, work group or organization to a group level (Chan, 1998). According to leading safety climate theorists (Zohar and Luria, 2005) because focal organizational facets may represent competing operational demands (e.g. safety climate vs. productivity) the best indicators of an organization's true priories are enacted policies, practices and procedures. Therefore, similar to common definitions of safety climate, PSC refers to PPPs but specifically in relation to the protection of worker psychological health and well-being (Dollard, 2007).

PSC has important implications for work stress interventions. Any intervention will be affected by extant PSC conditions. For example, getting interventions started and appropriately implemented within organizations and work groups will be largely affected by prevailing levels of management commitment and support. Similarly, ensuring intervention effects on outcomes will be somewhat dependent on existing communication procedures regarding the identification of psychosocial risk factors and the implementation of control procedures. We expect that readiness to change in both management and workers will be also be affected by levels of PSC.

Theoretically, PSC is largely driven by management, and given that managers have discretionary authority regarding whether or not policies will be enacted, we expect that PSC will vary between organizations and between work teams or units. We confirmed that PSC had group level properties in Dollard and Bakker (2010). We found that 22 per cent of the variance in PSC could be explained by differences between work units; there was acceptable homogeneity (i.e. mean rwg agreement index = .76; James, Demaree, and Wolf, 1984), and reliability (ICC(2) = .74) of PSC within units. Together these effects suggest PSC could be aggregated to the unit level. The important implication is that PSC is a feature of the organization and as a precursor to work stress, the "cause" of work stress may be identified outside of the individual. We expect that there is a "psychological climate" at play as well, whereby individual workers experience PSC differently, based on their experiences. For instance, a worker who has experienced bullying in the workplace may view PSC differently from a worker who has not experienced bullying because their experience of the enactment of relevant PPP would be salient. Workers who have submitted a compensation claim for psychological injury would view PSC differently from those who had not, due to their interaction with the PPP system post-injury. We are interested in decomposing the variance in PSC to identify multiple levels of the phenomenon and their possible levels of influence on work conditions and stress outcomes.

The domain composition of PSC initially comprised two areas; management commitment and support; and participation and consultation. In the first study these were assessed using a four item scale (Dollard and Bakker, 2010). In the interests of adequate domain sampling we expanded the scale to include two further domain areas (Dollard and Kang, 2007) following a review of the safety climate literature (e.g. Clarke, 2006; Cox and Cheyne, 2000; Gershon et al., 2000; Pronovost et al., 2003). We developed a 26-item PSC scale to reflect the four inter-related domains (Dollard and Kang, 2007):

1 *Senior management support and commitment* for stress prevention through involvement and commitment;
2 *Management priority* towards psychological health and safety versus productivity goals;

3   *Organizational communication*, i.e. the organization listens to contributions from employees;
4   Organizational *participation and involvement* in relation to psychological health and safety, e.g. participation and consultation occurs with employee representatives (i.e. unions) and occupational health and safety representatives.

Next, we developed a parsimonious scale with increased domain coverage, and reduced the 26 items to a concise 12-item scale, with three items reflecting each of the four domains (Hall, Dollard, and Coward, 2010). This latter study also substantiates the group level properties of PSC and affirms its status as an organizational phenomenon.

Step 2: Classification

This phase involves the classification of the phenomenon into categories (Carlile and Christensen, 2005). The construct was initially termed "best practice in OHS (occupational health and safety)", and subsequently as "OHS culture" in draft papers emerging from the study. After consideration of the safety climate literature, parallels emerged between the safety climate construct and its operations and the phenomenon under observation. The safety climate construct has been extremely useful for promoting OHS practice particularly in high risk industries (Cox and Cheyne, 2000), offering prediction of objective and subjective safety criteria (e.g. injury rates, accidents) (Christian et al., 2009). But in its 30 year history the safety climate construct (Zohar, 2010) has rarely been linked to the psychological health and well-being of workers. PSC is in the genre of safety climate but it is a high fidelity safety climate construct focusing on psychological health. From a social policy perspective, psychological health at work is an important occupational health and safety issue, so elaborating the PSC construct in safety climate terms could draw fresh attention to the issue from an OHS perspective, and enhance PPP for psychological health. Even though the empirical base linking psychosocial hazards to health is strong, in OHS practice much less attention is given to psychosocial hazards compared to physical hazards, by OHS legislation and OHS inspectors (Australian Productivity Commission, 2010). Yet theories matter and help to draw attention to an issue, shape public policy, and promote a research agenda in the area (Pfeffer, 2010). The term PSC was therefore created[1] and theory developed to situate the construct within an occupational health and safety framework.

Step 3: Defining relationships

In the descriptive phase of theory development relationships are proposed so that models may be developed. This completes the first descriptive stage of the inductive theory building process. Testing these hypotheses for example in different samples constitutes the deductive portion of the cycle (Carlile and Christensen, 2005). Here we present hypotheses or assumptions, empirical evidence, and tentative statements of causality-moving from descriptive to normative theory.

As a general proposition PSC is a contextual variable and is a lead indicator of work conditions identified in work stress models as work stressors or psychosocial risk factors (e.g. the Job Demand-Control model (JD-C), Karasek, 1979; or the Job Demand-Resources (JD-R) model, Demerouti et al., 2001). Here we extend JD-R theory with the following propositions or premises (see Figure 5.1).

JD-R theory classifies psychosocial risk factors as demands and resources. Job demands refer to psychological, social, physical or organizational aspects of the job (e.g. work pressure, emotional demands) that require sustained physical or psychological effort and are therefore associated with certain physiological or psychological costs. Job resources are psychological, social, or organizational aspects of the job (e.g. social support, autonomy) that are functional in achieving work goals; and/or reducing job demands and the associated psychological and physiological cost; and/or stimulate personal development, growth, and learning (Demerouti et al., 2001). The PSC propositions are:

1   PSC is a lead indicator of workplace demands (Dollard and Bakker, 2010; Hall et al., 2010; Idris, Dollard and Winefield, 2011). Theoretically, in high PSC contexts where managers are concerned about worker well-being, managers will ensure that the demands that workers face are manageable; there is a negative relationship between PSC and work demands.
2   PSC is a lead indicator of workplace resources (Dollard and Bakker, 2010; Idris et al., 2011). Theoretically in high PSC contexts where managers are concerned about worker well-being, managers will ensure that workers have

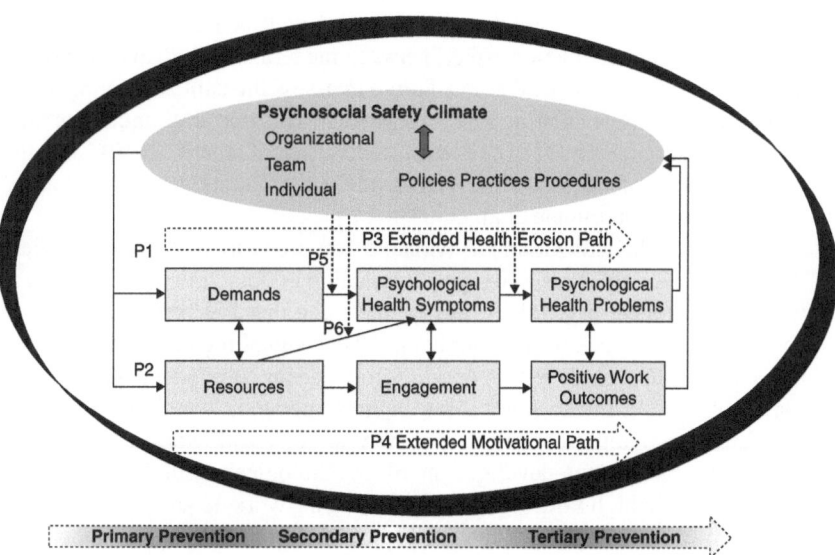

Figure 5.1 Psychosocial safety climate model.

enough resources to do the job; there is a positive relationship between PSC and resources.

3   JD-R theory proposes a health erosion process such that exposure to chronic demands leads to an erosion of personal resources and energy and in turn leads to burnout and health problems. We propose an *extended health erosion path* (P3 in Figure 5.1) whereby PSC is a precursor to this path. In a mediated path, PSC is negatively related to psychological health problems via its negative relationship to work demands (Dollard and Bakker, 2010; Law et al., under review).

4   The JD-R model also proposes a motivational process, whereby adequate job resources lead to work engagement. According to social exchange theory, workers will give back to the organization (i.e. engagement) what they receive (i.e. resources) (Blau, 1964). We propose an *extended motivational path* (P4 in Figure 5.1) whereby PSC is a precursor to this process; in a mediated path the effect of PSC on work engagement is carried by job resources (Dollard and Bakker, 2010; Idris et al., 2011).

5   PSC is a support laden contextual variable and moderates the negative impact of psychosocial risk factors on psychological health (Dollard and Bakker, 2010). In particular, consistent with many work stress theories (e.g. the JD-C, Karasek, 1979; the JD-R model, Demerouti et al., 2001), we expect that the detrimental effect of demands on psychological health will be moderated in high resource conditions (i.e. high PSC, high control, high support) (see also Xanthopoulou et al., 2007). Evidence from major reviews of the JD-C and DCS models, derived from 126 studies from 1979 to 2007 (van der Doef and Maes, 1999; Häusser et al., 2010) and 45 longitudinal studies (de Lange et al., 2003), generally supports a main effects model whereby demands, control and support additively predict psychological well-being. Yet the theories generally predict interaction effects between the terms. Failure to detect stress interaction effects may be due to a failure to assess the context or climate (de Lange et al., 2003) such as PSC. We have found that PSC moderates the effect of bullying on PTSD (Bond, Tuckey, and Dollard, 2010), and the adverse effect of negative customer behaviour on employee psychological well-being (Zimmermann et al., 2009).

6   PSC enables the utilization of available resources to build psychological health. This is evident as a two-way interaction effect. Drawing from behavioural and clinical psychology theory, we argue that PSC serves as a safety signal (Dollard, Tuckey, and Dormann, 2012), signalling action options that are available to provide respite or relief from danger cues (Lohr, Olatunji, and Sawchuk, 2007). In this sense safety signals operate as conditioned positive reinforcers providing cues regarding how to cope with adverse events and avoid the consequent development of psychological disorders (see Lohr et al., 2007). In high PSC organizations or teams, workers know that they can speak up without negative consequences, and may seek to utilise support to cope with demands without concern about negative consequences. They will feel reassured for example, that action will be taken when reporting bullying.

Evidence in support of the proposition is that Dollard and Karasek (2010) showed the positive benefits of decision latitude in reducing psychological distress only in high PSC work units. In other words high PSC enabled (signalled) the (safe) utilisation of the already available resource, decision latitude. Other interactions with PSC noted above could also be explained in terms of safety signal theory as well (i.e. Dollard and Bakker 2010; Zimmermann et al., 2009; Bond et al., 2010).

Additional propositions are:

7   PSC is negatively related to unsafe and hazardous behaviour, e.g. work place bullying, and harassment (Bond et al., 2010; Law et al., 2011). Theoretically in a strong PSC environment, clear PPP will be enacted that motivate desired worker behaviours, roles and responsibilities in delivering OHS obligations enshrined in duty of care or OHS legislation. However, due to leadership discretion considerable variation may exist in the delivery of PPP. For example, in organizations with a strong macho hierarchical culture, such as policing, there is likely to be a low concern for employee psychological health (Tuckey, Winwood, and Dollard, 2011). Such conditions are prone to give rise to bullying and due to management discretion other psychologically threatening behaviours.
8   PSC is a construct distinct from related constructs such as organizational social support, team psychological climate, and physical safety climate. This divergent validity is demonstrated in a study by Idris, Dollard, Coward, and Dormann, 2012. Theoretically, PSC is distinct from these measures as outlined in Dollard and Bakker (2010) and Hall et al. (2010).

In sum, descriptive theory and empirical evidence to-date suggest a model whereby PSC precedes work conditions that, in turn, relate to health and important industry outcomes (e.g. sickness absence, engagement), and moderates the relationship between work conditions and psychological health. PSC therefore serves both a *preventative* (proposition 1, 2) and an *ameliorative* (proposition 5, 6) function in the development of work stress and is therefore an optimal target for primary and secondary intervention. PSC also serves a tertiary intervention function, providing PPPs for when workers experience psychological problems (e.g. Employee Assistance Programs). PSC serves an integral component of a resilient organizational system that protects and builds individual resilience. Next, we explore organizational conditions, i.e. prevailing PSC levels, and their effects on intervention implementation.

Starting conditions and process of the intervention

Evaluation research on work stress interventions assesses whether specific interventions lead to improvements in worker health and/or to improvements in aspects of the work environment that affect worker health. Most research has a narrow focus on effectiveness alone and addresses the question "did the intervention have the desired effect?". However, Goldenhar et al. (2001) argue that all three phases of the intervention should be evaluated: development, implementation, as well as effectiveness. This is because the impact of an intervention first and foremost depends on how well the intervention was developed and implemented.

Research on the development and implementation phases of OHS interventions is emerging to answer the question of how and why observed changes were or were not achieved. This research is also referred to as formative evaluation, process evaluation, program monitoring and implementation assessment (Goldenhar et al., 2001). Process factors that may impact on an intervention's effectiveness include weak, inconsistent or nonexistent implementation (Goldenhar et al., p. 618) and contextual factors (Nytrø et al., 2000; Saksvik et al., 2002). So the question of *what works* (effectiveness evaluation) depends not only on the content of the intervention but also on the process (*how*) and context of implementation (Nytrø et al., 2000).

Researchers have outlined a number of process and contextual issues influencing the developmental stage of an organizational stress intervention (Biron, Gatrell, and Cooper, 2010; Dollard and Karasek, 2010). In particular, organizational capacity and resource issues (financial, human, expertise, and skills) are crucial, as is the political will of stakeholders (Dollard and Karasek, 2010).

An often studied feature in evaluation of the developmental stages of intervention is readiness for change and willingness to participate – both potentially limiting and facilitating factors in organizational change (DeJoy et al., 2010; Pasmore and Fagans, 1992). Individual contributors to readiness for organizational change may include personal coping strategies and self-efficacy, i.e. the perceived ability to manage change, successfully (Prochaska, Redding, and Evers, 1997). For example, those with a higher level of self-efficacy may be more confident about job change and adopt an active approach to change problem solving, have a higher level of readiness for change and be more likely to participate in organizational redesign activities (Cunningham et al., 2002). Workplace factors are also expected to affect readiness for organizational change. Cunningham et al. (2002) showed that readiness for change was best predicted by both individual factors and by jobs combining high demands and high control (i.e. active jobs as described by Karasek, 1979).

Even further upstream, organizational culture and climate, and group norms are key factors in readiness for change or resisting factors in organizational change (Heward, Hutchins, and Keleher, 2007). In a similar vein we assert that PSC is a feature of organization climate that affects readiness or capacity to change at an organizational, team and individual level. Any workplace intervention is introduced into a context, a pre-existing set of social, political and structural conditions that influence the intervention (Heward et al., 2007; Pettigrew, Ferlie, and McKee, 1992). Here we specifically focus on the influence of the prevailing PSC on intervention components, quality and progress, and change in stress outcomes.

Case study

In this study, we evaluate the early phases of a stress prevention pilot project and test two hypotheses:

1 PSC will be related to intervention components (workshop sessions attended and ratings of workshop quality), intervention quality, and intervention progress; and
2 PSC, along with intervention components, quality and progress, will be related to change in stress outcomes.

Stimulus for the intervention and approach

An OHS Roundtable comprising senior executives of Victorian Government Departments, Unions, and the Victorian WorkCover Authority[2] identified stress as a priority issue due to the rising incidence of stress claims. The Roundtable endorsed a tripartite study tour to Europe and the UK to explore models to address stress in the workplace. The study tour group endorsed the following best practice principles of a stress prevention model[3] (Giga et al., 2003; European Agency for Safety and Health at Work, 2002; Kompier and Cooper, 1999):

- Support of top management through involvement and commitment.
- Human resource management practices include leadership and management capability and accountability.
- A positive work group climate – colleague and manager support.
- Involvement of all layers of the organization.
- Integration into the day to day operations of the organization.
- Risk identification, risk assessment and implementation of risk controls at organizational and work unit levels.
- Participation and consultation with employees, unions and health and safety representatives – "A bottom up approach".

To enable employers to meet the requirements of OHS legislation, the tour group recommended a trial of the risk management model featuring the ingredients as outlined above (see Cox et al., 2000; Leka, Cox and Zwetsloot, 2008; Nytrø et al., 2000).

Research design and participants

Two government organizations from the Australian state of Victoria participated in the pilot.[4] Initially there were 41 naturally occurring workgroups randomly assigned to the intervention ($N = 20$) or control ($N = 21$) group. In this study, we focus on the intervention group (because only they can evaluate the intervention process), and groups with at least five responses at Time 1 ($N = 18$ intervention groups). Potential participants were initially asked to volunteer by signing a

Table 5.1 Sample demographics

	Mean age	Gender	% volunteers	Mean group size
Department 1 $N = 108$	45–54	79% female 21% male		
Department 2 $N = 73$	25–54	71% female 29% male		
T1 $N = 181$			76	10.05
T2 $N = 124$			52	7.77
T3 $N = 119$			50	6.61

consent form and full ethics approval for the project was obtained. For ethical and practical reasons participants were informed whether they were in an intervention or control group. The sample demographics are shown in Table 5.1 and were *representative* of the demographics of the respective departments.

Survey data were obtained before the intervention to establish baseline measures of health and stress factors and to identify risk factors (T1). Stage one comprised the implementation of the workshops over a six-week period. A second survey (T2) was conducted two weeks after the completion of the workshops, and a third survey (T3) was conducted 10 months later (12 months from commencement). Four additional measures were used to evaluate the workshops: a pre-workshop survey, a post-workshop survey immediately following the workshop completion, in-depth interviews with workshop participants ($N = 13$), and a four-hour focus group with workshop facilitators ($N = 6$).

Participatory design

A participatory action research design (plan, action, observe-data, feedback, reflect) was used (Dollard, le Blanc, and Cotton, 2008) with all key stakeholders involved in the design, implementation, and evaluation of the overall program.

Workshops

The intervention groups participated in a maximum of four interactive training and development group sessions for a maximum of 16 hours over a 4 week period (70 per cent participation rate). Units included education on using the stress risk management process. Within the workshops, employees and managers in their natural workgroups worked together to develop and implement a risk management process (interpreting risks, developing action plans, implementing, evaluating and improving interventions) to prevent occupational stress. With the assistance of an external facilitator the work groups triangulated data from a range of sources (e.g. aggregated grievance data, absentee records; the results of the T1 survey are available from the author) to assess local psychosocial risks and needs. They considered risks in the context of current management systems and employee support mechanisms to determine residual risk that required action. Action plans were developed to implement risk controls at a work group level, and actions beyond their resources or authority were referred to a Regional Steering Committee established for the project (see Dollard and Karasek, 2010).

Measures

Measures and their properties are shown in Table 5.2.

Table 5.2 Description of intervention measures and stress outcome measures

	N items	Example items and/or response options	Cronbach's α
Intervention measures			
Sessions attended (collected at T2)	1	Minimum 0, maximum 4	
Workshop quality (collected at T2)	3	"The facilitated workgroup sessions were worthwhile"	.91
		"My workgroup was able to discuss issues openly"	
		"My workgroup was able to determine actions to address stress factors in the workplace"	
		From 1 (*strongly disagree*) to 5 (*strongly agree*)	
Psychosocial safety climate (collected at T2)[5] (Dollard and Bakker, 2010)	4	"Senior management shows support for stress prevention through involvement and commitment"	.81
		From 1 (*strongly disagree*) to 5 (*strongly agree*)	
Intervention quality (collected at T3)	2	"To what extent have you been listened to in the stress prevention process (SPP)"	.86
		"To what extent has trust been built (e.g. from feeling supported, listened to within the organization) as a result of the SPP?"	
		From 0 (*to a very small extent*) to 4 (*to a large extent*)	
Intervention progress (collected at T3)	2	"To what extent are actions from your workgroups action plans being addressed?" "Based on your experience to what extent has the SPP led to change in the work environment?"	.84
		From 0 (*not at all*) to 5 (*to a large extent*)	

(Continued overleaf)

88 *Challenges and methodological issues*

Table 5.2 Continued

	N items	Example items and/or response options	Cronbach's α
Stress outcome measures (all collected at T2 and T3)			
Psychological distress (Goldberg, 1978)	12	"Have you recently (last two weeks) felt constantly under strain?" From 0 (*not at all*) to 3 (*much more than usual*)	.90
Emotional exhaustion (Schaufeli et al., 1996)	2	"I feel emotionally drained from my work" From 0 (*very rarely/never*) to 4 (*very often, always*)	.85
Engagement (Schaufeli and Bakker, 2003)	2	"I am enthusiastic about my job" From 0 (*very rarely/never*) to 4 (*very often, always*)	.80
Intention to resign	1	"Have you ever seriously considered resigning from your work?" From 0 (*very rarely/never*) to 4 (*very often, always*)	
Job satisfaction (Warr, Cook, and Wall, 1979)	1	"Taking everything into consideration how do you feel about your job as a whole?" From 1 (*extremely dissatisfied*) to 7 (*extremely satisfied*)	
Sickness absence		Data were obtained from organizational data bases and matched to individual data	

Results

Relationship of psychosocial safety climate to intervention components

In relation to Hypothesis 1 that PSC will be related to intervention components (workshop sessions attended and ratings of workshop quality), intervention quality, and intervention progress, as shown in Table 5.3, we found strong correlations between PSC and all aspects of the intervention except workshop quality.

Next, we assessed the relative importance of PSC, sessions attended, and workshop quality on intervention quality at T3 (see Table 5.4). The model accounted for 67 per cent of the variance in the intervention process ($F(3, 14) = 9.28$, $p < .01$). PSC and number of workshop sessions attended were most important.

Additionally we regressed intervention progress on PSC, sessions attended, and quality of the workshop (see Table 5.4). The model accounted for 38 per cent of the variance ($F(3, 14) = 2.91$, $p < .10$). Only PSC was significantly related to the

Table 5.3 Pearson correlations between PSC, sessions attended, workshop quality, intervention quality, and intervention progress

	PSC (T2)	Sessions attended	Workshop quality	Intervention quality
Sessions attended (T2)	.63**			
Workshop quality (T2)	.21	.32		
Intervention quality (T3)	.68**	.72**	.46^	
Intervention progress (T3)	.60**	.37	.26	.73**

Note. Intervention groups $N = 18$ groups; $^\wedge p < .1$, $* p < .05$, $** p < .01$.

Table 5.4 Regression of intervention quality and progress at T3 on PSC, sessions attended, and workshop quality

	Intervention Quality T3		Intervention Progress T3	
	Beta	p	Beta	p
Psychosocial safety climate (T2)	.37	.08	.61	.04
Sessions attended (T2)	.41	.07	−.05	.85
Workshop quality (T2)	.25	.15	.14	.53

Note. Intervention groups $N = 18$.

progress of the intervention. In sum, PSC is related to the success of intervention components, i.e. involvement of participants in workshop sessions, and is an important predictor of both intervention progress and quality, over and above workshop participation and quality.

Relationship of psychosocial safety climate and intervention components, to stress outcomes

To test Hypothesis 2 that PSC, intervention components, intervention quality, and intervention progress are related to change in stress outcomes, we used hierarchical linear modelling (HLM) and HLM 6.06 software (Raudenbush et al., 2005). The data were nested, within individuals (T2, T3), within units ($N = 18$), and within departments ($N = 2$). Initial analyses showed that for some of the outcome measures random effects at Level 3 (department) were significant. Therefore, we continued with a three level model controlling for random Level 3 effects. We regressed T3 stress outcome measures on group level measures beginning with PSC,[6] controlling for baseline T2 stress measures, age, gender and individual perception of PSC. All variables were group mean-centered.

Table 5.5 shows that in each case PSC added significant variance to the model (M1). Intervention progress added variance to some models (M2), but workshop sessions attended, quality of workshops, and intervention quality, did

Table 5.5 Random Coefficient Modelling regressing stress outcomes at T3 on PSC (group and individual), and intervention progress (intervention groups only)

	Psychological distress T3		Exhaustion T3		Engagement T3		Intention to Resign T3		Satisfaction T3		Sickness Absence T3	
	M1	M2	M1	M2	M1	M2	M1	M2	M1	M2	M1	M2
Grand Mean	11.89	12.03	4.14	4.11	5.37	5.38	1.13	1.14	5.34	5.32	7.69	7.70
Individual level predictors												
Gender	2.80	2.77^	0.65	0.65	−0.15	−0.14	−0.07	−0.07	0.43	0.43	0.68	0.68
Age	0.33	0.32	−0.63**	−0.63	0.20	0.20	0.04	0.04	0.00	0.00	−1.11	−1.11
Time 2 outcome	0.45**	0.44**	0.64***	0.64***	0.46***	0.43**	0.67***	0.67***	0.36***	0.36***	0.29**	0.28**
PSC	0.18	0.18	0.02	0.02	0.03	0.02	0.02	0.02	0.09*	0.09*	−0.34	−0.34
Group level predictors												
PSC	−0.66^	−0.29	−0.59***	−0.42**	0.39**	0.26^	−0.25***	−0.22**	0.29***	0.21**	−1.02^	−1.01
Intervention progress		−0.96^		−0.21		0.29^		−0.07		0.17^		−0.02
χ2(1)	3.24^	3.03^	15.16**	2.96	8.25**	2.93^	17.27***	0.97	13.97**	4.06*	3.54^	0.00

Note. ^, p < .1, *p < .05, **p < .01, ***p < .001. Groups, N = 18; Individuals, T2 N = 124; T3 N = 119. Unstandardized regression parameters. M1 adds PSC; M2 adds intervention progress. Chi-square refers to the significance of the change in deviance as M1 adds PSC, and M2 adds intervention progress.

not add any further variance to the models. Results showed that PSC is predictive of change in a number of stress outcomes T2-T3 including psychological health consequences (i.e. psychological distress and emotional exhaustion) and important organizational outcomes (i.e. work engagement, job satisfaction, intention to resign, sickness absence). Hypothesis 2 is supported with PSC as the predictor. If PSC was grand-mean centred at the lower level, results were similar, but for satisfaction in M2 neither PSC or intervention progress was significant.

Discussion of case study

The aim of this study was to determine the effect of starting conditions on components, quality, and progress of an organizational stress risk management intervention, focusing on the role of PSC. It is the first to attempt to link elements of context (prevailing PSC) to the success of an intervention implementation. Understanding the context of organizational interventions provides valuable information to industry, researchers, and practitioners regarding when intervention strategies are likely to be successful, and how to better prepare the context in applied organizational interventions.

We found strong support for the proposition that the starting condition, PSC estimated at a group level, predicts the success of the intervention implementation. It was associated with the extent to which the intervention groups attended the capacity building sessions, the quality of the intervention, and the progress (quantity) of the intervention, and predicted health and organizational stress-related outcomes over time.

Importance of psychosocial safety climate as grounding for the intervention

PSC was operationalized here in terms of best practice principles in stress prevention intervention initially identified by key stakeholders (i.e. the study tour group), based on inputs from the literature and international experts in the area. This construct was associated with reduced stress over time and, importantly, gave rise to better quality interventions (i.e. better ratings of trust being developed, and being listened to) and better intervention implementation (i.e. better ratings of the extent of the intervention and observed change).

These results provide strong support for the notion that prevailing levels of organizational PSC are extremely important (for instance, participation, consultation and communication) to consider in the development and preparation stage of the project and can potentially explain inconsistent implementation of stress interventions, and weak implementation of interventions found in other studies. In our study if groups began with low levels of PSC, participants not only attended fewer workshop sessions on average, but they also had poorer intervention implementation, poorer progress and poorer stress outcomes. The observation that group PSC had stronger effects than individual PSC on stress outcomes is consistent with research on safety climate; because group PSC reflects shared perceptions of individuals, it is likely to be more potent and more influential than when divergent individual perceptions prevail (cf. Christian et al., 2009).

Further, PSC is important in its own right explaining unique variance in a range of stress-related outcomes such as job attitudes (i.e. job satisfaction) and organizational outcomes (i.e. sickness absence) over and above the actual intervention for some outcomes. These results confirm Taris et al.'s (2003) argument that these best practice qualities of the implementation process are important to evaluate in order to identify specific aspects of the intervention most closely *linked* to positive/favourable impacts. For example we find that stronger participation and consultation with employee groups and representatives (i.e. unions) earlier on could lead to future stronger interventions and improved outcomes. We argue that these best practice elements in sum constitute organizational PSC.

Even though workshops were evaluated favourably by participants, the quality of the workshops was not related to stress outcomes, nor to the quality or progress of the intervention. Participants rated the workshops very favourably in terms of capacity building (i.e. participation improved my understanding of stress and the risk management process) and also in terms of belief about the relevance of intervention (i.e. it is wise to use the risk management process). This evidence accords well with reports from facilitators, and provides qualitative support that "the proposed intervention is suitable for the problem" (Nytrø et al., 2000). Competency-building in the intervention process appears to be an essential step; alone, however, it is not sufficient to lead to necessary changes.

Other starting conditions

Implementation of a well-controlled study in a "real life" context has many challenges including the timing of project events within dynamic organizational schedules. Preparation and planning required:

a   communication about the project to potential participants (Kompier and Kristensen, 2001), and
b   providing adequate lead time for work scheduling, and back filling resources to ensure that volunteers can participate. A strategy we used in this study to address additional demands and to provide flexibility, was to ask participants to represent other workgroup members in the workshops and discuss decisions and actions back in the workplace. Nevertheless, adequate resourcing was a key issue in starting conditions for some participants.

Limitations and future research

Participation and attrition of the sample were affected by workload, timing, scheduling problems, and staff turnover. Nevertheless, the demographics of the participating samples seemed representative of the respective departments, such that conclusions from the study can be generalized to the organizations themselves, and to other large government agencies. Theoretical and empirical developments in relation to PSC suggest that repeated measures of PSC should be used to assess whether the intervention itself improved perceptions of PSC overtime. In this regard the reciprocal effects of the intervention (e.g. improved psychological health) on evolving PSC through the course of the intervention would be of much interest. We concur with Saksvik et al. (2002) regarding the importance of undertaking of pilot work to investigate the cultural preparedness of the organizations, in this case PSC. A final limitation is the absence of a control group; however, the impact of the intervention relative to a comparison condition was beyond the scope of this chapter and is reported elsewhere.

Building psychosocial safety climate

Taken together, theoretical, empirical, and case study evidence suggests that PSC itself is an efficient target for a stress intervention. In particular, PSC is a good predictor of work conditions: at high levels it will ameliorate the relationship between demands and stress outcomes, it will facilitate the uptake and utilisation of available resources, and as demonstrated here PSC affects the success of an intervention that is largely conducted at a workgroup level. In sum, actions to build PSC should lead to many positive flow-on effects for worker conditions, worker health, and engagement. We concur with Heward et al. (2007) in their reflections on health promotion within organizations: "far from being peripheral, organizational change is a crucial element to effective action" (p. 176). In essence PSC is what an intervention should aim to improve. Every component of an intervention, the participatory process, the communication process, the support from management, and the priority of the project as action foci together should build PSC. But our research illuminates a conundrum; it is more likely to be difficult to implement interventions in organizations or groups with low PSC starting conditions, precisely where the intervention is probably required.

In the Healthy Conducive Production Model, Dollard and Karasek (2010) describe how PSC may be built within organizations. Karasek 2008 argue that support and stability outside an organization are being eroded or undermined by global socio-economic phenomena (e.g. deregulation, competition, breakdown of the family). Therefore internal organizational level regulatory structures are required to produce order in the environment, through the coordination of responses to random, unexpected and uncontrollable demands, and thus decrease entropy in the environment (Dollard and Karasek, 2010).

At a higher level within an organization, a lack of coordination and resourcing of incoming demands could in turn lead to threats to individual workers' stable self-regulation – interfering with work performance, coordination of tasks, personal development, personal resilience, job stability, and work/family life at the individual level (Dollard and Karasek, 2010). They propose a social collective structure at the organization level that will enable internal control and co-ordination of incoming demands that will build external social control, and theoretically enable workers' internal capacity for self-regulation without being overwhelmed (Dollard and Karasek, 2010).

In the intervention case study presented here, there was extensive social coordination of stakeholders (described in greater detail in Dollard and Karasek, 2010). Within an organization this social structure could be a tripartite committee: a social dialogue structure necessary for the development, implementation, and sustainability of stress prevention initiatives as required on a "business as usual" basis (Leka, Cox, and Zwetsloot, 2008). As depicted in Figure 5.2, the necessary ingredients for the operation of the committee include management and political will, employee good will, union support, financial and/or capacity surplus, trust, and other resources (Dollard and Karasek, 2010). These ingredients parallel the contextual ingredients identified by others that affect the developmental stage of an organizational stress intervention (Biron, Gatrell, and Cooper, 2010). Without careful consideration of each of these ingredients the operation of the committee may fail, along with the development of PSC.

The goal of the committee is to increase order (decrease entropy), and design well-coordinated PPP across levels and between departments, including PPP that protect the psychological well-being of workers (e.g. return to work policies and practices).

Successful social coordination depends on strong communication and participation processes, upwards and downwards. Risk assessment of psychosocial risks operating at both group and organizational levels is necessary information for the operation of the committee, and in turn the organization and the work group.

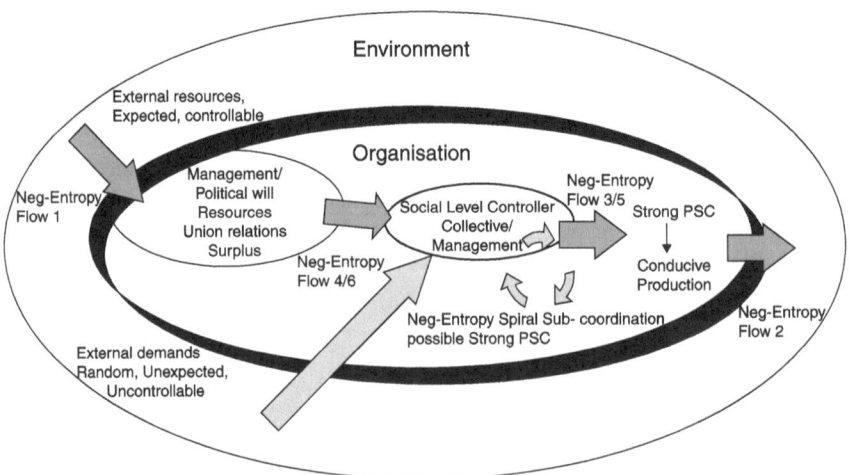

Figure 5.2 The Healthy Conducive Production Model (adapted from Dollard & Karasek, 2008).

Cooperative action plans developed and implemented at a local level using risk assessment data are shown to be effective in improving work conditions and reducing psychological distress (Le Blanc et al., 2007).

Capacity building in terms of management/leadership development and group skills development will assist in building PSC. Although policies, practices and procedures in relation to psychological health may be in place, their enactment is somewhat discretionary depending on top management, and local management preferences and values. Therefore, PSC is expected to vary by work unit and by organization. It is important to build an awareness of the work stress process, the detrimental effects on worker health and performance, and the economic costs of work stress to "unfreeze" unfriendly health attitudes and practices. In low PSC contexts where management priority is on productivity and profitability to the detriment of worker psychological health, we expect higher work demands, low supply of resources such as job control, and rewards, lower worker engagement, higher emotional exhaustion, psychological distress, physical health injury, sickness absence, and workers compensation claims. Development of group skills is required interpret to risks, develop action plans, implement, evaluate and improve interventions, as part of a risk management cycle is required. Common risk themes will emerge from local work groups, and this information along with possible actions and resource requirements need to be communicated upwards. A regional committee may be required for this.

The expected outputs from an effective social level controller system are: a strong PSC; primary, secondary and tertiary levels of prevention; reduced workplace injury; production conducive with health; risk assessment; policy and practice coordination; and a resilient system with resilient workers. The social coordination in stress prevention model (Figure 5.3) brings these elements together in a practical intervention design. Note that control workgroups within organizations may be difficult to sustain when implementing organizational level interventions because of expected wide ranging effects across groups.

Conclusions

This discussion and illustration of PSC makes an important contribution to knowledge, practice and theory in relation to the process of developing and implementing stress prevention interventions. Evaluation of stress prevention intervention starting conditions are relatively rare in the research literature. The results indicated that the prevailing PSC is extremely important in predicting the success of the intervention: high PSC begets intervention success (a positive spiral), whereas low PSC begets intervention failure (a negative spiral). Building PSC as a continuous organizational development strategy should bring many positive benefits including a positive context for future stress intervention strategies.

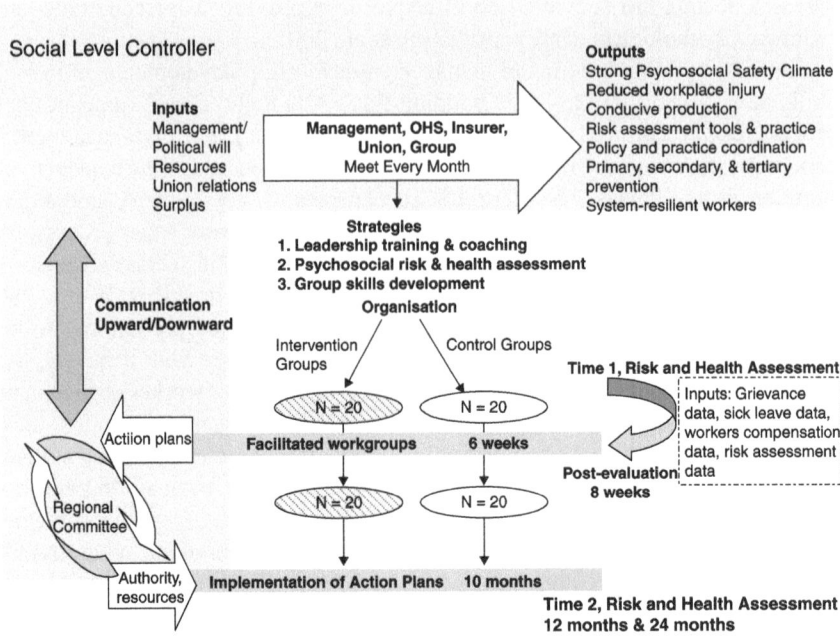

Figure 5.3 Social Coordination in Stress Prevention Model.

Notes

1 The term was first conceived by Maureen Dollard (in July 2007; Dollard, 2007), presented publicly as psychosocial safety culture at the national Human Factors and Ergonomics Conference (November 2007, Perth) and the terms *psychosocial safety climate* were first published in a press release to *The Advertiser* (December 2007) and described as "the organization's values and practices towards protection of the psychological health of employees".
2 Victorian WorkCover Authority (VWA), a government agency primarily responsible for workplace health and safety, public safety, and workers compensation within the State of Victoria.
3 Several concepts relate to the UK Health and Safety Executive's report on best practices (Jordan et al., 2003). Given that PSC is based on some of these components, PSC relates/allows to test that model as well.
4 The Department of Education and Training plans, regulates, manages, resources and delivers primary and secondary education. The Department of Human Services provides social services to the most disadvantaged including child protection, secure juvenile care, primary health, and school nursing.
5 We compared the survey levels of PSC at T2 with a measure taken from workshop participants prior to the workshop and found no significant difference in levels. We continued to use this survey PSC measure as we had greater responses to it (i.e. not all people attended the workshop).

6   The percentage of variance in PSC due to workgroup effects was significant: 15.76 per cent due to workgroup, 84.20 per cent due to individual, and .04 per cent due to organization. The ICC(2), the reliability of the group level PSC measure, was reasonable (.56) given the small number of higher level units. Based on these metrics aggregation of PSC to the group level was justifiable.

Author's note

Correspondence concerning the chapter should be addressed to M. F. Dollard, Work & Stress Research Group, Centre for Applied Psychological Research, School of Psychology, University of South Australia, Adelaide, South Australia. Phone +61 8 83022775, Fax+61 8 83022956, email, Maureen.dollard@unisa.edu.au

The author was funded to conduct the evaluation by Vic WorkCover Authority. Very sincere thanks to all who participated in the pilot project.

References

Australian Productivity Commission (2010). *Performance Benchmarking of Australian Business Regulation: Occupational Health & Safety*, Research Report, Canberra.

Biron, C., Gatrell, C., & Cooper, C. L. (2010). Autopsy of a failure: Evaluating process and contextual issues in an organizational-level work stress intervention. *International Journal of Stress Management, 17*, 135–158.

Blau, P. M. (1964). *Exchange and power in social life*. New York: Wiley.

Bond, S. A., Tuckey, M. R., & Dollard, M. F. (2010). Psychosocial safety climate, workplace bullying, and symptoms of posttraumatic stress. *Organization Development Journal, 28*, 37–56.

Carlile, P. R., & Christensen, C. M. (2005). The cycles of theory building in management research. www.innosight.com/documents/Theory%20Building.pdf (accessed January 2012).

Caulfield, N., Chang, D., Dollard, M. F., & Elshaug, C. (2004). A review of occupational stress interventions in Australia. *International Journal of Stress Management, 11*, 149–166.

Chan, D. (1998). Functional relations among constructs in the same content domain at different levels of analysis: A typology of composition models. *Journal of Applied Psychology, 83*, 234–246.

Christian, M. S., Bradley, J. C., Wallace, J. C., & Burke, M. J. (2009). Workplace safety: A meta-analysis of the roles of person and situation factors. *Journal of Applied Psychology, 94*, 1103–1127.

Clarke, S. (2006). The relationship between safety climate and safety performance: A meta-analytic review. *Journal of Occupational Health Psychology, 4*, 315–327.

Cox, S. J., & Cheyne, A. J. T. (2000). Assessing safety culture in offshore environments. *Safety Science, 34*, 111–129.

Cox, T., Griffiths, A., Barlowe, C., Randall, R., Thomson, L., & Rial-González, E. (2000). *Organisational interventions for work stress: A risk management approach*. Sudbury: HSE Books.

Cox, T., Karanika, M., Griffiths, A., & Houdmont, J. (2007). Evaluating organisational-level stress work interventions: Beyond traditional methods. *Work and Stress, 21*, 348–362.

Cox, T., Randall, R., & Griffiths, A. (2002). *Interventions to control stress at work in hospital staff*. Sudbury: HSE Books.

Cunningham, C. E., Woodward, C. A., Shannon, H. S., Macintosh, J., Lendrum, B., Rosenbloom, D., & Brown, J. (2002). Readiness for organizational change: A longitudinal study of workplace, psychological and behavioural correlates *Journal of Occupational and Organizational Psychology, 75*, 377–392.

DeJoy, D. M., Wilson, M. G., Vandenberg, R. J., McGrath-Higgins, A. L., & Griffin-Blake, C. S. (2010). Assessing the impact of health work organization intervention. *Journal of Occupational and Organizational Psychology, 83*, 139–165.

de Lange, A. H., Taris, T. W., Kompier, M. A. J., Houtman, I. L. D., & Bongers, P. M. (2003). "The very best of the millennium": Longitudinal research and the Demand-Control-Support model. *Journal of Occupational Health Psychology, 8*, 282–305.

Demerouti E., Bakker, A. B., Nachreiner, F., & Schaufeli, W. B. (2001). The Job Demands-Resources model of burnout. *Journal of Applied Psychology, 86*, 499–512.

Dollard, M. F., & Kang, S. (2007). *Psychosocial safety culture and climate survey*. Work & Stress Research Group, University of South Australia, Adelaide.

Dollard, M. F. (2007). *Psychosocial safety culture and climate; Definition of a new construct*. Adelaide: Work and Stress Research Group, University of South Australia.

Dollard, M. F., & Bakker, A. B. (2010). Psychosocial safety climate as a precursor to conducive work environments, psychological health problems, and employee engagement. *Journal of Occupational and Organizational Psychology, 83*, 579–599.

Dollard, M. F., & Karasek, R. (2010). Building psychosocial safety climate: Evaluation of a socially coordinated PAR risk management stress prevention study. In J. Houdmont & S. Leka (Eds), *Contemporary occupational health psychology: Global perspectives on research and practice* (pp. 208–234). Chichester: Wiley Blackwell.

Dollard, M. F., le Blanc, P., & Cotton, S. (2008). Participatory action research as work stress intervention. In Ed Katharina Näswall, Johnny Hellgren, & Magnus Sverke. *Balancing work and well-being: The individual in the changing working life* (pp 351–353), Cambridge: Cambridge University Press.

Dollard, M. F., Tuckey, M. R., & Dormann, C. (in press). Psychosocial safety climate moderates the demand-resource interaction in predicting work stress, *Accident Analysis & Prevention*.

European Agency for Safety and Health at Work. (2002). *Systems and programmes: How to tackle psychosocial issues and reduce work-related stress*. Luxembourg: Office for the Official Publications of the European Communities.

Gershon, R. R. M., Karkashian, C. D., Grosch, J. W., Murphy, L. R., Escamilla-Cejudo, A., Flanagan, P. A., & Martin, L. (2000). Hospital safety climate and its relationship with safe work practices and workplace exposure incidents. *American Journal of Infection Control, 28*, 211–221.

Giga, S. I., Noblet, A. J., Faragher, B., & Cooper, C. L. (2003). The UK perspective: A review of research on organisational stress management interventions. *Australian Psychologist, 38*, 158–164.

Glick, W. H. (1985). Conceptualising and measuring organizational and psychological climate: Pitfalls in multi-level research. *Academy of Management Review, 10*, 601–616.

Goldberg, D., & Williams, P. (1991). *A User's Guide to the General Health Questionnaire*. London: NFER–Nelson.

Goldenhar, L. M., LaMontagne, A. D., Katz, T., Heaney, C., & Landsbergis, P. (2001). The intervention research process in occupational safety and health: An overview from the National Occupational Research Agenda Intervention Effectiveness Research Team. *Journal of Occupational and Environmental Medicine, 43*, 616–622.

Griffiths, A. (1999). Organizational interventions: Facing the limits of the natural science paradigm. *Scandinavian Journal of Work Environment and Health*, *25*, 589–596.

Hall, G. B., Dollard, M. F., & Coward, J. (2010). Psychosocial safety climate: Development of the PSC-12. *International Journal of Stress Management*, *4*, 353–383.

Häusser, J. A., Mojzisch, A., Niesel, M., & Schulz-Hardt, S. (2010). Ten years on: A review of recent research on the Job Demand-Control-Support model and psychological well-being. *Work & Stress*, *24*, 1–35.

Heward, S., Hutchins, C., & Keleher, H. (2007). Organisational change–key to capacity building and effective health promotion. *Health Promotion International*, *22*, 170–178.

Idris, M. A., Dollard, M. F., Coward, J., & Dormann, C. (2012). Psychosocial safety climate: Conceptual distinctiveness and effect on job demands and worker psychological well-being. *Safety Science*, *50*, 19–28.

Idris, M. A., Dollard, M. F., and Winefield, A. H. (2011). Integrating psychosocial safety climate in the JD-R Model: A study among Malaysian workers. *South African Journal of Industrial Psychology*, *37*, 1–11.

James, L. R., Demaree, R. G., & Wolf, G. (1984). Estimating within-group inter-rater reliability with and without response bias. *Journal of Applied Psychology*, *69*, 85–98.

Jordan, J., Gurr, E., Tinline, G., Giga, S. I., Faragher, B., & Cooper, C. L. (2003). *Beacons of excellence in stress prevention*. HSE Books, www.hse.gov.uk/RESEARCH/rrhtm/rr133.htm (accessed January 2012).

Kang, S. Y., Staniford, A., Dollard, M. F., & Kompier, M. (2008). Knowledge development and content in occupational health psychology: A systematic Analysis of the *Journal of Occupational Health Psychology, and Work & Stress*, 1996–2006, pp. 27–63. In J. Houdmont & S. Leka (Eds.), *Occupational Health Psychology: European Perspectives on Research, Education and Practice*, Vol. 3, Maia, Portugal: ISMAI Publishers.

Karasek, R. A. (1979), Job demands, job decision latitude, and mental strain: Implications for job redesign. *Administration Science Quarterly*, *224*, 285–307.

Karasek, R. A. (2008). Low social control and physiological deregulation – the stress-disequilibrium theory, towards a new demand-control model. *Scandinavian Journal of Work Environment and Health*, *6*, 117–135.

Kompier, M., & Cooper, C. (1999). *Preventing stress, improving productivity: European case studies in the workplace*. London: Routledge.

Kompier, M., & Kristensen, T. S. (2001). Organizational work stress interventions in a theoretical, methodological and practical context. In J. Dunham (Ed.), *Stress in the Workplace: Past, Present and Future* (pp. 164–190), London: Whurr.

Landsbergis, P. A., & Vivona-Vaughan, E. (1995). Evaluation of an occupational intervention in a public agency. *Journal of Organizational Behavior*, *16*, 29–48.

Law, R., Dollard, M. F., Tuckey, M. R., & Dormann, C. (2011). Psychosocial safety climate as a lead indicator of workplace psychosocial hazards, psychological health and employee engagement. *Accident Analysis & Prevention*, *43*, 1782–1793.

Le Blanc, P. M., Hox, J. J., Schaufeli, W. B., Taris, T. (2007). Take care! The evaluation of a team based burnout intervention program for oncology providers. *Journal of Applied Psychology*, *92*, 213–227.

Leka S., Cox, T., & Zwetsloot, G. (2008). The European Framework for Psychosocial Risk management (PRIMA-EF). In S. Leka, & T. Cox (Eds). *The European Framework for Psychosocial Risk Management*, Nottingham; I-WHO, pp. 1–16.

Lohr, J. M., Olatunji, B. O., & Sawchuk, C. N. (2007). A functional analysis of danger and safety signals in anxiety disorders. *Clinical Psychology Review*, *27*, 114–126.

Nytrø, K., Mikkelsen, A., Bohle, P., & Quinlan, M. (2000). An appraisal of key factors in the implementation of occupational stress interventions. *Work & Stress, 14*, 213–225.

Pasmore, W. A., & Fagans, M. R. (1992). Participation, individual development, and organizational change: A review and synthesis. *Journal of Management, 18*, 375–397.

Pettigrew, A., Ferlie, E., McKee, L. (1992). *Shaping Strategic Change*. London: Sage.

Pfeffer, J. (2010). Building sustainable organizations: The human factor. *The Academy of Management Perspectives*, 24, 34–45.

Prochaska, J. O., Redding. C. A., & Evers, K. (1997), The transtheoretical model of change. In K. Glanz, E. M, Lewis & B. K. Rimer (Eds), *Health behavior and health education: Theory, research, and practice* (pp. 60–84), San Francisco: Jossey-Bass.

Pronovost, P. J., Weast, B., Holzmueller, C. G., Rosebstein, B. J., Kidwell, R. P., & Haller, K. B. (2003). Evaluation of the culture of safety: Survey of clinicians and managers in an academic medical center. *Quality and Safety in Health Care, 12*, 405–410.

Raudenbush, S. W., Bryk, A. S., Cheong, Y. F., & Congdon, R. (2005). *HLM6.01: Hierarchical Linear and Nonlinear Modeling* [software]. Lincolnwood, Ill: Scientific Software International.

Saksvik, P. Ø., Nytrø, K., Dahl-Jorgensen, C., & Mikkelsen, A. (2002). A process evaluation of individual and organizational occupational stress and health interventions. *Work & Stress, 16*, 37–57.

Schaufeli, W., & Bakker, A. (2003). *Utrecht Work Engagement Scale. Preliminary Manual*. Utrecht: Occupational Health Psychology Unit, Utrecht University.

Schaufeli, W. B., Leiter, M. P., Maslach, C., & Jackson, S. E. (1996). The MBI-general survey. In C. Maslach, S. E. Jackson, & M. P. Leiter (Eds.), Maslach Burnout Inventory manual (3rd ed., pp. 19–26). Palo Alto, CA: Consulting Psychologists Press.

Taris, T. W., Kompier, M. A. J., Geurts, S. A. E., Scheurs, P. J. G., Schaufeli, W. B., de Boer, E., Sepmeijer, K. J., & Wattez, C. (2003). Stress management interventions in the Dutch domiciliary care sector: Findings from 81 organizations. *International Journal of Stress Management, 10*, 297–325.

Tuckey, M. R., Winwood, P. C., & Dollard, M. F. (2011, in press). "No-one to talk to": Psychosocial culture and psychological injury within policing. *Police Practice and Research: An International Journal*.

van Den Bossche, S., Smulders, P., & Houtman, I. (2006). *Trends and risk groups in working conditions*. Hoofddrop, TNO Work & Employment.

van der Doef, M., & Maes, S. (1999). The Job Demand-Control (-Support) model and psychological well-being: A review of 20 years of empirical research. *Work and Stress, 13*, 87–114.

Warr, P., Cook, J. D., & Wall, T. D. (1979). Scales for the measurement of some work attitudes and aspects of psychological well-being. *Journal of Occupational Psychology, 52*, 129–148.

Xanthopoulou, D., Bakker, A. B., Dollard, M. F., Demerouti, E., Schaufeli, W. B., Taris, T. W., & Schreurs, P. J. G. (2007). When do job demands particularly predict burnout?: The moderating role of job resources. *Journal of Managerial Psychology, 22*, 766–786.

Zimmermann, B. K., Haun, S., Dormann, C., & Dollard, M. (2009). Stress and reciprocity in service interactions: Main and moderating effects of psychosocial safety climate. In P. H. Langford, N. J. Reynolds, & J. E. Kehoe (Eds), *Meeting the future: Promoting sustainable organisational growth*, 8th Industrial and Organisational Psychology Conference proceedings: 150–155. Sydney: Australian Psychological Society.

Zohar, D. (2010). Thirty years of safety climate research: Reflections and future directions. *Accident Analysis and Prevention, 42,* 1517–1522.

Zohar, D., & Luria, G. (2005). A multilevel model of safety climate: Cross-level relationships between organization and group-level climates. *Journal of Applied Psychology, 90,* 616–628.

6 Perspectives on the intervention process as a special case of organizational change

Sturle D. Tvedt and Per Øysten Saksvik

Introduction

The present chapter presents a perspective on the intervention process as a special case of organizational change processes. Specifically, it takes as its point of departure the context of the Nordic participative work life tradition. The approach of healthy organizational change originates from within this tradition and key issues from this research are discussed as to whether they make transferable knowledge to the field of interventions; the importance of the process, the individual variable impact of interventions, and individual differences in personal resources for coping with interventions. Possible threats and benefits from implementation processes in themselves are also discussed. As discussed by Tetrick, Quick, and Gilmore (this volume) these issues might be informed by the field of organizational change.

Process issues of stress interventions as a specific case of organizational change processes

In our own research we moved from research on interventions (initiated from outside or from the enterprise itself) to research on different forms of organizational change. Some differences between organizational interventions for work-related health and organizational change activities should be mentioned. One such difference is who is initiating the project. Researchers seldom have the power or take the initiative to restructure a whole business, whereas organizational change tends to be initiated from within the organization. Interventions are often suggested from outside (researchers) or from the plant floor level (from the employees' representatives themselves or with the help of ergonomists or organizational psychologists) who then have to promote the project among senior managers. Even the employees sometimes have to be convinced that it is in their own interest to take part in the interventions when they perceive they have too little time to participate. One advantage of studying organizational change is that the decision is almost always anchored at board level. Although it could be argued that organizational change tends to benefit organizations, whereas interventions aim to benefit employees, we have seldom seen this as a problem in practice. Most interventions will benefit the enterprise as well as the employees although not all organizational changes (such as downsizing) will benefit (all) employees. We have elaborated on these dilemmas in Dahl-Jørgensen and Saksvik (2005).

Organizational change as a psychosocial risk

Organizational change can be a difficult endeavour. The probability of any change project realizing planned financial and strategic objectives has been found to be 25–50 per cent (Clegg andWalsh, 2004; Kramer, Dougherty, and Pierce, 2004). Nguyen and Kleiner (2003) conclude that, in the world of increasing mergers and acquisitions, 75–83 per cent of organizations fail to reach targeted objectives. Negative occupational health effects are part of this equation. Individual reactions to ongoing change efforts are one of the reasons why change does not move as smoothly as intended. These reactions include uncertainty for the future job situation that the change creates (Blau, 2003; Bordia et al., 2004; Furnham, 1997; Nguyen and Kleiner, 2003), a loss of control (Proctor and Dukakis, 2003; Worrall and Cooper, 1998), reduced role clarity (Korunka et al., 2003), or a change in group climate when colleagues are discharged or when well-established organizational structures change or are dissolved (Kivimäki et al., 2003; Noel, 1998). This, in turn, may contribute to different short- and long-term outcomes at the individual and organizational level, such as psychological morbidity (Virtanen et al., 2005), early retirement (Saksvik and Gustafsson, 2004), increased job strain (Korunka et al., 2003), sickness absenteeism (Nguyen and Kleiner, 2003), and injuries (Quinlan, Mayhew, and Bohle, 2001; Virtanen et al., 2005). This academic interest in the association between psychosocial risks and organizational change is reflected in a focus on the management of organizational change by both the UK's Health & Safety Executive (Cousins et al., 2004) and the Norwegian Labour Inspectorate (NLI) (Saksvik et al., 2007). However, it is still only sparsely reflected in current national surveillance surveys of psychosocial risk factors in the work place (Dollard et al., 2007).

Intervention process

Learning from organizational change has led us to focus on the intervention (or change) process. Early in our research we stated that the intervention process had a greater impact on the results than the actual content of the intervention. The content, or what to change, has to do with the purpose or mission of the organization. The process, the *how*, concerns implementation and adoption; how the change is planned, launched and carried out. We defined the process as "individual, collective or management perceptions or actions in implementing any intervention and their influence on the overall result of the intervention" (Nytrø, et al., 2000, p. 214). Our subsequent empirical research identified certain process criteria to be fulfilled for success in interventions:

1 the ability to learn from failure and to motivate participants,
2 multi-level participation and negotiation, and differences in organizational perception,

3 insight into tacit and informal organizational behaviour,
4 clarification of roles and responsibilities, especially the role of middle management, and
5 simultaneously competing projects (Saksvik et al., 2002).

A participative approach to guiding the change process

The context of the Nordic participative tradition

The approach of healthy organizational change processes originates in the context of the Nordic participative tradition. Parallel to our research effort, the revised Working Environment Act of Norway was launched on 1st January 2006 (Arbeidstilsynet, 2006). This is, to our knowledge, the first time the concept of organizational change processes has been presented in any legislation. The act provided us with a ready frame of reference to our understanding of how to run a good organizational change or intervention. The following excerpt from the law told us what the law defined as a good process (Arbeidstilsynet, 2006):

> § 4–2. *Requirements regarding arrangement, participation and development.*
> (3) During reorganization processes that involve changes of significance for the employees' working situation, the employer shall ensure the necessary information, participation and competence development to meet the requirements of this Act regarding a fully satisfactory working environment.
> [authors' translation]

What is new here is the explicit mention of the change process, and not specifically the concepts "information", "participation", and "competence". These latter concepts have been present in Norwegian legislation in the work environment since the then pioneering *Psychosocial Act* of 1977 (Norge, 1996). The problem remains that these concepts lack the concrete operationalization needed for practitioners in the field. We tried to dig a bit deeper to find success factors (Saksvik et al., 2002), but in retrospect we are not sure we achieved our aim. Our findings, the five process criteria mentioned above, are still of a general nature. The lack of concrete action-oriented guidance to healthy change was exactly the reason why the NLI, prompted by the new legislation, sought information from us relevant to giving shop floor level advice about change processes.

The proliferation of radical reorganizations in the Norwegian public sector in recent years must be regarded as an important context for this legislation, and there is reason to suspect that public sector managers are hit especially hard because decisions of change are most likely to be made outside the organization by politicians and based on international trends like New Public Management. Additionally, public sector enterprises are often more bound than commercial enterprises in the way they operate, giving their local managers less room for manoeuvre.

Scandinavian work life has been steeped in a zeitgeist of participation and self-governing work groups since the 1960s (see e.g. Emery and Thorsrud, 1976). This can be seen in connection with a Scandinavian egalitarian culture described in terms of Hofstede's (2001) cultural dimensions as having short power distances. Thus, employees expect to participate in decision-making; which makes it difficult for middle managers to sustain a distanced position to their employees and concentrate on selling goals outlined by top management. Second, this short power distance is probably also a prerequisite for the remarkably high degree of agreement and shared perspective between the management and their co-workers. It has been demonstrated elsewhere that disparity between employers and employees in work environment risk estimates is symptomatic of unhealthy organizations (Warren et al., 1999).

Change and interventions can also be studied favourably from an open systems theory perspective (Katz and Kahn, 1978), where decisions to change are usually seen as leader responses on behalf of the organization to perturbations in the environment. However, it also follows from open system theory that different departments in an organization form subsystems for which the larger organization constitutes part of the environment. Earlier research on organizational change has also pointed out differences in organizational perception of the change (see e.g. Fay and Lührman, 2004; Saksvik et al., 2002). This supports our own experiences that employees present different "changes" both within and between different departments. In line with open systems theory, we further expect that there will be real differences between the departments – or subsystems – of an organization in the nature and extent of change impact. It follows that the changes take on a very local nature in reaching the actual level of implementation when the change decision meets the actual organization members who will carry out and embody the change or the intervention (Dahl-Jørgensen and Saksvik, 2005).

The challenge of putting participation in concrete terms

A study was initiated in 2004 to find behavioural criteria behind the broad concepts "information", "participation", and "competence development" that we find in the new Work Environment Act of 2006 (Saksvik et al., 2007). In this study, five categories or criteria appeared as underlying or conveying these concepts in the 90 organizations undergoing very different types of changes. These categories emerged through extensive interviewing and an expanded grounded theory analysis. It was found that organizational change processes were better managed by:

1   more attention to awareness of local norms,
2   and to diversity among employees in the perception and reactions to change efforts,
3   early role clarification,
4   manager availability, and
5   using conflicts constructively to deal with change.

Manager availability was found to be the main content of "information" in the Working Environment Act. The point here is not the contents of information, but how information is passed on, and also whether there is a possibility that employees can communicate freely based on that information. Early clarification of roles was found to be a prerequisite for the development of qualifications ("competence development"), according to the Act. It is only when roles and new purposes are clarified that the employees are able to improve their qualifications further, or maybe develop their new roles in the new organization based on their existing qualifications. Finally, it was found that conflict management was a central feature in the Working Environment Act that defines "participation". Change can increase the level of conflict, and capable conflict management increases the probability of active participation during a change process, which in turn creates a ground for constructive dialogue. This is considered important if proper employee participation is to be achieved.

Further development of process factors

Later quantitative studies have supported the importance of these factors for process healthiness through the statistical testing of the Healthy Change Process Index (HCPI) (Tvedt and Saksvik, 2008a; Tvedt, Saksvik, and Nytrø, 2009b). A factor analysis revealed the following factors:

1. awareness of diversity,
2. leader availability,
3. role clarification, and
4. destructive conflicts (Tvedt et al., 2009a).

As expected, the factors were substantially inter-correlated, with r coefficients ranging between .39 and .56. The analysis showed that the two original dimensions of "awareness" and "diversity" were best treated as one factor reflecting the open treatment of differing views and perspectives in a considerate and non-threatening manner. The former factor "constructive conflicts" appears to be unduly inspired by the trend of positive psychology and can best be understood as the absence of destructive, alienating conflict management. Lastly, the dimensions of leader availability and role clarification are refined and focused, but keep their central content as they materialize as distinct factors. These four factors form a reasonably focused instrument. Although the number of items forming each factor is a limitation to the content validity and robustness of the instrument, the concise nature simplifies its use and can prevent missing data. Further research is needed to test the factor structure in subsamples across a range of sociodemographic variables, organizational structures, and the contents of organizational change. In addition, the convergent and predictive validity of the instrument should be established.

When the intervention meets the individual worker

Different impacts of organizational changes

Official intervention programs aside – the reality of the change is local, concrete, individually diverse and sometimes unexpected. There is a growing interest in cascading change through the hierarchical levels in organizations and ultimately to individual employees (Judge et al., 1999). That is to say that interventions at the organizational level will have different impacts on different departments and work groups and even the individuals that inhabit these (Mohrman, Mohrman, and Ledford, 1989). Nielsen, Randall, and Albertsen (2007) have shown the importance of individuals' perceptions and participation in the change process. As employees actively appraise the intervention, their perceptions mediate exposure to interventions and outcome measures. Similarly, Lau and Woodman (1995) concluded that employees are more focused on their immediate work environment than on the meaning or significance of the larger organizational-level change. Models such as that of Burke and Litwin (1992) portray individuals as assessing the personal impact of larger organizational changes by considering the more immediate aspects of the changes, such as adjustments in work processes or routines.

Ashford (1988) also argues that extensive transitions in employees' immediate work settings represent more difficult transitions, greater adaptation demands, and potentially more threat and uncertainty.

Such observations prompted Caldwell, Herold, and Fedor (2004) to include the Individual Job Impact Index as a control variable when researching the relationships between the extent of change in the work unit and process quality. This construct was designed to capture the personal-level impact of the change concerning detrimental changes in job demands, expectations, and responsibilities. Fedor, Caldwell, and Herold (2006) even found individual job impact to be one of the determining factors for employee perception of change and the change process. Here we are aligned with Fedor, Caldwell, and Herold (2006), however, we wanted a neutral measure instead of one capturing the negative impacts of change. Hence, we developed a scale termed the Change Impact Factor (CIF), which measures nine different ways in which employees may be affected by the change. Preliminary studies (Saksvik and Tvedt, 2009; Tvedt and Saksvik, 2008c) indicate that employees appraise the changes differently and that CIF shows substantial explanatory power in health and stress outcomes, as well as the importance put on the process. Yet, the scale was uncorrelated with individual differences predisposing people for change, suggesting there are real differences in how people are impacted by organization change and interventions.

Naturally, just like the content of the change, the process is not appraised alike among employees. The quality of a process is more adequately viewed as a composite appraisal of different aspects of the change process. This appraisal is nuanced rather than absolute, and is undertaken individually and collectively in the immediate work group. As such, we should expect individual variation in process appraisals, although with some collective anchoring. Thus, when it comes to measuring process quality, we need continuous rather than dichotomous parameters. It is because of this kind of reasoning we have developed our own HCPI (Tvedt et al., 2009b; Tvedt, Saksvik, and Nytrø, 2009a) as a multidimensional and continuous measurement and would advocate this strategy on the evaluation of process in intervention research.

Individual differences in change resilience and aptitude

As shown in a recent review (Heuvel et al., 2010), interest in the influence of individuals' characteristics on their reactions to organizational change is recent. Correspondingly, the current Norwegian legislation does not explicitly recognize the importance of individual differences related to organizational change. However, there is an ever-present theme of the differing change resilience amongst employees in our interviews about change processes, indicating that individual co-workers and different workforces show widely differing aptitudes for change, creating different contexts for process work. This has led us to develop measures relating to individual change resilience. We conceptualized this resilience as stabile trait-like dispositions within individuals pertaining to their aptitude for change. Specifically, we have focused on three scales that are independently established in the research literature:

1 Readiness to change (Oreg, 2003), which specifically measures a disposition for life changes
2 Optimism (Scheier, Carver, and Bridges, 1994) which basically measures a tendency to expect good things in general, and
3 Self-efficacy (Schwarzer, 1993) which measures a person's general belief that the person is able to master the challenges they are faced with.

Preliminary research based on this operationalization (Saksvik and Tvedt, 2009; Tvedt and Saksvik, 2008c) indicates that together these three dimensions form a measure of resilience that substantially predicts stress and health outcomes as well as the demands employees put on the quality of the process.

Other similar conceptualizations for research on individual differences relating to organizational change include Heuvel et al. (2009), who connect themselves with the positive psychology project of Luthans, Youssef, and Avolio (2007). Heuvel et al. (2009) propose the following working definition for what they label "personal resources in organizational settings":

> Personal resources are lower-order, cognitive-affective aspects of personality; developable systems of positive beliefs about one's "self" (e.g. self-esteem, self-efficacy, mastery) and the world (e.g. optimism, faith) which motivate and facilitate goal-attainment, even in the face of adversity or challenge.
>
> (p. 129)

The only substantial difference between this definition of personal resources and our own suggested change resilience, is the inclusion in the latter of readiness to change (Oreg, 2003). Heuvel et al. (2009) suggest the inclusion of readiness *for* change (see Armenakis, Harris, and Mossholder, 1993; Weiner, Amick, and Lee, 2008) as an attitudinal concept to be included together with personal resources in their research framework. At the attitudinal level, the importance of employee readiness for the intervention of relevance has been argued by several researchers (Biron, Cooper, and Bond, 2009; Nielsen, Taris, and Cox, 2010). Far from disputing its importance, we simply argue for the distinctiveness and importance of the dispositional concept of readiness to change (Oreg, 2003) in terms of assessing change resilience.

Cognitive adaptation theory (Taylor, 1983) can readily explain why we would expect people high on optimism, self-esteem, and personal control to adjust well to stressful life events like organizational change. Also, there is increasing empirical evidence supporting the claims of the theory in the context of organizational change (see, for example, Ashford, 1988; Campbell, 2006; Caldwell et al., 2004; Jimmieson, Terry and Callan, 2004; Martin, Jones, and Callan, 2005; Wanberg and Banas, 2000).

Acknowledging differing levels of change resilience is recommendable for interventions; requiring interventionists to assess workers' resilience in order to tailor process management.

The meta-perspective: stress interventions as potential stressors

Is it useful to evaluate the potential for harmfulness of interventions using the same criteria for evaluating process quality and change impact, as well as change resilience? Although interventions are done with the best of intentions, it is justified to evaluate the psychosocial work environment hazards of stress interventions like any other kind of organizational change. Given the importance of the process quality for the psychological health of employees with any organizational change, there is a risk that well-intended interventions have some serious adverse health effects. The importance of process issues in intervention research is increasingly being stressed (Egan et al., 2009; Murta, Sanderson, and Oldenburg, 2007; Nielsen et al., 2010; Semmer, 2003; Semmer, 2006) and we have already been involved with such adaptation in the evaluation of stress interventions (Randall, Nielsen, and Tvedt, 2009). However, we are now going even further in saying that the risk that intervention efforts can represent—potential stressors in themselves, should be considered.

We thus suggest our own measures may be adapted to the *risk* evaluation of interventions. There is a conceptual distinction between effective change processes and processes that are healthy to employees. In the same line, we believe there is a parallel conceptual distinction within intervention research. Hence there might be a unique contribution from the HCPI and the CIF to the field of interventions.

As existing work on intervention processes seems preoccupied with the successfulness of interventions, certain aspects of the HCPI approach will not overlap but complement existing work on intervention process. The active encouragement of the expression of diverse ideas and views on the intervention (awareness of diversity); the close dialogue between manager and employee prescribed to clarify the meaning of intervention-related information for each employee (manager availability); the clarification of any changes in roles and competence requirements that might arise – or be seen to arise from the intervention (role clarification); and finally the willingness to constructively discuss conflict material whether it is old conflicts that are "unearthed" through the intervention or new ones inadvertently created by the intervention (constructive conflicts); are all aspects that go beyond what one would normally look for when evaluating the successfulness of an intervention process.

Thus the chief requirement for adapting the HCPI to evaluation of intervention processes would be to substitute "organizational change" for "intervention" in questionnaires. The same goes for the CIF and to us it is self-evident that the level of perceived impact from the intervention thus measured should determine if precaution is needed as to whether or not a high individual impact is predicted by the intervention program. Finally, keeping with the perspective of intervention as an organizational change, we suggest that the change resilience of the workforce subject to intervention might very well co-determine the reception of an intervention. Our perspective is that any change or intervention will be demanding, and so the process must be healthy in order to achieve any kind of healthy intervention outcome.

The positive potential of processes

Limitations and opportunities of process

Caldwell et al. (2004) warned about focusing solely on change implementation processes, as they found support in their study for several moderating effects on the relationship between process and important outcomes, by characteristics of the change and the workforce. Also, there are limitations in terms of contextual features of organization and intervention as to how much can be achieved by managing change processes in a healthy way (Saksvik and Tvedt, 2009; Tvedt and Saksvik, 2008c). Hence the sheer magnitude and extent of an intervention is likely to be intrinsically problematic independent of the purpose of the intervention and its implementation. Likewise, conducting interventions with workforces that struggle with changes will mean increased challenges with regard to the interventions, even if the workforce may need the intervention.

On the other hand, serious attention to change process management is exactly what is needed when the process is challenged. Our preliminary research (Saksvik and Tvedt, 2009; Tvedt and Saksvik, 2008c) shows a substantial mediation of both the impact of change and the resilience of employees through the healthiness of the process. This means that the impact of change and resilience has an effect through the perceived healthiness of the process, such that the quality of the process can dampen or increase their effects.

Process as an engagement- and commitment-building tool

Modern culture and history are permeated with anecdotal evidence for the idea of positive group effects through shared event, journeys, or processes. For example, Nielsen et al. (2010) suggest that the use of participative strategies for the implementation of interventions becomes an intervention in its own right. Although our focus until now has been on the dangers of change and intervention and the resulting need for a redeeming process, we ought to point out that there is some indication that process in itself may produce some positive outcomes. Indeed, commitment has been studied earlier in connection to process and impact of change (e.g. Fedor et al., 2006; Hui and Lee, 2000; Judge et al., 1999). However, it is then most often studied as an eroding commitment, feeding into absenteeism, as well as satisfaction and positive well-being in general (see e.g. Mikkelsen, Saksvik, and Landsbergis, 2000).

Heuvel et al. (2009) argue for the importance of studying the sustainability of engagement during change. In line with this, our own case studies of organizational change (Tvedt and Buvik, 2009) and experimental simulation of organizational change (Tvedt, 2010; Tvedt and Saksvik, 2011) give an indication that the process shows substantial contributions to measures like engagement (Schaufeli, Bakker, and Salanova, 2006) and affective commitment (Meyer and Allen, 1991). Thus, if managed healthily, the process might be a positive intervention factor in its own right, much like the increased attention to employees so famously produced by the Hawthorne effect (Landsberger, 1958). In fact, from within the existing intervention literature, the "special attention effect" implied in the Hawthorne effect is already being suggested as a viable working mechanism supporting interventions (Semmer, 2011).

The complexity of process

Our studies support a view of the change process and intervention implementation as broader and more complex than before. The majority of the organizational change models offer a normative or prescriptive approach that identifies what managers and consultants should and should not do and how they should implement a specific change initiative. Doyle, Claydon, and Buchanan (2000) report that managers and change leaders often found change methodologies too "pre-packaged" and not able to address the contradictions, tensions, and linkages in the change. Whelan-Berry, Gordon, and Hinings (2003) extend this critique by requesting a more comprehensive process model of change that includes the relationships between change processes at the organizational level and change processes at the individual and group levels.

Whelan-Berry, Gordon, and Hinings (2003) argue that major organizational change cannot occur without specific groups and individuals changing, that is, without teams and individual workers adopting altered work routines and different models, structures, or standards to direct their action. Hence, appreciating the group and the individual change processes that occur as part of organizational change processes is fundamental. Whelan-Berry, Gordon, and Hinings (2003) found few models in the organizational change literature that specifically described the change process at the group or team level of analysis as well as a lack of models that address how individual change occurs within the organizational context. From a broad case study, they proposed a detailed and more complete change process model that includes the interplay between the individual, group, and organizational levels.

Pettigrew (1985) also demonstrates the limitations of theories that view *change* either as a single event or as a sequence of episodes that are regarded in isolation from the context. Since the 1980s, there has been a theoretical and empirical movement toward more fluid, dynamic, and process-oriented explanations of change. Pettigrew (1985) proposed a processual-contextual perspective to organizational change, further developed by Dawson (2003), which argues that change has to be understood in terms of multi-level connections between change substance, contexts, implementation process, and organization politics over time. The processual framework is concerned with understanding the political activities of consultation, negotiation, conflict, and resistance, which occur at different levels outside and within the organization during the process of organizational change (Dawson, 2003). This framework refers to the extent to which the change is viewed as being crucial to the continued existence of the organization. This perspective contrasts sharply with the models presented as steps and guides typical of the oversimplified, practitioner-focused managerial literature of change (Buchanan, 2003).

An organization is a complex network of individuals, groups, and systems that operate within a culture and structure. A change model that appeals to many managers is that of Kotter's (1990) "eight steps to transforming your organization". This model highlights eight lessons based on an analysis of organizations going through change and it addresses some of the issues associated with making change happen and emphasizes the particular need to communicate the vision. Despite its appeal, it is criticized for being too linear and for disregarding the change process as a continuous cycle (Cameron and Green, 2004). As Dawson (2003) illustrates:

> Change does not occur in a neat linear fashion, but is messy, murky and complicated. It involves twists and loops, turns and returns, omissions and revisions, the foreseen and the unforeseen, and is marked by the achievement of planned targets, failures, resistance, celebration, ambivalence, fatigue, conflict and political manoeuvring.
>
> (p. 144)

According to Jimmieson, Terry, and Callan (2004), one of the managerial challenges facing organizations today is the effective implementation of organizational change programs that minimize feelings of uncertainty and the negative consequences that follow. Increased job insecurity is a consequence of almost all types of organizational change, including interventions, which consequently will be subject to resistance. Worrall and Cooper (1998), for instance, found that job insecurity resulted in less commitment as well as lowered degrees of motivation and morale. Similar results were found by King (2000): job insecurity has a negative effect on support for organizational goals, quality of work, and turnover. A related concept is control at work. Considered an important motivational force and necessarily negatively related to increased job insecurity, control at work is in a positive relationship with job satisfaction, commitment, involvement, and performance (Sparks, Faragher, and Cooper, 2001). Cunningham et al. (2002) suggest that, since job insecurity is negatively related to readiness for change, there must be an evident need for change and at the same time employees must be given the feeling that they are self-efficient; they must feel that they are able to carry out the change and be given the possibility to participate. An empowering work environment offers people abilities, attitudes, and possibilities beneficial for the handling of change, as this means an increased level of self-efficacy.

Intervention on change – the development of a therapeutic instrument

Our long research into the change process of organizational interventions has now led to the need to develop interventions that are directly related to the change process. Based on what we said above about individual reactions on change we now wish to guide individuals who tend to perceive most changes negatively (Saksvik and Hetland, 2009; Oreg, 2003). It is important that clinicians and change management specialists can provide an evidence- and theory-based tool to meet these individuals who react negatively to change. The objective of our effort has been to develop a therapeutic tool for helping individuals experiencing resistance to organizational change. The tool has been developed based on our research presented above and inspiration from the organization psychology literature.

We have identified three steps to be followed in order to treat resistance to change (Saksvik, Saksvik and Tvedt, 2011):

Step 1: Individual assessment of dispositional resistance with the test developed by Oreg (2003) and further extended by us including self-efficacy and optimism labelled resilience. Resilience refers to the ability to recover from adverse events, or cope successfully (Rutter, 1985). According to Luthans et al. (2007) the process of resilience can be built using cognitive coaching interventions (Luthans et al., 2006). This is the assessment of individual reactions to change on the emotional, attitudinal and behavioural level.

Step 2: Workplace assessment with the CIF model among those with dispositional resistance to change. Impact of change is about the meaning of change for the person, how it is interpreted and acted upon. Resistance and impact identify individual attitudes and reactions towards change.

Step 3: Therapy for vulnerable individuals (those with high scores at Steps 1 and 2) with the "Left hand column method" inspired by Argyris and Schön (1996). The therapist first interviews the client about their perceived change status today (right column). Next the therapist challenges the client on expectations about what changes are already manifesting, and what changes are to come (left column). This finally leads to acceptance of a new reality orientation of the client.

This therapeutic tool could empower individuals who may feel insecure and defensive in times of organizational change, and increase their psychological flexibility (Bond, Flaxman and Bunce, 2008). This could help them restore perceived control and promote job security. The tool has not yet been tested out in practice, and further research is needed to validate the tool. This is an intervention that ultimately takes us back to our origin and links together our intervention and change research. Interventions on change processes are as difficult as most interventions, but represent a test of important aspects of our HCPI model.

Conclusions

Starting out from the Nordic participative work life tradition, we have now deliberated upon the various points of learning that might have transfer value from the field of organizational change, to the field of intervention research. Fortunately, lessons and experiences are already being exchanged between these fields, as many scholars straddle them both. However, we conclude that there are certain issues within intervention research that might be enriched by viewing it as a special case of organizational change:

1. the process of intervention in itself contains both threats and potential benefits, not simply a means for more or less effective intervention;
2. interventions are highly individualized by the time they meet with actual employees;
3. employees show varying degrees of resilience in terms of the stress of interventions.

Ultimately, a new form of individual-level intervention may be called for in order to help individuals less resilient to change and stress cope with both change and interventions.

References

Arbeidstilsynet [Norwegian Labour Inspection]. (2006). www.arbeidstilsynet.no

Argyris, C. & Schön, D. A., (1996). *Organizational learning II*. Reading, MA: Addison-Wesley.

Armenakis, A. A., Harris, S. G., & Mossholder, K. W. (1993). Creating readiness for organizational change. *Human Relations*, 46, 681–703.

Ashford, S. J. (1988). Individual strategies for coping with stress during organizational transitions. *Journal of Applied Behavioral Science*, 24, 19–36.

Biron, C., Cooper, C. L., & Bond, F. (2009). Mediators and moderators of organizational interventions to prevent occupational stress. In S. Cartwright & C. L. Cooper (Eds.), *The Oxford handbook of organizational well-being* (pp. 441–465). Oxford: Oxford University Press.

Blau, G. (2003). Testing for a four-dimensional structure of occupational commitment. *Journal of Occupational and Organizational Psychology*, *76*, 469–488.

Bond, F. W., Flaxman, P. E., & Bunce, D. (2008). The influence of psychological flexibility on work redesign: Mediated moderation of a work reorganization intervention. *Journal of Applied Psychology*, *93*, 645–654

Bordia, P., Hobman, E., Jones, E., Gallois, C., & Callan, V. J. (2004). Uncertainty during organizational change: Types, consequences, and management strategies. *Journal of Business and Psychology*, *18*, 507–532.

Buchanan, D. A. (2003). Getting the story straight: Illusions and delusions in the organizational change process. *TAMARA: Journal of Critical Postmodern Organizational Science*, *2*, 7–21.

Burke, W., & Litwin, G. (1992). A causal model of organizational performance and change. *Journal of Management*, *18*, 523–545.

Caldwell, S. D., Herold, D. M., & Fedor, D. B. (2004). Towards an understanding of the relationships between organizational change, individual differences, and changes in person-environment fit: Across-level study. *Journal of Applied Psychology*, *89*, 868–882.

Cameron, E., & Green, M. (2004). *Making sense of change management: A complete guide to the models, tools and technologies of organizational change*. London: VA Kogan Page.

Campbell, D. J. (2006). Embracing change: Examination of a "capabilities and benevolence beliefs model" in a sample of military cadets. *Military Psychology*, *18*, 131–148.

Clegg, C., & Walsh, S. (2004). Change management: Time for a change! *European Journal of Work and Organizational Psychology*, *13*, 217–239.

Cousins, R., Mackay, C. J., Clarke, S. D., Kelly, C., Kelly, P. J., & McCaig, R. H. (2004). 'Management Standards' and work-related stress in the UK: Practical development. *Work & Stress*, *18*, 113–136.

Cunningham, C. E., Woodward, C. A., Shannon, H. S., Mcintosh, J., Lendrum, B. Rosenbloom, D. & Brown, J. (2002). Readiness for organizational change: A longitudinal study of workplace, psychological and behavioural correlates. *Journal of Occupational and Organizational Psychology*, *75*, 377–392.

Dahl-Jørgensen, C., & Saksvik, P. Ø. (2005). An evaluation of the impact of two workplace interventions on the health of service workers. *International Journal of Health Services*, *35*, 529–549.

Dawson, P. (2003). *Reshaping change: A processual perspective*. London: Routledge.

Dollard, M., Skinner, N., Tuckey, M. R., & Bailey, T. (2007). National surveillance of psychosocial risk factors in the workplace: An international overview. *Work & Stress*, *21*, 1–29.

Doyle, M., Claydon, T., & Buchanan, D. (2000). Mixed results, lousy process: The management experience of organizational change. *British Journal of Management*, *11*, 59–80.

Egan, M., Bambra, C., Petticrew, M., & Whitehead, M. (2009). Reviewing evidence on complex social interventions: appraising implementation in systemic reviews of the health effects of organisational-level workplace interventions. *Journal of Epidemiology & Community Health*, *63*, 4–11.

Emery, F., & Thorsrud, E. (1976). *Democracy at work: The report of the Norwegian industrial democracy program*. Martinus Nijhoff: Leiden.

Fay, D., & Lührman, H. (2004). Current themes in organizational change. *European Journal of Work and Organizational Psychology*, *13*, 113–119.

Fedor, D. B., Caldwell, S., & Herold, D. M. (2006) The effects of organizational changes on employee commitment: A multilevel investigation. *Personnel Psychology*, *59*, 1–29.

Furnham, A. (1997). *The psychology of behavior at work. The individual in the organization.* New York: Psychology Press.

Heuvel, M. van den, Demerouti, E., Bakker, A. B., & Schaufeli, W. B. (2010). Personal resources and work engagement in the face of change. In J. Houdmont & S. Leka (Eds.), *Contemporary occupational health psychology: Global perspectives on research and practice* (pp. 124–150) Chichester, UK: John Wiley & Sons Ltd.

Hofstede, G. (2001). *Culture's consequences: Comparing values, behaviors, institutions, and organizations across nations* (2nd ed.). Thousand Oaks, CA: Sage.

Hui, C. & Lee, C. (2000). Moderating effects of organizational-based self-esteem on organization uncertainty: Employee response relationships. *Journal of Management, 26,* 215–232.

Jimmieson, N. L., Terry, D. J., & Callan, V. J. (2004). A longitudinal study of employee adaptation to organizational change: The role of change-related information and change-related self-efficacy. *Journal of Occupational Health Psychology, 9,* 11–27.

Judge, T. A., Thoresen, C. J., Pucik, V., & Welbourne, T. M. (1999). Managerial coping with organizational change: A dispositional perspective. *Journal of Applied Psychology, 84,* 107–122.

Katz, D. & Kahn, R. L. (1978). *The social psychology of organizations* (2nd ed.). New York, NY: Wiley.

Kets de Vries, M. F. R., & Balazs, K. (1997). The downside of downsizing. *Human Relations, 50,* 11–50.

King, J. E. (2000). White collar reactions to job insecurity and the role of the psychological contract: Implications for human resource management. *Human Resource Management, 39,* 79–91.

Kivimäki, M., Vahtera, J., Elovainio, M., Pentti, J., & Virtanen, M. (2003). Human costs of organizational downsizing. Comparing health trends between leavers and stayers. *American Journal of Community Psychology, 32,* 57–67.

Korunka, C., Scharitzer, D., Carayon, P., & Sainfort, F. (2003). Employee strain and job satisfaction related to an implementation of quality in a public service organization: a longitudinal study. *Work & Stress, 17,* 52–72.

Kotter, J. (1990). *A force for change: How leadership differs from management.* London: Sage.

Kramer, M. W., Dougherty, D. S., & Pierce, T. A. (2004). Managing uncertainty during a corporate acquisition: A longitudinal study of communication during airline acquisition. *Human Communication Research, 30,* 71–101.

Landsberger, H. A. (1958). *Hawthorne revisited: Management and the worker, its critics, and developments in human relations in industry.* Ithaca, New York: Cornell University.

Lau, C. M., & Woodman, R. W. (1995). Understanding organizational change: A schematic perspective. *Academy of Management Journal, 38,* 537–554

Luthans, F., Avey, J. B., Avolio, B. J., Norman, S. M., & Combs, G. J. (2006). Psychological capital development: towards a micro-intervention. *Journal of Organizational Behavior, 27,* 387–393.

Luthans, F., Youssef, C. M., & Avolio, B. J. (2007). Psychological capital: Investing and developing positive organizational behaviour. In D. L. Nelson & C. L. Cooper (Eds.), *Positive organizational behaviour* (pp. 9–24). London: Sage.

Martin, A. J., Jones, E. S., & Callan, V. J. (2005). The role of psychological climate in facilitating employee adjustment during organizational change. *European Journal of Work and Organizational Psychology, 14,* 263–289.

Meyer, J. P., & Allen, N. J. (1991). A three-component conceptualization of organizational commitment. *Human Resource Management Review, 1*, 61–89.

Mikkelsen, A., Saksvik, P. Ø., & Landsbergis, P. (2000). The impact of a participatory organizational intervention on job stress in community health care institutions. *Work & Stress*, 14, 156–170.

Mohrman, S. A., Mohrman, A. M., & Ledford, G. E. (1989). Interventions that change organizations. In A. M. Mohrman, S. A. Mohrman, G. E. Ledford, T. G. Cummings, E. E. Lawler & associates (Eds.), *Large-Scale Organizational Change* (pp. 145–153). San Fransisco: Jossey Bass.

Murta, S. G., Sanderson, K., & Oldenburg, B. (2007). Process evaluation in occupational stress management programs: A systematic review. *American Journal of Health Promotion, 21*, 248–254.

Nguyen, H., & Kleiner, B. H. (2003). The effective management of mergers. *Leadership & Organizational Development Journal, 24*, 447–454.

Nielsen, K., Randall, R., & Albertsen, K. (2007). Participants' appraisals of process issues and the effects of stress management interventions. *Journal of Organizational Behavior, 28*, 793–810.

Nielsen, K., Randall, R., Holten, A., & Rial-Gonzáles, E. (2010). Conducting organizational-level occupational health interventions: What works? *Work & Stress, 24*, 234–259.

Nielsen, K., Taris, T., & Cox, T. (2010). The future of organizational interventions: Addressing the challenges of today's organizations. *Work & Stress, 24*, 219–233.

Noel, D. (1998). Layoff survivor sickness: What it is and what to do about it. In M. K. Gowing, J. D. Kraft, & J. C. Quick (Eds.), *The new organizational reality – downsizing, restructuring and revitalization*, Washington, DC: American Psychology Association.

Norge (1996). *Lov av 4. februar 1977 nr. 4 om arbeidervern og arbeidsmiljø m.v. (arbeidsmljøloven): med endringer.* [Law of February 4th on workers' protection and work environment (the work environment act), with subsequent ammendments.] Oslo: Grøndahl.

Nytrø, K., Saksvik, P. Ø., Mikkelsen, A., Quinlan, M., & Bohle, P. (2000). An appraisal of key factors in the implementation of occupational stress interventions. *Work & Stress, 13*, 213–225.

Oreg, S. (2003). Resistance to change: Developing an individual difference measure. *Journal of Applied Psychology, 88*, 680–706.

Pettigrew, A. (1985). *The awakening giant: Continuity and change in Imperial Chemical Industries.* Oxford: Basil Blackwell.

Proctor, T., & Dukakis, I. (2003). Change management: The role of internal communication and employee development. *Corporate Communications, 8*, 268–276.

Quinlan, M., Mayhew, C., & Bohle, P. (2001). The global expansion of precarious employment, work disorganization and occupational health: a review of recent research. *International Journal of Health Services, 31*, 335–414.

Randall, R., Nielsen, K. M., & Tvedt, S. D. (2009). The development of five scales to measure employees' appraisals of organizational-level stress management interventions. *Work & Stress, 23*, 1–23.

Rutter, M. (1985). Resilience in the face of adversity: protective factors and resistance to psychiatric disorder. *British Journal of Psychiatry, 147*, 598–611.

Saksvik, I. & Hetland, H. (2009). Exploring dispositional resistance to change. *Journal of Leadership & Organizational Studies, 16*, 175–183.

Saksvik, I. B., Saksvik, P. Ø., & Tvedt, S. D. (2011). Surviving change: Developing a therapeutic tool for overcoming dispositional resistance to change. Poster presented at EAWOP, Maastricht, Netherlands.

Saksvik, P. Ø., & Gustafsson, O. (2004). Early retirement from work. A longitudinal study of the impact of organizational change in a public enterprise. *Policy and Practice in Health and Safety, 2*, 43–55.

Saksvik, P. Ø., Nytrø, K., Dahl-Jørgensen, C., & Mikkelsen, A. (2002). A process evaluation of individual and organizational occupational stress and health interventions. *Work & Stress, 16*, 37–57.

Saksvik, P. Ø., & Tvedt, S. D. (2009). Employee change resilience and impact of change: Simultaneously limiting and highlighting healthy change processes' importance. Paper presented at the APA Work, Stress, and Health Conference, Puerto Rico, US.

Saksvik, P. Ø., Tvedt, S. D., Nytrø, K., Buvik, M. P., Andersen, G. R., Andersen, T. K., & Torvatn, H. (2007). Developing criteria for healthy organizational change. *Work & Stress, 21*, 243–263.

Schaufeli, W. B., Bakker, A. B. & Salanova, M. (2006). The measurement of work engagement with a short questionnaire: A cross-national study. *Educational and Psychological Measurement, 66*, 701–716.

Scheier, M. F., Carver, C. S., & Bridges, M. W. (1994). Distinguishing optimism from neuroticism (and trait anxiety, self-mastery, and self-esteem): A reevaluation of the life orientation test. *Journal of Personality and Social Psychology, 67*, 1063–1078.

Schwarzer, R. (1993). *Measurement of perceived self-efficacy: Psychometric scales for cross cultural research*. Berlin: Freien Universitat, Institut fur Psychologie.

Semmer, N. (2003). Job stress interventions and organization of work. In L. E. Tetrick & J. C. Quick (Eds.), *Handbook of occupational health psychology* (pp. 325–353). Washington, DC: APA.

Semmer, N. K. (2006). Job stress interventions and the organization of work. *Scandinavian Journal of Work and Environmental Health, 32*, 515–527.

Semmer, N. K. (2011). Job stress interventions and organization of work. In J. C. Quick & L. E. Tetrick (Eds.), *Handbook of occupational health psychology* (2nd ed., pp. 299–318). Washington, DC: American Psychological Association.

Sparks, K., Faragher, B., & Cooper, C. L. (2001). Well-being and occupational health in the 21st century workplace. *Journal of Occupational and Organizational Psychology, 73*, 389–509.

Taylor, S. (1983). Adjustment to threatening events: a theory of cognitive adaptation. *American Psychologist, 38*, 1161–1173.

Tvedt, S. D. (2010). Experimentally manipulated organizational change: Importance of healthy processes and change resilience for engagement, commitment, stress, and health complaints. Poster presented at the EA-OHP Annual conference, Rome Italy.

Tvedt, S. D., & Buvik, M. P. (2009). Building engagement and commitment through healthy change processes: a case study examining possibilities and limitations. Paper presented at the APA Work, Stress, and Health Conference, Puerto Rico, US.

Tvedt, S. D., & Saksvik, P. Ø. (2008a). Organizational change and employee health. Paper presented at the APA Work, Stress, and Health Conference. Washington DC, US.

Tvedt, S. D., & Saksvik, P. Ø. (2008b). Embracing change. Paper presented at the OPEN workshop, Innsbruck, Austria.

Tvedt, S. D., & Saksvik, P. Ø. (2008c). Individual disposition, personal impact, and healthy processes as moderators for stress and health complaints connected with organizational change. Paper presented at the EA-OHP Annual Conference, Valencia, Spain.

Tvedt, S. D., & Saksvik, P. Ø. (2009). Employee change resilience and impact of change: Simultaneously limiting and highlighting healthy change processes' importance. Paper presented at the APA Work, Stress, and Health Conference, Puerto Rico, US.

Tvedt, S. D., & Saksvik, P. Ø. (2011). Experimental manipulation of organizational change process healthiness: Triangulating correlational findings and exploring the power of workplace simulations. Paper presented at the EAWOP, Maastricht, Netherlands.

Tvedt, S. D., Saksvik, P. Ø., Lau, B., Finne, L. B., Buvik, M. P., Ingstad, G. Z., & Vaag, J. R. (2009a). Developing the Healthy Change Process Index: a new measure assessing management of organizational change processes. Paper presented at the APA Work, Stress, and Health Conference, Puerto Rico, US.

Tvedt, S. D., Saksvik, P. Ø., & Nytrø, K. (2009b). Does change process healthiness reduce the negative effects of organizational change on the psychosocial work environment? *Work & Stress, 23*, 80–98.

Virtanen, M., Kivimäki, M., Joensuu, M., Virtanen, P., Elovainio, M., & Vahtera, J. (2005). Temporary employment and health: a review. *International Journal of Epidemiology, 34*, 610–622.

Wanberg, C. R. & Banas, J. T. (2000). Predictors and outcomes of openness to changes in a reorganizing workplace. *Journal of Applied Psychology, 85*, 132–142.

Warren, N., Karasek, R., Punnett, L., & Houtman, I. (1999). Worker-employer risk estimate disparity: Identifying the unhealthy organization. In APA (American Psychological Association), *Work, Stress, and Health Conference*. Baltimore, ML, US, 11–13 March 1999.

Weiner, B. J., Amick, H., & Lee, S.-Y. D. (2008). Conceptualization and measurement of organizational readiness for change: A review of the literature in health services research and other fields. *Medical Care Research and Review, 65*, 379–436.

Whelan-Berry, K. S., Gordon, J. R., & Hinings, C. R. (2003). Strengthening organizational change processes: Recommendations and implications from a multilevel analysis. *Journal of Applied Behavioral Science, 39*, 186–207.

Worrall, L., & Cooper, C. L. (1998). *Quality of working life: 1998 survey of managers' changing experiences*. London: Institute of Management.

7 Does the intervention fit?
An explanatory model of intervention success and failure in complex organizational environments

Raymond Randall and Karina M. Nielsen

There is much debate about the effectiveness of organizational-level interventions. Many interventions based on sound theory have been shown to have inconsistent or small effects. In this chapter we argue that intervention outcomes are heavily influenced by:

1. the degree of fit between the intervention and the individual employee (person-intervention fit), and
2. the degree of fit between the intervention and the environment within which it is implemented (environment-intervention fit).

The factors that contribute to these fit dimensions are examined and we discuss how the analysis of fit can be used to enhance intervention process evaluation.

Introduction

The implementation of organizational-level interventions presents many challenges for researchers and practitioners. Much of the research on interventions reflects a desire for simple answers to questions such as "what works?" At present this question remains unanswered because organizational-level intervention research is scarce and what there is has produced inconsistent results (Richardson and Rothstein, 2008). The search for simple answers to simple questions about intervention effectiveness is not an easy one because interventions are complex sequences of events taking place in complex social systems (Semmer, 2006). In this chapter we will argue that this search for simple, linear and universal intervention effects does not fit with the complexity of intervention processes and outcomes (Cox et al., 2007; Nielsen et al., 2010). Simply put, for interventions to be effective it seems logical that they need to fit the problem as it is perceived by employees and the context within which it occurs.

Implicit in the existing research literature is a search for interventions that are "guaranteed" to be effective. As many chapters in this book show, this approach has highlighted the need for models of intervention processes that help us to explain inconsistencies in intervention outcomes. When evaluating organizational-level interventions the question "what works?" is too simple. We also need to know *why* interventions succeed or fail. In this chapter we describe how this concept of intervention fit could be used to help to answer such questions.

We argue that the degree of fit between the *active ingredients of the intervention* and the *required remedy for a specific presenting problem in a specific context* shapes the intervention process and as a consequence intervention outcomes. We discuss both the extent to which the intervention fits with individual employees (person-intervention fit) and the extent to which the intervention fits with organizational contexts (environment-intervention fit). We propose that inconsistencies in intervention outcomes are linked to inconsistent fit across employee populations and intervention contexts. Applying the concept of fit to organizational-level interventions means that in some situations an intervention might be appropriate and powerful (because of good fit), but the same intervention might be inappropriate and weak in others (because of poor fit). The evidence of inconsistent intervention implementation and associated inconsistent intervention effects seems to indicate that this is a distinct possibility (Egan et al., 2009; Murta, Sanderson, and Oldenburg, 2007).

We present a model of intervention fit that is intended to offer researchers and practitioners a framework for understanding better the factors that influence the success (or otherwise) of intervention processes. This requires a more complex model of intervention design, implementation and outcomes variables than is typically used in intervention research. In describing this model we also discuss how fit can be evaluated and improved during intervention processes to maximise the chances of intervention success.

The importance of fit in work psychology

The notion of person-environment fit is found in many theories and practical interventions that have been used successfully by work psychologists. The concept of fit can be seen (either implicitly or explicitly) in many theories of work stress (see Cooper, 1998; Mark and Smith, 2008). Transactional theories of stress (e.g. Lazarus and Folkman, 1992) describe a process where there is an exchange between the worker and their environment that alters the degree of perceived person-environment fit. In these theories fit is the result of the coming together of many factors (such as individual differences, the work environment, coping strategies, cognitive appraisal, etc.) that are in a constant state of flux. Therefore fit is a function rather than a simple one-dimensional measurable variable. In most stress theories a good fit is hypothesized to be healthier than a poor fit. For example, a worker who perceives low control in their job might initiate or welcome changes to their job to make their perceived situation better (Bond and Bunce, 2001). Theories of stress also indicate that there is tremendous potential for heterogeneity among a group of employees: some may suffer while others thrive in the same situation (Daniels et al., 2009). Transactional theories of stress are designed to capture the complexity and diversity of the exchanges that take place between individuals and the organizational environment (Mark and Smith, 2008). This is in sharp contrast to prevailing models of intervention evaluation that are built around linear stimulus-response frameworks (Semmer, 2006).

A rather different example of fit can be found throughout the personnel selection literature. Selection processes designed around the requirements of a well-specified job role have the best chance of being effective (Sackett and Lievens, 2008). This is because the demands of the job and competencies of the employee are well-matched (i.e. there is good person-environment fit). In this situation the organizational context means that the job may change (e.g. if new work equipment becomes available or the company's objectives change) and with it change the fit between the job and the organizational environment. This might mean that although there is a good fit between the person and the job role, fit becomes irrelevant because the job role no longer fits with the wider organizational context, thus weakening the effectiveness of the selection process. In a similar way it can be that promising organizational-level interventions no longer fit with the organizational context.

Risk management approaches to work-related stress emphasise the need to tailor problem diagnosis and subsequent interventions to the needs of employees in order to help ensure that problem diagnosis and intervention planning have good fit (Cox, Randall, and Griffiths, 2000). Therefore, it is surprising that the concept of fit has not yet been applied to understand better the various effects of organizational-level interventions. These interventions are often delivered to diverse working populations across complex, different and fluctuating organizational contexts. For example it is not uncommon for the same intervention to be planned for different teams or for people working in different locations (Egan et al., 2009).

Examining the fit of an intervention invites us to look differently at intervention effectiveness. Quasi-experimental intervention research has often been built around the assumption that interventions have an inherent potency (or lack thereof) that needs to be established. One possible explanation for the inconsistency of outcomes in intervention research is that we might be observing good effects when an intervention has good fit and weak effects when it does not. By using the concept of fit, we could examine the potency of an intervention for dealing with a presenting problem in the context within which the problem occurs while simultaneously taking into account the diversity of the recipient population. This changes the key evaluation questions from "is an intervention effective?" to "did the intervention fit?" and "how powerful is the intervention when it fits the organizational context and the needs of the individual employee?"

A model of intervention fit

Inspection of the literature on organizational-level interventions reveals that when interventions fail many of the reasons given for disappointing outcomes relate to the fit of the intervention to the organizational context (Parkes and Sparkes, 1998; Semmer, 2006). This may be a re-structuring of the organization (Landsbergis and Vivona-Vaughan, 1995), the threat of downsizing or redundancies (Nielsen, Randall and Christensen, 2010), a lack of resources to allow proper implementation or any number of other chronic or acute problems facing the organization

(Randall, Cox, and Griffiths, 2007). There are also numerous examples of studies that cite the support of senior management as maintaining a favourable organizational context, good communication and intervention activities that are faithful to the intervention plan (Kompier, Cooper, and Geurts, 2000). In a similar way when interventions succeed there is often discussion of the importance of participatory intervention process and careful assessment of presenting problems that allow tailored interventions to be developed and delivered (LaMontagne et al., 2007).

Therefore, the model shown in Figure 7.1 is built around two dimensions of fit. The first is the fit between the intervention and the organizational environment. This is the extent to which the intervention is suited to the various constraints and opportunities found in the organizational setting: environment-intervention fit. The second is the fit between an intervention and those who receive it: person-intervention fit.

This model emerged from a review of the published intervention research. In contrast to other recent reviews we followed the recommendations of Semmer (2006) and sought to extract useful information from less-than-perfect studies. We took the position that problems with study design are rarely the result of poor researcher competence but, rather, they are often an inevitable consequence of attempting to intervene in complex functioning organizations. Our analysis indicated that the intervention fit was a common latent theme in accounts of intervention processes and qualitative explanations of intervention effectiveness.

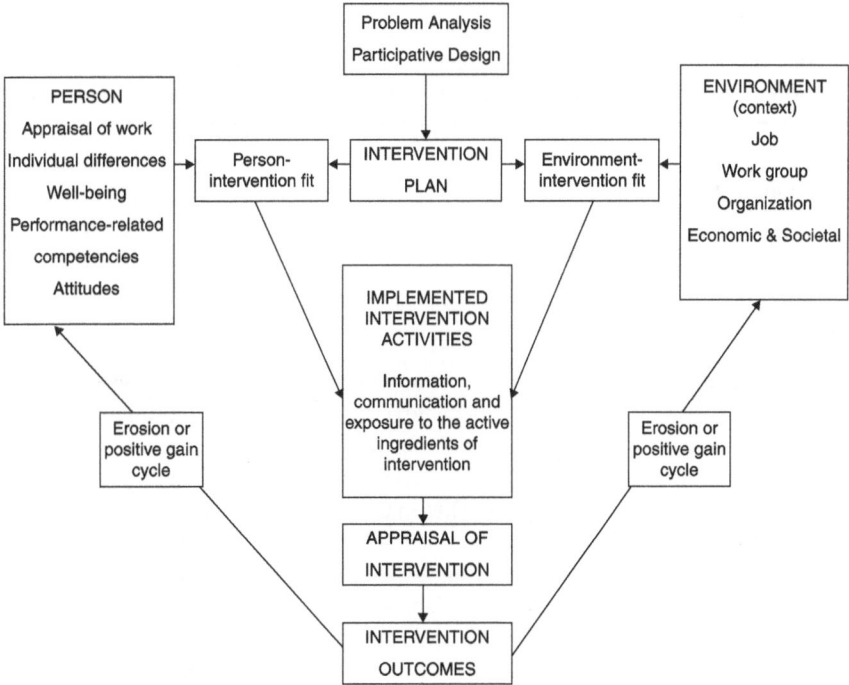

Figure 7.1 Intervention fit model.

Complex functions of fit are used in many contemporary theories of work stress to describe the processes through which stress develops (see Mark and Smith, 2008). We extend this notion into the problem-solving process and propose that each dimension of intervention fit is a complex function that results from the coming together of many factors. Both dimensions of fit (see next section) are subject to the influence of intervention planning activities (such as risk assessment and participatory intervention design processes). In the model we also include two mediating steps between the planned intervention and intervention outcomes. These are from the intervention plan, through fit, to implemented intervention activities (Step 1) and from the intervention activities to the appraisal of intervention activities (Step 2). Recent accounts of intervention research that include process evaluation have indicated that good fit will result in intervention plans that are enacted (Step 1) and enacted plans being appraised positively by employees (Step 2) (see Egan et al., 2009; Murta, Sanderson, and Oldenburg, 2007; Nielsen, Randall, and Albertsen, 2007). We also argue that many intervention studies describe worker-environment transactions that occur during the intervention process (Nielsen et al., 2010). This suggests that feedback loops exist in intervention process just as they do in transactional models (see Mark and Smith, 2008) and cybernetic theories of work stress (Edwards, 1992): intervention outcomes can change the characteristics of the intervention contexts and modify individual workers' psychological processes thereby changing the fit of an intervention.

In the following sections we deconstruct person-intervention and environment-intervention fit to describe the various elements of these dimensions. We also discuss the antecedents and consequences of fit. In doing this we attempt to develop a comprehensive explanatory framework linking intervention processes and intervention outcomes. An important aim of this chapter is to offer suggestions for how fit can be improved through intervention planning, implementation and monitoring activities.

Dimensions of fit

Environment-intervention fit

Environment-intervention fit refers to the appropriateness of the intervention in its setting. Good fit is characterized by a context that facilitates the initial implementation of the intervention and does not disrupt, displace or dilute ongoing intervention activities. Many examples of poor intervention-context fit can be found in the discussion sections of intervention studies that identified small intervention effects (see Landsbergis and Vivona-Vaughan, 1995, for a specific example and LaMontagne et al., 2007, and Parkes and Sparkes, 1998, for summaries of these challenges presented by contextual variables in intervention research). Poor intervention fit is often linked to concurrent changes in the team or organizational context (e.g. restructuring, downsizing or re-organization) or to pre-intervention job design (e.g. a lack of opportunities for workers to participate in intervention activities as they cannot set aside other work demands).

Context is a complex construct with there being multiple contexts that are subject to change over time (see Randall, Cox, and Griffiths, 2007). Our review of intervention research suggests that at least four levels of context need to be considered in the evaluation of environment-intervention fit: individual employees' immediate working environment, the environment of the team they work in, the wider organizational environment and the economic and social contexts within which the organization functions. These distinctions are important because it may be that an intervention can fit well within the immediate context of an individual's work but could be damaging to the organization of work tasks within the wider team context. In the same way an intervention that fitted well with the tasks carried out by the team may not fit well with the problems being faced by the whole organization that originate in the social or economic context within which the organization operates (e.g. economic recession or shortages of skilled employees).

Person-intervention fit

Various theories of work stress indicate that an intervention is more likely to be effective for those workers for whom it has a good "fit" for them (i.e. when it changes for the better a working condition that has been appraised as a problem). Most organizational-level interventions are based on the premise that certain working conditions (e.g. high job control, low job demands and high social support) are almost universally healthy. This assumption appears to be relatively sound (Bond, Flaxman, and Loivette, 2006).

Diversity of appraised working conditions within an intervention group can influence the person-intervention fit. Kompier and Kristensen (2000) have argued that interventions to improve working conditions may not have the same effects for every employee because not every employee reports a problem that needs to be addressed. For example, improving opportunities for control by re-designing work would be unlikely to yield strong effects for a person who already perceived their job control to be high (a ceiling effect). Figure 7.1 indicates that this aspect of fit should also therefore include a consideration of the factors that influence the appraisal of working conditions and employees' readiness to benefit from an intervention. These include individual differences (e.g. psychological flexibility, locus of control, personality, self-efficacy and coping styles), attitudes (e.g. job satisfaction and organizational commitment), well-being and job-related competencies (i.e. knowledge, skills and abilities). Saksvik and Tvedt (2009) have argued that identifying and taking account of this potential for diversity within intervention groups is an important part of successful intervention processes.

The principle that an individual difference (psychological flexibility) can moderate intervention outcomes has been elegantly demonstrated by Bond, Flaxman and Bunce (2008) in their study of an intervention to improve job control. In their intervention study those with higher levels of psychological flexibility benefitted the most from an intervention to increase control. This appeared to be because the individual difference meant that employees accessed better the benefits of improved control at work.

Other individual differences could operate in a similar way. It has been demonstrated that several organizational-level interventions place additional demands on employees (Mikkelsen, Saksvik and Landsbergis, 2000). A person's belief that they are capable (self-efficacy) may put them in a better position to engage in such demanding intervention activities. For example, participatory intervention processes may be more appropriate for employees who are high in self-efficacy because they are more likely to believe that taking responsibility will have a successful outcome (Nielsen and Randall, 2009). Employees with an internal locus of control are also likely to be motivated to do something themselves about their situation. This may mean that interventions that give them more autonomy to solve their own problems have a better fit for them than interventions requiring that others (e.g. managers or external consultants) take responsibility for change.

Some employees may be more anxious about change (e.g. those high on the personality dimension of extraversion or high on negative affect) than others (McCrae and Costa, 1997). Other personality dimensions may also influence person-intervention fit. For example, for those high on openness creative changes may provide a good fit, while those low on agreeableness may not be worried by the prospect of conflict that accompanies some interventions. An important aspect of well-being is a high level of arousal and energy (Daniels, 2000; van Horn et al., 2004). Therefore, those employees high in well-being possess the resources to "make the most" of any intervention.

Organizational-level interventions are designed to alter significantly the demands of a job and some interventions may even reduce person-job fit for those struggling to meet pre-intervention job demands. Guth and Macmillan (1986) have argued that employees resist change when they fear they do not have the competencies to work in changed structures. The finding that cognitive ability is most strongly predictive of performance in highly complex job roles (Ones, Viswesvaran, and Dilchert, 2007) suggests that changes such as job enrichment might be most beneficial to those with high levels of this ability. In a similar way workload reduction may have the best "fit" for employees with insufficient pre-intervention knowledge and skills. It is therefore unfortunate that selection and training interventions are not more widely investigated as organizational-level interventions that could be used to improve employee well-being through better person-job fit.

Finally, employees who are skilled in using active coping strategies may have less need for intervention as they are more likely to engage in job crafting activities. Indeed it is possible that an organizational-level intervention may disrupt their own self-initiated problem-solving activities.

In this section we have presented quite a complex account of how a range of "worker variables" might influence person-intervention fit. Targeted research into the effects of these variables on intervention outcomes is needed. However, the measurement of such a wide range of variables is rarely practicable. Therefore, it may be that various influences of these variables can be measured by capturing employee attitudes to the intervention plan. Attitudes are well-documented as correlates of individual differences, well-being and work competencies (Arnold et al., 2010). Measures of attitudes indicating readiness for change may be particularly suitable for capturing reliable and valid data on person-intervention fit. Readiness for change encompasses the various attitudes that employees have about the intervention plan (i.e. the extent which the employee is looking forward to the intervention or believes it will be useful). Such direct appraisals of intervention plans have already been shown to be an important factor in determining whether employees are willing to engage in important intervention activities (Cunningham et al., 2002).

It may also be that the importance of person-intervention fit is also likely to vary across interventions. This is because some interventions require more active involvement of employees than do others. A level of active involvement by intervention recipients is typical of many participative change processes (Cox, Randall, and Griffiths, 2002). It could be that positive attitudes to the intervention are particularly strongly related to the outcomes of such intervention.

Research indicates that workload reduction and increasing levels of control are consistently related to improvements in employee health and well-being (Bond, Flaxman, and Loivette, 2006). This might indicate that interventions targeting these factors are likely to provide a good person-intervention fit irrespective of individual differences. For such interventions less detailed assessments of the person-intervention fit may be required because of the consistent effects of these job design variables across the employee populations: steps would still need to be taken to ensure good environment-intervention fit.

In this section we have argued that each dimension of fit could influence intervention outcomes independent of the effects of the other. However, intervention outcomes may be better understood by considering the interactions between the two dimensions of fit. In intervention research there is some evidence to suggest that person-intervention fit and environment-intervention fit are interdependent. Nielsen, Randall, and Christensen (2010) found that intervention plans were perceived as inappropriate by employees (poor person-intervention fit) because the intervention was planned when the organization was facing severed and pressing difficulties (poor environment-intervention fit).

Antecedents of intervention fit

Existing research has highlighted two processes that appear to be particularly important in determining intervention fit (Nielsen et al., 2010). These are:

1 a thorough pre-intervention assessment of presenting problems in context, and
2 intervention planning activities that involve key stakeholders (e.g. participative action research).

Conducting a thorough assessment of the problem before interventions are implemented is widely recommended in the literature on organizational interventions (see Cooper and Cartwright, 1997; Cooper, Dewe, and O'Driscoll, 2001; Cox et al., 2000; Kompier et al., 1998). The focus of the problem analysis is often on the immediate work environment as perceived by employees. This can help to design interventions that have good person-intervention fit but may neglect important elements of environment-intervention fit. There is evidence of this problem in several studies in which promising interventions have been de-railed or disrupted by contextual events (Parkes and Sparkes, 1998). The concept of fit implies that problem identification activities should also capture information about the wider context within which work takes place. Contextual analysis can be used to capture data about the issues and constraints facing work groups (e.g. those carrying out a particular role, those in a particular location) and other organizational contexts that could impact on viability of any intervention plans. This may be achieved through audits of existing management and employee support systems (e.g. interviews with senior managers, analysis of organizational records, etc.) the results of which can then be considered during intervention planning (Nielsen et al., 2010).

Participatory approaches have been recommended as ways of increasing the relevance and uptake of interventions (LaMontagne et al., 2007). Employee participation in the development and implementation of interventions has been shown to help with the integration of intervention activities into existing organizational structures (Tsutsumi et al., 2009) thus securing sustainable changes (Daltuva et al., 2009). Participatory approaches encourage good intervention fit because of the gathering of rich data about intervention feasibility and potential intervention impact (Rosskam, 2009). Participatory approaches also bring together the various parties who will be involved in the implementation of the intervention with the intended beneficiaries of it (LaMontagne et al., 2007; Pasmore and Friedlander, 1982). Fit is likely to occur because of a critical analysis of change options that means that it is only those interventions that have a good chance of survival that are retained within the intervention plan (Mikkelsen and Gundersen, 2005; Rosskam, 2009).

The consequences of fit

Figure 7.1 is an initial working model of the concept of fit and how it could be related to intervention outcomes. The model is designed to illustrate how fit could underpin and feed into the mechanisms which determine intervention outcomes.

In Figure 7.1 we make the distinction between intervention plans and implemented interventions. A key outcome of fit is the translation (or otherwise) of intervention plans into enacted activities: good fit is likely to result in intervention plans being enacted (Steckler and Linnan, 2002). In many intervention studies poor intervention fidelity is often attributed to problems in the organizational environment that disrupted intervention plans as a result of poor intervention-environment fit. There is often less discussion of person-intervention fit: the assumption is often that the intervention will fit equally all of those targeted by the intervention and all contexts within which the intervention is delivered. Results of intervention studies such as that by Bond, Flaxman, and Bunce (2008) indicate that this assumption needs to be more widely tested in intervention research.

Recent research has highlighted the importance of measuring intervention exposure as it is by no means guaranteed even during comprehensive and well-resourced intervention strategies (Randall, Griffiths, and Cox, 2005; Nielsen et al., 2006). An apparent consequence of poor fit is a large gap between the intervention plan and the enacted activities. For example, when resources are scarce it may be that cost-effective interventions are most likely to have good environment-intervention fit because expensive interventions are unlikely to be initiated or maintained. Interventions that require time and effort from employees may not survive in high workload environments. Employees are less likely to seek out and engage in intervention activities that are not communicated well partly because this leaves them uncertain about how the intervention will fit with their own circumstances and preferences (Nielsen, Randall, and Albertsen, 2007).

Low levels of exposure found in specific defined locations or work groups are likely to indicate poor intervention-environment fit. This is because the environment is hostile or unstable and the active ingredients of the intervention cannot be delivered or sustained. If low levels of exposure are more widely dispersed then it may be that individual-intervention fit is the problem: the environment might be supportive of the intervention activities but interventions activities do not meet the requirements of individual employees.

It is also likely that poor intervention fit will result in negative appraisals of interventions that are enacted. Employees' direct appraisals of interventions have received little research attention (Randall, Nielsen, and Tvedt, 2009). This is surprising given the central importance of appraisal and perception in various well-validated theories that describe the links between work and health (see Mark and Smith, 2008). The research that does exist shows that these perceptions of the intervention can have a dramatic impact on how employees engage with and benefit from interventions. Nielsen, Randall, and Albertsen (2007) found that employees' direct appraisal of the sustainability of interventions predicted their perceptions of changes in the working environment. Many of these perceptions were likely consequences of person-intervention fit (e.g. intervention relevance and suitability) while others were recognisable as outcomes of environment-intervention fit (e.g. sustainability).

Good intervention outcomes include improved working conditions, employee attitudes, motivation, self-efficacy and well-being (Richardson and Rothstein, 2008). These outcomes can place employees in a better position to engage in the intervention thus improving person-intervention fit. Intervention outcomes may also stimulate adjustments in the environment to allow for further intervention activities to be supported (i.e. better environment-intervention fit). This pattern of transactions could establish a positive gain cycle. Naturally, the opposite could occur if interventions have poor outcomes (an erosion cycle). This model may help to explain why some intervention activities decay in the face of difficult intervention environments, while others flourish after some initial successes.

It has often been argued that organizational-level interventions have the best prospects of securing long-term improvements in employee well-being and satisfaction but, as yet, there is little evidence to support such claims (Richardson and

Rothstein, 2008). Long-term effects are only likely if intervention fit remains good up to the census point for evaluation. This makes the timing of evaluation crucial in determining the validity of intervention effects (Semmer, 2003). Lengthy evaluation periods mean that employees' appraisals of problem and the organizational context and individual employees' psychological states may change many times during the intervention period. In other words, even if the intervention was suitable at the start of the intervention period its suitability may change in the period before evaluation occurs. If intervention fit decays over the period before evaluation, any initial positive effects of the intervention may be missed. Poor environment-intervention fit may mean that the effects of the intervention have been underestimated because the intervention was not implemented as intended or implemented without the conditions needed to maximise its effects. This evaluation problem has been described as a Type III error (Dobson and Cook, 1980). Estimates of intervention fit could be used to help avoid Type III errors.

Conclusions: Using intervention fit

The concept of fit can be used to shape the search for the conditions that can influence strongly intervention effectiveness. For it to be used effectively, researchers will need to document carefully the environment and features of the working population during intervention studies. Expanding the nexus of data collection during interventions in this way will help researchers to identify the conditions under which the effects of the intervention are likely to generalise. The methods and tools described throughout this volume will aid the collection of these data.

The model presented in Figure 7.1 is designed to also help researchers and practitioners to carry out targeted analysis of some of the factors linked to fit. It is not our intention to over-complicate the evaluation process but rather to provide a different and, we argue, a more appropriate lens through which to examine the factors that may influence intervention outcomes. The body of intervention research suggests that there are many such factors residing outside of the intervention itself. We have developed the concepts of environment-intervention fit and person-intervention fit in an attempt to produce heuristics that can be used to assess and manage these factors in a way that supports effective intervention.

Problems that might lead to poor fit are often seen as confounding variables that obstruct proper evaluation rather than as important elements of the mechanisms of change (see Parkes and Sparkes, 1998). The prevalence of such confounds in published evaluation studies suggests researchers will find few stable intervention environments containing heterogeneous working populations to support carefully controlled quasi-experiments. By examining fit such variables are *integrated* into the evaluation process (rather than designed out of it). This can help to make evaluation more realistic and accurate in complex and unstable organizational settings.

The concept of fit could provide a useful way of monitoring intervention suitability throughout intervention processes. Documented analysis of fit would provide early warning signs of interventions at risk of failure and point to ways of increasing the chances of success. In much intervention research the "goodness of fit" of an intervention is assumed but not measured. For example, an intervention designed to increase social support at work is assumed to have a positive outcome (i.e. it has a universal good fit). By posing some relatively simple questions we can see that these assumptions about fit should be tested as part of process evaluations. We could better assess the person-intervention fit of the intervention by asking each employee if they wanted more support or if they felt ready for it. This could be achieved by asking each employee if they saw the increased support as relevant to solving their particular problems or difficulties at work. We could also monitor whether their appraisal of the intervention changed over time and examine whether the intervention placed additional demands on employees (e.g. a requirement to attend more meetings) that offset some of its positive effects (see Saksvik and Tvedt, 2009). We might also consider whether the intervention or its impact was altered by concurrent events of changes in the intervention environment (e.g. we might need to know whether there had been sufficient resources to implement the intervention as intended or as planned). We might also want to consider whether major structural changes within the company had interfered with or prevented implementation.

These are just some of the questions that could be asked. However, the concept of fit is not intended to be constrained by such suggestions. Indeed, just as transactional theories of stress have helped to broaden the field of enquiry into sources of stress, we hope that the concept of intervention fit provides stimulus for intervention researchers to consider capturing data on more contextual and psychological variables as part of the intervention evaluation process.

Information about fit gathered throughout the intervention process could also be used to make decisions about how intervention plans could be modified to make them more effective. An emerging problem in the intervention literature (see Nielsen et al., 2010) is that those employees who need the intervention the most are least "ready" to receive it because of poor person-intervention fit (e.g. low self-efficacy) or poor environment-intervention fit (e.g. poor leadership within the group). This "intervention paradox" could be tackled by taking steps to improve the resources available to employees to place them in a better state to benefit from the interventions. For example, steps could be taken to develop employees' skills or competence or to educate them about the possible benefits of involvement in intervention activities. Similarly, pre-intervention contextual modifications could be made in intervention groups for whom environment-intervention fit might be poor (e.g. the allocation of additional intervention resources to support intervention delivery, postponement of other planned changes, delivering leadership training, etc.).

We accept that the model of fit and the hypotheses we have developed in this chapter need to be tested fully. As part of this testing process we are examining how information about fit can be integrated with outcome evaluation data to provide more information about the key determinants of intervention success. The model will also need to be refined as the body of process evaluation research grows through the application of methods such as those described throughout this book.

References

Arnold, J., Randall, R., Patterson, F., Silvester, J., Robertson, I., Cooper, C., Burnes, B., Swailes, S., Harris, D., Axtell, C., & Den Hartog, D. (2010). *Work psychology: Understanding human behaviour in the workplace (5th Ed.)*. Harlow: Pearson.

Bond, F., & Bunce, D. (2001). Job control mediates change in a work reorganization intervention for stress reduction. *Journal of Occupational Health Psychology, 6*, 290–302.

Bond, F. W., Flaxman, P. E., & Bunce, D. (2008). The influence of psychological flexibility on work redesign: Mediated moderation of a work reorganization intervention. *Journal of Applied Psychology, 93*, 645–654.

Bond, F. W., Flaxman, P. E., & Loivette, S. (2006). *A business case for the Management Standards for Stress*. Norwich, UK: Her Majesty's Stationery Office, Research Reports.

Cooper, C. L. (1998). *Theories of organizational stress*. Oxford: Oxford University Press.

Cooper, C., & Cartwright, S. (1997). An intervention strategy for workplace stress. *Journal of Psychosomatic Research, 43*, 7–16.

Cooper, C., Dewe, P., & O'Driscoll, M. (2001). Organizational Interventions. In C. Cooper, P. Dewe, & M. O'Driscoll (Eds.), *Organizational Stress: A review and critique of theory, research, and applications* (1st ed., pp. 187–232). Thousand Oaks: Sage Publications.

Cox, T., Griffiths, A., Barlow, C., Randall, R., Thomson, L., & Rial-González, E. (2000). *Organisational interventions for work stress*. Sudbury UK: HSE Books.

Cox, T., Karanika, M., Griffiths, A., & Houdmont, J. (2007). Evaluating organisational-level work stress interventions: Beyond traditional methods. *Work & Stress, 21*, 348–368.

Cox, T., Randall, R., & Griffiths, A. (2002). *Interventions to control stress at work in hospital staff*. Sudbury, UK: HSE Books.

Cunningham, C. E., Woodward, C. A., Shannon, H. S., MacIntosh, J., Lendrum, B., Rosenbloom, D., and Brown, J. (2002). Readiness for organizational change: A longitudinal study of workplace, psychological and behavioural correlates. *Journal of Occupational and Organizational Psychology, 75*, 377–392.

Daltuva, J. A., King, K. R., Williams, M. K., & Robins, T. G. (2009). Building a strong foundation for occupational health and safety: Action research in the workplace. *American Journal of Industrial Medicine, 52*, 614–624.

Daniels, K. (2000). Measures of five aspects of affective well-being at work. *Human Relations, 53*, 275–294.

Daniels, K., Boocock, G., Glover, J., Hartley, R., & Holland, J. (2009). An Experience Sampling Study of Learning, Affect, and the Demands Control Support Model. *Journal of Applied Psychology, 94*, 1003–1017.

Dobson, L. D., & Cook, T. J. (1980). Avoiding type III error in program evaluation: Results from a field experiment. *Evaluation and Program Planning, 3*, 269–276.

Edwards, J. R. (1992). A cybernetic theory of stress, coping, and well-being in organizations. *Academy of Management Review, 17*, 238–274.

Egan, M., Bambra, C., Petticrew, M., & Whitehead, M. (2009). Reviewing evidence on complex social interventions: appraising implementation in systemic reviews of the health effects of organisational-level workplace interventions. *Journal of Epidemiology & Community Health, 63*, 4–11.

Guth, W. D., & Macmillan, I. C. (1986). Strategy implementation versus middle manager self-interest. *Strategic Management Journal, 7*, 313–327.

Kompier, M., Cooper, C., & Geurts, S. (2000). A multiple case study approach to work stress prevention in Europe. *European Journal of Work and Organizational Psychology, 9*, 371–400.

Kompier, M., Geurts, S., Grundemann, R., Vink, P., & Smulders, P. (1998). Cases in stress prevention: The success of a participative and stepwise approach. *Stress Medicine, 14*, 155–168.

Kompier, M. A. J., & Kristensen, T. S. (2001). Organizational work stress interventions in a theoretical, methodological and practical context. In J. Dunham (Ed.), *Stress in the workplace: Past, present and future* (pp. 164–190). London: Whurr.

LaMontagne, A. D., Keegel, T., Louie, A. M., Ostry, A., & Landsbergis, P. A. (2007). A systematic review of the job-stress intervention evaluation literature, 1990–2005. *International Journal of Occupational and Environmental Medicine, 13*, 268–280.

Landsbergis, P., & Vivona-Vaughan, E. (1995). Evaluation of an occupational stress intervention in a public agency. *Journal of Organizational Behavior, 16*, 29–48.

Lazarus, R., & Folkman, S. (1992). *Stress, appraisal and coping*. New York: Springer Publications.

Mark, G. M., & Smith, A. P. (2008). Stress Models: A review and suggested new direction. In J. Houdmont & S. Leka (Eds.), *Occupational health psychology: European perspectives on research, education and practice* (pp. 111–144). Nottingham: Nottingham University Press.

McCrae, R. R., & Costa, P. T. Jr., (1997). Personality trait structure as a human universal. *American Psychologist, 52*, 509–516.

Mikkelsen, A., & Gundersen, M. (2003). The effect of participatory organizational intervention on work environment, job stress, and subjective health complaints. *International Journal of Stress Management, 10*, 91–110.

Mikkelsen, A., Saksvik, P. Ø., & Landsbergis, P. (2000). The impact of a participatory organizational intervention on job stress in community health care institutions. *Work & Stress, 14*, 156–170.

Murta, S. G., Sanderson, K., & Oldenburg, B. (2007). Process evaluation in occupational stress management programs: A systematic review. *American Journal of Health Promotion, 21*, 248–254.

Nielsen, K., Fredslund, H., Christensen, K. B., & Albertsen, K. (2006). Success or failure? Interpreting and understanding impact of interventions in four similar worksites. *Work & Stress, 20*(3), 272–287.

Nielsen, K., & Randall, R. (2009). Managers' active support when implementing teams: The impact on employee well-being. *Applied Psychology: Health and Well-being, 1*, 374–390.

Nielsen, K., Randall, R., & Albertsen, K. (2007). Participants' appraisals of process issues and the effects of stress management interventions. *Journal of Organizational Behavior, 28*, 793–810.

Nielsen, K., Randall, R., & Christensen, K. B. (2010). Does training managers enhance the effects of implementing teamworking? A longitudinal, mixed methods field study. *Human Relations, 63*, 1719–1742.

Nielsen, K., Randall, R., Holten, A-L., & Rial-González, E. (2010). Conducting organizational-level occupational health interventions: What works?. *Work & Stress, 24*, 234–259.

Ones, D. S., Viswesvaran, C., & Dilchert, S. (2007). Cognitive ability in personnel selection decisions. In A. Evers, O. Voskuijl and N. Anderson (Eds.), *Handbook of Selection*. Oxford: Blackwell.

Parkes, K. R., & Sparkes, T. J. (1998). *Organizational interventions to reduce work stress: Are they effective? A review of the literature*. Sudbury: HSE Books.

Pasmore, W., & Friedlander, F. (1982). An action research programme for increasing employee involvement in problem solving. *American Science Quarterly, 27*, 343–362.

Randall, R., Cox, T., & Griffiths, A. (2007). Participants' accounts of a stress management intervention. *Human Relations, 60,* 1181–1209.

Randall, R., Nielsen, K., & Tvedt, S. D. (2009). The development of scales to measure participants' appraisals of organizational-level stress management interventions. *Work & Stress, 1,* 1–23.

Richardson, K. M., & Rothstein H. R. (2008). The effects of worksite stress management intervention programs: A systematic review. *Journal of Occupational Health Psychology, 13,* 69–93.

Rosskam, E. (2009). Using participatory action research methodology to improve worker health. In P. Schnall, M. Dobson, & E. Rosskam (Eds.), *Unhealthy work: Causes, consequences, cures* (pp. 211–229). NY: Baywood Publishing Company.

Sackett, P. R., & Lievens, F. (2008). Personnel selection. *Annual Review of Psychology, 59,* 419–450.

Saksvik, P. Ø., & Tvedt, S. D. (2009). Leading change in a healthy way. *Scandinavian Journal of Organizational Psychology, 1,* 20–28.

Semmer, N. K. (2003). Job stress interventions and organization of work. In J. C. Quick & L. E. Tetris (Eds.), *Handbook of Occupational Health Psychology*. Washington, DC: American Psychological Association.

Semmer, N. K. (2006). Job stress interventions and the organization of work. *Scandinavian Journal of Work and Environmental Health, 32,* 515–527.

Steckler A., & Linnan, L. (Eds.) (2002). *Process Evaluation for Public health Interventions and Research*. San Francisco: Jossey-Bass.

Tsutsumi, A., Nagami, M., Yoshikawa, T., Nogi, K., & Kawakami, N. (2009). Participatory intervention for workplace improvements on mental health and job performance among blue-collar workers: A cluster randomized controlled trial. *Journal of Occupational and Environmental Medicine, 51,* 554–563.

van Horn, J. E., Taris, T. W., Schaufeli, W. B., & Schreurs, P. C. (2004). The structure of occupational well-being: A study among Dutch teachers. *Journal of Occupational and Organizational Psychology, 77,* 365–375.

8 How can qualitative studies help explain the role of context and process of interventions on occupational safety and health and on mental health at work?

Geneviève Baril-Gingras, Marie Bellemare, and Chantal Brisson

Introduction

In the field of occupational health and safety (OHS) intervention research, as for mental health at work, there is a growing concern over the effectiveness of interventions (e.g. Goldenhar and Schulte, 1994; Zwerling et al., 1997). This highlights the scientific, social and public policy issues around health and safety in the workplace (see Goldenhar et al., 2001). In OHS and ergonomics, this preoccupation led to literature reviews (e.g. Cohen and Colligan, 1998; Cole et al., 2005; Denis et al., 2005; Denis et al., 2008; Karsh, Moro, and Smith, 2001; Goldenhar and Schulte, 1994; Guastello, 1993; Zwerling et al., 1997). Methodological recommendations regarding evaluation designs were also made (e.g. Goldenhar et al., 2001; Heacock, Koehoorn, and Tan, 1997; Robson et al., 2001; Shannon, Robson, and Guastello, 1999).

Research has mainly focused on evaluating the effects of interventions using quantitative methods, and our comprehension of how interventions produce their effects is rather limited. Some authors recommended adding qualitative information to the data collected in studies of OHS interventions (Robson et al., 2001; Shannon et al., 1999; Zwerling et al., 1997). Needleman and Needleman (1996) provided rich insights on the use of qualitative methods for intervention. Baril-Gingras, Bellemare, and Brun (2006a) argued for the contribution of qualitative analysis of intervention to a better understanding of the factors influencing their implementation and effects.

A growing number of articles reporting interventions on OHS, such as musculoskeletal disorders (MSD), include thorough descriptions of the interventions' context and process, and insightful reflections on change mechanisms (e.g. Garrety and Badham, 1999; Polanyi et al., 2005; Theberge et al., 2006; Trudel and Montreuil, 1999). The same pattern is observed for interventions aimed at improving the psychosocial environment of work (e.g. Bourbonnais et al., 2006a; Bourbonnais, Brisson, and Vézina, 2010; Nielsen et al., 2006; Park et al., 2004; Saksvik et al., 2002). In both fields, literature reviews included information on the context, on the interventions' process and on the actual changes proposed and implemented.

This kind of description is not yet the rule. Completing a systematic review of MSD preventive interventions, Denis et al. (2005, p. 2) observed that in previous literature reviews, nothing was said about the intervention approach. In the field of mental health at work interventions, a recent examination of studies on health effects of organizational-level workplace interventions reported that "many studies referred to implementation, but reporting was generally poor and anecdotal in form. A minority of studies described how implementation may have influenced outcomes" (Egan et al., 2009, p. 4). LaMontagne et al. (2007, p. 277) stressed that "the published literature tends to focus more on evaluation and often provides only limited descriptions of interventions." This is in line with the conclusions made by Murta, Sanderson, and Oldenburg (2007) in their systematic review regarding the information on process issues in occupational stress management programs.

As pointed out by Kristensen (2005), systematically documenting the intervention process allows the researcher to distinguish between theory failure (intervention is implemented but is not effective), and program failure (also termed Type III error, which refers to the conclusion that an intervention is ineffective, when in fact it was not adequately implemented) (Cox et al., 2007, p. 354). Semmer (2006) suggest that "this type of evaluation is more than a narrative account of things. It requires the careful documentation of as much information as possible in a systematic fashion" (p. 523). The question is thus to define what is relevant information.

Objectives

This chapter aims to illustrate the contribution of qualitative studies to answer the following questions:

> How can the context and process of interventions, as well as the changes proposed and implemented (on OHS and on mental health at work) be described?
> How can the context and process of interventions, as well as the changes proposed and implemented (on OHS and on mental health at work) influence the intermediate and final outcomes of these interventions?

We focus on a qualitative multiple case study of OHS interventions, whose precise aim was to understand the role of context and process (Baril-Gingras et al., 2006a, 2006b; 2007). First, we briefly expose the methods used in this multiple case study and the theoretical model drawn from it. Second, we present factors related to context, process and nature of proposed/implemented changes which were identified as relevant for OHS interventions. We then examine recent intervention studies on mental health at work and report on what they say on each of these factors.

We do not pretend the descriptive model exposed here and the potential relationship between the factors it includes and the interventions' effects are definitive. This chapter is actually a step in our own ongoing multidisciplinary reflection, combining ergonomics, industrial relations and epidemiology. Our endeavour is to contribute to the reflection on how the utilization of qualitative research methods could support comparisons between interventions.

Methods

Multiple case study of OHS interventions

The descriptive model proposed by Baril-Gingras et al. (2004, 2006a, 2006b, 2007) was elaborated on the basis of in-depth analysis of OHS interventions. The seven interventions were conducted by OHS advisors (generalists, ergonomists, hygienists, etc.) employed by non-profit joint (unions and employers associations) OHS sector-based associations. These interventions took place in six different workplaces with very diverse characteristics (size, type of economic activity, etc.). They were chosen based on the heterogeneity of OHS activities to perform and changes to be implemented: training on specific risks (e.g. confined spaces entry), ergonomic analysis to reduce manual handling risks, updating of a prevention program including safety procedures, drafting of a program to prevent aggressions from clients, integration of OHS in an architectural, technological and organizational change, creation of an OHS committee. The effects that were looked at were not final ones – like the reduction of occupational injuries and illnesses – but intermediate changes, i.e. the implementation of change proposals formulated during the intervention.

The following methods were used: observations (43 events where an external advisor was interacting with workplace actors); direct (46) and telephone interviews (57) with the external advisors and different workplace actors (OHS coordinators, directors, supervisors, workers' and unions' representatives, workers, etc.). Depending on each case, the data collection extended between 10 and 19 months. Inter and intra-case analysis were completed (Huberman and Miles, 1991). From the analysis of a first case, hypotheses were formulated with the following format: "if this condition (context, process, type of proposed change) is present, then a specific change proposal is (or is not) implemented". In a process of analytical induction, each of these hypotheses was then applied to the cases, and then accepted, rejected or modified. The descriptive model was constructed with an approach combining both induction and deduction (Eisenhardt, 1989) taking into account the results of a literature review (Baril-Gingras, 2003) on the factors influencing OHS prevention organization and OHS interventions' effects.[1] It was further validated by a collaborative study with 14 OHS practitioners, based on 27 interventions, aiming at transforming this model into tools for their analysis of the social and organizational dimensions of interventions (Baril-Gingras et al., 2010).

Sources for the comparison with mental health at work interventions' studies

In order to compare the way interventions are described in the fields of OHS and mental health at work, and what is said about the influence of context and process on intermediate and final outcomes, the sources used here are scientific articles and book chapters that can be classified in three categories.[2]

1. Literature reviews on mental health at work interventions studies (including stress management or intervention on psychosocial work environment) which addressed some process issues: Egan et al. (2009); LaMontagne et al. (2007); Murta et al. (2007).
2. Qualitative studies of the context, process or nature of changes proposed/ implemented in interventions on psychosocial health at work, or more generally on health at work, including mental health: Biron et al. (2009); Bourbonnais et al. (2006a); Brisson et al. (2006); Gilbert-Ouimet et al. (2010 and forthcoming); Nielsen et al. (2006); Saksvik et al. (2002 and 2007).
3. Articles reporting a reflection on the role of context, process or nature of changes proposed/implemented or on the methodology for the study of these interventions: Biron et al. (2009); Cox et al. (2007); Goldenhar et al. (2001), Kristensen (2005); Levenstein (1996); Murphy and Sauter (2004); Nytrø et al. (2000); Randall, Nielsen, and Tvedt (2009); Semmer (2006).

We believe that this sample of studies and conceptual papers provides a relevant basis to describe the process and contextual issues and their potential influences on outcomes targeted by the intervention.

Theoretical model

The theoretical model developed following the multiple case study is illustrated by Figure 8.1. In this model, the intermediate and final outcomes of the intervention depend on its process (participation, activities and change proposals) and on the changes that are actually implemented. The intervention and its outcomes are modulated by the workplace internal and external context. Some of the implemented changes (the first intermediate outcomes of the intervention) are modifying the workplace context, and the work place actors' preventative capacities.

Context

Context here refers to both external and internal factors. Both are embeded in the political and economical context of a given historical period (see Bambra, 2011; Benach, Muntaner, and Santana, 2007; Nichols, 1997; Quinlan, Mayhew, and Bohle, 2001). External factors considered here correspond to the characteristics of the OHS regime that apply to this particular workplace (in Québec, some sectors are covered by systematic OHS management and Joint OHS committees requirements, others not), and by the way these requirements are enforced by the inspectorate. Internal factors refer to structural characteristics of the workplace and to the actors' "capacity" to be active in prevention, to their "dispositions" (willingness) to do so. These two concepts of capacities and dispositions are inspired by Dawson et al. (1988). The present model adds the social interactions between

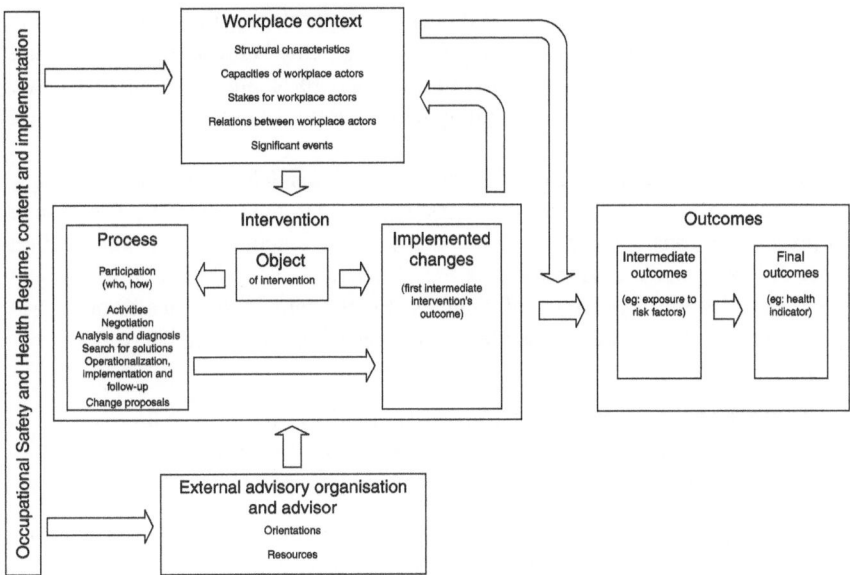

Figure 8.1 Model used for analyzing interventions context and process.
(Adapted from Baril-Gingras et al., 2010)

actors, e.g. high management, first line supervisors, groups of workers, workers' unions to better describe the way OHS is treated, and the intervention embedded.

Intervention and process

In this model, the intervention is defined by its object, its process and the nature of the implemented changes, which represent a first level of intervention outcome.

The intervention process itself is described using three concepts:

1   the participation device that organizes the interactions between the OHS external advisor and the workplace actors;
2   the characteristics of the successive phases and activities (e.g. negotiation of the intervention, its object, objectives and conditions, analysis and diagnosis, search for solutions, operationalization of solutions, implementation and follow-up); and
3   the content of the proposals for change emerging during the intervention, which may or may not be implemented.

Implemented changes

Among all the changes proposed during the intervention process, the ones that are implemented could be considered as the first intervention outcomes and may take

140 *Challenges and methodological issues*

place at two different levels: prevention activities and structures, or working conditions themselves. Activities and change proposals together thus qualify the nature of the intervention, which may vary, for a same "object" like "preventing back injuries", from one intervention to the other: for example training for "safe handling techniques" versus undergoing an ergonomic analysis to modify layout, equipment or work organization.

The implemented changes correspond to "what was delivered", and other steps would have to be taken to assess workers' exposure to these changes. The change proposals are considered here as an intermediate outcome of the intervention process. The process thus modulates the translation of the intervention's object and objectives into implemented changes. Final changes, like health outcomes were not measured in the case studies whose results are presented here.

An examination of the literature on mental health at work interventions with the help of a model drawn from a qualitative analysis of OHS interventions

In this section, we successively compare the concepts of the model just exposed, which was developed on the basis of OHS interventions, to those used in the selected mental health at work intervention literature. For each of them, we briefly expose the main conclusions drawn from the iterative comparison of multiple case study of OHS interventions[3] and discuss the similarities and differences found with what is said on the same topics in the selected literature on mental health at work intervention.

Intervention's context

Various authors stress the need to describe diverse features of the context of the workplace. For example, Goldenhar et al.'s (2001) proposal of an intervention research process in OHS includes the need to document "What is the context for the intervention?" and "What is the perspective of target population(s)?" in the intervention research development phase. They include amongst intervention research tasks to "gather background information to help characterize the problem and its history and the sociopolitical work and evaluation settings (e.g. labour-management relations, different evaluation agendas among stakeholders)." Egan's et al. (2009, p. 6) and Murta et al. (2007, p. 250) reviews do include this type of factors. Cox et al. (2007, p. 355) recall that "variability in macro processes in the wider organizational, social and socio-economic, and political contexts may explain why some interventions are successful while others are not (Goldenhar and Schulte, 1994)."

When conducting intervention studies, context and process are difficult to separate. This may explain why some authors include both in the way they define "process", as is the case with the definition by Nytrø et al. (2000, p. 214), also used by Saksvik et al. (2007, p. 245) where perceptions of the different actors are considered as well as their actions. What Cox et al. (2007, p. 357) call the "implementation process" and the "macro processes" (social, organizational, socio-economic, and political) – in which the intervention process is imbedded – correspond to what is defined here as the context.

The multiple case study of OHS interventions suggests that the following contextual factors influence the process and the outcomes of an intervention:

- Legal requirements and their enforcement;
- Significant events affecting the workplace during the intervention;
- Workplace structural characteristics;
- Capacities related to prevention, in particular the degree of development of structures and activities in prevention;
- Convergence or divergence of interests between stakeholders on OHS issues;
- Relations between these collective actors, which result in their relative power over a specific question.

All of these factors vary in each intervention and can be classified as favourable or unfavourable to the intervention as they contribute to define how OHS and change proposals are dealt with.

Legal requirements and their enforcement

One of the external context factors is the legal requirements and the OHS compensation regime. Problems of recognition for compensation and the possibility to reduce costs by other ways than prevention are barriers to successful interventions. Results from interventions made on a voluntary basis may not be generalized to the whole range of employers. For example, Biron et al. (2010) observed that legal requirements were not sufficient to elicit ownership of the intervention by the managers who were responsible for the prevention of psychosocial risks. The question is thus what is needed to avoid paper compliance and make prevention effective.

The actual enforcement of legal requirements is also to be considered. Levenstein (1996, p. 359) stresses the negative influence of a context marked by "a government committed to deregulation, an industry faced with declining or highly competitive markets, and a union reeling from unemployment and the loss of membership."

Workplace structural characteristics

In our case studies, workplaces were chosen to cover an array of sizes, levels of qualification of their workforce, ethnic and gender composition, levels of wages, levels of turnover, etc. All these characteristics create conditions that tend to be more or less favourable for prevention. Some authors (e.g. Dorman, 2000; Graham and Sakow, 1990; Simard and Marchand, 1997) regroup them according to the characteristics of primary (large firms, unionized, white qualified men workers, neither "young" nor "old", with stable employment) and secondary (the opposite) labour markets and report how it relates to the quality of OHS conditions. Our multiple case study suggested that these structural characteristics influenced the degree of development of prevention activities and structures at the beginning of the intervention and the fate of change proposals. Relations between actors appeared to be profoundly influenced (but not "determined") by these workplaces' structural characteristics. In the same way, Kristensen (2005, p. 208) suggested that:

> Health promotion programmes, for example, should not only be tested in middle class samples, which so far has been the most common setting, but also among blue collar workers, immigrants, young workers, obese workers, workers in small and medium size enterprises, or temporary workers.

Capacities

Our results indicate that changes most easily implemented corresponded to actors' existing capacities. In other words, changes implemented were less demanding, i.e. not requiring the development of new competencies, structures, etc. This suggests the need to identify the "development stage" of the workplace, in relation to prevention. These stages will be presented later in relation to the fate of change proposals.

The multiple case study also suggests that external actors conducting an intervention cannot substitute themselves to internal actors. It is true in particular for all changes that go beyond technical "one shot" modifications, like those requiring the integration of new activities, new knowledge and to implement and maintain preventive actions of a recurring nature (e.g. a safety procedure, a new work organization). One context variable to consider is the presence of one or more internal actors that will relay the intervention. Some authors call it an internal change agent (e.g. Nytrø et al., 2000; Saksvik et al., 2007). If the capacities already exist, they have to be utilised, otherwise to be developed. Both the process and the effects of the intervention seem to be greatly influenced by the presence of these relay agents and their capacities. In the proposed model, a distinction is made between technical (e.g. OHS knowledge and experience, time) and political capacities (e.g. access to financial resources, influence over decision making). These will be examined below in relation to the construction of the participation device of the intervention (how these capacities are recruited/found) and the activities (how these capacities are used and developed) of the intervention.

Stakes

Relations between workplace actors are dependent upon the issues (stakes) that the object of the intervention represents for each of them, in the broader context of social relations of production (division of work, power over means of labour, etc.). The model assumes a basic antagonism, but leaves open the question of convergence or divergence of interest around a particular problem and specific change proposal: stakes may converge on a particular solution but diverge on another.

In the multiple case study, workplace actors' issues were examined first in relation to the situation at the origin of the intervention, and second in relation to change proposals. In their study on ergonomic training, Trudel and Montreuil (1999) unveil what had generated the demand for the intervention, the acceptation of an offer or the obligation made to the employer. They uncover the interaction and "negotiation" process – between workplace actors – that preceded the intervention. The chain of interaction may have been initiated by workers, their representatives or by a first-line supervisor. Who relayed these demands? Who blocked them? What were the motives for the employer? Was it to comply with an obligation from an inspector, to reduce compensation cost, to reduce production problems? This creates a different context, more or less favourable to prevention. These preoccupations are coherent with the view expressed by Biron et al. (2009, p. 455), whose model suggests considering the motives (legal, economical, political) and the expectations about the intervention and the readiness to change of workplace actors.

Employers' support for the intervention is almost systematically mentioned in the literature (e.g. Cox et al., 2007, p. 355; Egan et al., 2009, p. 6; Semmer, 2006, p. 522). In their review, Egan et al. (2009, p. 7) observed that negative health outcomes were more likely to result from interventions that were motivated for business reasons (managerial efficiency, productivity, cost) rather than by employee health concerns. Semmer (2006, p. 522) indicates that "many authors report that [management support] was not given strongly, or consistently enough." One issue is thus to define appropriate indicators of management support, in particular at the beginning of an intervention, as pointed out by Brun, Biron, and Ivers (2008).

In the stress intervention studies, workers' perspectives on the intervention are also examined (Egan et al., 2009, p. 6) via questions about what are employees' motivations for change (Cox et al., 2007, p. 355) and their readiness for change (e.g. Cox et al., 2007, p. 355). Randall et al. (2009) demonstrated that this last factor is significantly correlated with post-intervention outcomes. We will come back to this when examining what may explain the fate of change proposals.

Relations between "subjects" as social actors

In OHS as in mental health intervention studies, socio-demographic data are often collected about workers as subjects of the intervention. Researchers may be interested in the number of participants, their occupational group (Murta et al., 2007, p. 250) and diverse health variables.

Another way of looking at populations or samples is as sociological subjects, i.e. actors, in their social, economic and political interrelations. Some broad characteristics of these actors transcend the specificity of each workplace, due to the basic power imbalance between employers and workers and the generally lower status attributed to women's work; but the logic of each workplace needs to be understood, as it contributes to shape the actors' issues and thus the relations between them around OHS matters. Appraising the key factors in the implementation of occupational stress interventions, Nytrø et al. (2000, p. 216) stress "the importance of taking into account the influence exerted by powerful coalitions, informal group processes and the bargaining positions adopted by various stakeholders". In the same way, Cox et al. (2007, p. 355) report that the quality of social relations and trust within the organization is among a number of possible process variables that need to be considered in organizational-level interventions for work-related stress. Nielsen et al. (2006, p. 280) also illustrate that conflicts may exist or arise among workers themselves, which limit the results of an intervention.

Another issue is what Biron et al. (2009, p. 455) called the alignment between the intervention and the organizational culture amongst the contextual issues likely to mediate or moderate the impact of an intervention on well-being at work. As raised by Randall and Nielsen (this volume), the question may thus be how to "adapt" the intervention to the prevailing context, without abdicating on key principles, like the need for workers' participation and representation.

In the multiple case study on OHS intervention, we observed that the way OHS matters were "dealt with" through the relations between workplace actors could be situated on a axis that partially mirrors the one opposing the secondary and primary labour markets. On the one hand, there is high turnover, either because "exit" (see Hirshman, 1972 and Fairris, 1997, 1998) of workers is associated with difficult working conditions or because of the employer's strategy to maintain low labour costs. On the other hand, we observed the expression of workers' voice around OHS matters, either informally, (between workers and supervisors) or more formally (OHS committee or negotiations between union's and employer's representatives). This voice is more easily expressed in situations that tend toward the primary labour market characteristics.

Significant events affecting the workplace during the intervention process

The multiple case study stresses the interest of documenting how the trajectory of the intervention meets the trajectory of other changes or events in the workplace context, like variations in the economic situation, technological changes, changes respecting work organization and human resources management and labour relations, or more "micro" events like the departure of a manager or the arrival of a new one. These events are mentioned amongst implementation barriers to OH change by Goldenhar et al. (2001) and also underlined in many mental health at work intervention studies (Biron et al., 2009, p. 455; Cox et al., 2007, p. 353; Nielsen et al., 2006, p. 277; Semmer 2006, p. 519 and 523). These events can also be potential levers (like the arrival of a new interlocutor). For example, a context of organizational downsizing could explain negative health outcomes of interventions (Egan et al., 2009, p. 7). Some authors talk of project fatigue (Nielsen et al., 2006) and Nytrø et al. (2000, p. 219) to describe the influence of the accumulation of previous organizational changes.

Intervention

Aside from contextual issues, the model developed from the multiple case study of OHS interventions uses four concepts to describe the intervention itself:

1 the intervention's object,
2 its participation device,
3 the activities conducted for the intervention, and
4 the changes proposals that emerge from these activities, even if they will not be retained and implemented.

The way the process is defined here thus fits with the general definition given by Saksvik et al. (2007, p. 244), for whom "the process, the how, concerns implementation and adoption; how the change is planned, launched, and carried out." It also fits with the definition by Cox et al. (2007, p. 353) for whom "process (...) refers to the flow of activities; essentially who did what, when, why, and to what effect". However, process here describes what is done to identify change proposal and to implement it. Implemented changes are considered as another element of the intervention. The effects included here in the intervention are thus only a first level of intermediate changes. Changes in the exposure to risk factors, or accidents or health variables are considered as part of the outcomes.

Two elements of the process (participation device and activities) are now examined. A third one (change proposals) will be looked at later, in comparison to the changes actually implemented, to understand how the stakes they represent, in the relations between workplace actors, may explain their trajectory.

Participation device

In OHS intervention studies, like in those devoted to stress, the concepts of "participation" and "participant" are understood in diverse ways. For example, in their systematic review of the job-stress intervention evaluation literature, LaMontagne et al. (2007, p. 277) noted that there was "likely to be a wide range of degrees of participation".

The question of recruitment, i.e. the "sources and procedures used to approach and attract potential participants to become effective participants" (Murta et al., 2007, p. 250) is present in stress interventions studies as well as in OHS ones, as this has to be addressed by every intervention study. However, participation goes beyond "recruitment". In this way, Goldenhar et al. (2001) include the development of partnerships in the intervention research tasks. In mental health at work interventions, "There is (...) a widespread agreement that a participative approach is the most promising" (Semmer, 2006, p. 522). One of its contributions is that it allows for matching content expertise of OHS professionals or researchers with context expertise of participants (LaMontagne et al., 2007, p. 270).

There appears to be two levels of participation in both OHS and mental health organizational-level interventions: participation as subjects that are expected to be exposed to the changes, and participation in the "device" (i.e. in defining the intervention, analyzing the diagnosis, or searching for solutions). In this model, the influence of each actor (management, supervisors, workers' representatives, workers) is modulated by their political capacities.

In the field of stress intervention studies, the participation device is identified as an important feature by Biron et al. (2009, p. 455) both for the assessment of risks and for the design of an action plan and its implementation. Nytrø et al. (2000, p. 221) stress another aspect of this dimension saying that "choosing competent persons with outspoken faith in the change project is advantageous, if not crucial, for intended change to take place." We referred to this in the "context" section when stressing the importance of the presence of "relay actors". This also relates to what Nielsen et al. (2006, p. 277) call "ownership" of the intervention project and activities. Another dimension of the participation device is related to the collaboration between workplace actors, which is mentioned by Nytrø et al. (2000, p. 217) and by Egan et al. (2009, p. 6). Randall et al. (2009, p. 18) illustrate the role of the participation device by suggesting that "the effects of this intervention were driven more by the behaviour of the line manager and the history of the team than by exposure to the content of the intervention itself." We now successively examine two dimensions of the participation device: who participates, and who interacts with whom.

Who participates?
In Baril-Gingras et al. (2004, 2006a, 2006b, 2007), we report that employers' representatives were always present when the intervention object, participation device, and activities were negotiated. The same was not true for the other steps of the processs. In the absence of an employer representative when a diagnosis of the situation is made and change proposals are discussed, changes are limited to what is within workers' reach, which is often limited. This often happens in informal discussion during training sessions during which workers reflect on what they need to support changes (enabling factors according to Green and Kreuter, 1991). When a direct contact occurred between the person(s) in charge of the intervention and a management representative on these intervention conditions, it contributed to resources allocation. However, the fulfilment of these conditions was not systematic, and was more probable when there was an association between OHS problems and production or service dysfunctions.

The presence of supervisors was quasi systematically asked for, but their workload often limited it. Supervisors may bring their own knowledge of the factors that shape working conditions, which is necessary for changes to be relevant and compatible (Trudel and Montreuil, 1999). They can endorse the decision for some of the changes and relay other change proposals to their own hierarchy. Their presence at the "diagnosis" and "search for solutions" phases positively influenced their views about proposals for change (their representation of the "stakes"). Their presence played an important role in the actual implementation of changes, even if it was not a condition sufficient in itself. However, their contribution depended on their political capacities, as many changes exceeded their own decision latitude.

Some observations from stress or health intervention studies also underline the role of manager's availability (Saksvik et al., 2007, p. 243), their competing activities versus the time allowed to work on the implementation of interventions (Biron et al., 2009, p. 455), or their "leadership style, way of communicating with subordinates and attitude towards change" (Nytrø et al., 2000, pp. 217–218). Randall et al. (2009) showed that line manager attitudes and actions had a significant correlation with post-intervention outcomes (job satisfaction, well-being, and self-efficacy). In this volume, Lewis, Yarker and Emma Donaldson-Feilder also discuss the vital role of line managers and how they can influence the success of interventions through their impact on process and contextual issues, as well as their role in causing or reducing psychosocial risks.

As for workers' participation, our study distinguished technical (ensuring the change proposals relevant and compatible with the context) and political dimensions (employee representatives giving a collective voice to problems and to change proposals). Observations and interviews also highlighted the role of collective (instead of individual) direct participation, i.e. where workers can exchange their views, build a collective point of view, influence their peers. It contributed to initiatives by workers individually or as a group.

In organizational-level stress management interventions, workers' participation is also mentioned as an important factor to explain the fate of interventions. Randall et al. (2009) observed that employee involvement showed significant correlation with post-intervention outcomes (job satisfaction, well-being, and self-efficacy). Interestingly, there are, in this field too, signs of recognition of the "political" dimension of workers' participation. For example, Cox et al. (2007, p. 355) report, amongst a number of possible process variables for organizational-level interventions for work-related stress, not only the "employees' willingness and ability to participate", but also the "employees' role in the decision-making process". This is in line with Nielsen, Randall, and Albertsen (2007) who demonstrate that employees are actively involved in shaping the intervention, which in turn has an influence on the outcomes. This political dimension inherent to workers participation is also illustrated by Nytrø et al. (2000, p. 219) who highlight that management can't call for participation and at the same time keep unilateral control over the change. Finally, the potential change mechanism that collective (instead of individual) workers' participation may represent is also underlined by Nytrø et al. (2000, pp. 218–219), who argue for "participation of collective employee voice in the change process".

Who interacts with whom?
The second dimension of the participation device is the interactions it organizes between actors who play different roles in the technical and social division of work, and for whom different stakes may be at play. These interactions also have a technical and a political dimension. On the technical dimension, the interactions contribute to the production of changes in part by transforming "representations" (a concept used by French ergonomics, see Teiger and Montreuil, 1995) or "understanding" of the real work activity of workers by supervisors and by those who conceive layouts, equipment and work organization. These interactions may be opportunities (but not guarantees) to conceive solutions which are relevant and compatible with the work activities, while also meeting the prevention objectives. The external OHS advisors were acting as the guardians of this last criterion.

148 *Challenges and methodological issues*

As for the political dimension, these interactions between actors do not per se lead to the implementation of change proposals, nor do they ensure the convergence of the actors' interests. When the interactions did contribute to changes, the following processes were observed: the integration of new prevention criteria into decision-making, formal or informal agreements between some actors, tacit alliances between some actors to convince another one in position to take a decision, pressure exerted on one actor by another one.

In the stress or health interventions studies we consulted, some authors explicitly mention the need "to provide opportunities for multi-level participation and negotiation in the design of interventions" (Nytrø et al., 2000, p. 213). An example is given by Bourbonnais et al. (2006a) intervention study in a hospital setting:

> The principles of these health circles are: (a) operating in small groups; (b) group members not of the same hierarchical level; (c) regularly scheduled work meetings; (d) preferably 8–10 meetings; (e) meetings led by an external moderator; and (f) individual knowledge of team members used as input for finding solutions to adverse factors.
>
> (p. 328)

Other authors explicitly include the use of "constructive conflicts to deal with change as an element of healthy organizational change" (Saksvik et al., 2007, p. 243).

Activities

Another dimension of the process documented by the exposed model is the activities that are realised to complete the intervention. Even if interventions seldom take place in a linear fashion, distinctions were made between five phases to enable comparisons: negotiation, analysis and diagnosis, search for solutions, operationalization, implementation (including formal or informal decision process) and follow-up. This helps when answering one of the questions posed by Goldenhar et al. (2001): "What are the components (e.g. activities, materials, technology) of the intervention and how were they provided to the target audience?"

Some authors describe these intervention components in terms of their format, i.e. "frequency of sessions, length of session, length of intervention, number of follow-ups, and time lapse between training and follow-up" (Murta et al., 2007, p. 250) or the project organization (Nielsen et al., 2006, p. 277). These intervention activities also cover the definition of "roles and responsibilities before and during the intervention period" (Nytrø et al., 2000, p. 222), which is identified by Saksvik et al. (2007, p. 243) as an element of healthy organizational change.

Negotiation of the intervention, its object, objectives, and conditions
In the cases studied, a negotiation of the conditions in which the intervention would take place was observed, even if informal. It could cover the definition of the object

and objectives of the intervention, the participation device (e.g. supervisor's presence), the activities to be held, the time schedule, the access to workplace "capacities" (time, budget, etc.), and various conditions e.g. availability, a training session of the required safety equipment. External OHS advisors tried to negotiate conditions that would enable changes. The presence or absence of the conditions asked for clearly influenced the scope of the change proposals that could be formulated and then implemented (e.g. equipment, layout and work organization, not only training). In relation to this "negotiation" phase, Kristensen (2005, p. 209) stressed that:

> It is necessary to distinguish between the three stages of contact with the workplace: (1) accept (the company has accepted the study); (2) access (the researchers have access to the workplace, the employees, and to written material); (3) commitment (the relevant parties at the workplace consider the intervention as their own and give it high priority).

In the field of mental health intervention studies, Egan et al. (2009, p. 6) included, in their review of the implementation of organizational-level workplace interventions on health issues, the item "planning consultations", asking "is there a report of consultation/collaboration processes between managers, employees and any other relevant parties during the planning stage?". Biron et al. (2009, p. 455) also stress the need for careful planning. In future research, it may be interesting to understand what influences the results of these consultations, preparations or "negotiations". Resources for intervention are also mentioned as important features (e.g. Biron et al., 2009; Cox et al., 2007, p. 355; Egan et al., 2009, p. 6). Biron et al. (2009, p. 455) also included the "feasibility of project given organizational constraints". Future research could help better understand what are the type and level of resources needed to conduct interventions and further our under— standing on what needs to be done when these resources are not made available–a major issue for public policy.

Analysis and diagnosis
Based on their analysis of musculoskeletal disorders preventive interventions' studies, Denis et al. (2008, p. 3) introduced a distinction between three types of processes of interventions:

1 the "complete" ones, which used the three steps of "preliminary analysis" (to define the scope of the problem), the diagnostic step (to find causes or determinants of problems) and the solution development step;
2 the shortened type of intervention process, where the diagnostic step was less important;
3 the "turnkey" process, which lacked diagnosis and solution development.

Even if the model exposed here did not use the same distinctions, Denis et al.'s work illustrates that the analysis and diagnosis steps may be partly or completely evacuated. According to these authors, these process characteristics influenced the changes that were produced by the interventions:

> Proportionately, "complete" interventions allowed more changes to be implemented, as well as more diversified changes (...). The "complete" type was the only intervention category interested in work organization (...). Two-thirds of the changes implemented in "turnkey" interventions involved worker training. In general, this was the only change introduced into the workplace.
> (Denis et al., 2008, p. 10)

In our multiple case study, OHS practitioners were often confronted with demands already formulated as solutions (like workers' training) instead of problems to solve (like musculoskeletal disorders). Reintroducing an analysis and diagnosis step had to be negotiated. Observations showed that analysis and diagnosis activities served a technical, social, and political function. The technical function refers to the understanding of causes of health and safety problems. The social function refers to what explains the presence of these risks, what are the levers and obstacles to change, and what are the actors' willingness and capacities to deal with them, i.e. an understanding of the context. Moreover, the analysis and diagnosis may have a political function for workplace actors in terms of convincing those in position to decide of changes.

These observations well relate to some of the questions Goldenhar et al. (2001) suggest should be asked during the developmental phase of the research:

- What changes are needed to enhance the health of the target population?
- What barriers hinder the desired changes from happening?
- To what extent does the target audience understand and buy into the need for the changes?

To answer these questions, Bourbonnais et al. (2006a, p. 328) used an interview grid to identify "favourable and unfavourable conditions for intervention: organizational constraints, communication problems, quality of social interactions, and conflicting needs and priorities." In the same way, Brisson et al. (2006, p. 243) used both a quantitative risk assessment and focus groups with employees "to obtain a more in-depth understanding of the main problems identified through the prior risk evaluation" (p. 250) and follow-up meetings with managers and union representatives, as complementary tools for the development of well-adapted interventions. Finally, our multiple case study suggested that when workers and supervisors were involved during the diagnosis and analysis, the solutions selected were perceived by workers as more relevant.

Search for solutions
Once the problems have been identified and solutions proposed, the change proposals have to be accepted and viewed favourably by the stakeholders. Our analysis of the multiple cases of OHS interventions showed that certain characteristics

of the change proposals influence the degree to which they were viewed favourably. First, the solutions have to make use of the capacities of the stakeholders, in particular of workers and supervisors. Another characteristic relates the possibility for experienced workers to use various strategies to limit or control risks, and to validate or modify some of the solutions when needed. This is in line with the central role given to the understanding of "real work activity" by French ergonomics (Guérin et al., 2006; Lamonde and Montreuil, 1995).

Nielsen et al. (2006, p. 274) and LaMontagne et al. (2007, p. 270) also observed that solutions to prevent occupational stress are likely to be unique, as they have to be specifically tailored and adapted to the needs of the work unit. However, the literature consulted did not provide information on the characteristics of the search for solutions activities and on their influence on the relevance, compatibility and actual implementation of these solutions.

Operationalization of solutions, implementation and follow-up
Our case studies suggested it is important to distinguish a specific intervention step called "operationalization", i.e. the passage from objectives or performance criteria or even general concepts about solutions to implement, to concrete solutions, "ready" to implement. Cox et al. (2000) labelled this step as "translation". By systematically examining the fate of change proposals, we observed the operationalization of each change proposal was positively related to how they were perceived and to which extent they were implemented.

In the studies on mental health at work interventions, this question of "operationalization" did not appear to be addressed in much detail. Nielsen et al. (2006) highlight the need to design changes targeted to the specific problems of the organization, and Cox et al. (2006) suggest that during this translation phase, underlying problems affecting most workers should be identified. This may however be related to the choice of participative approaches where the actual changes to be implemented are defined by workplace actors themselves, with the support of the external advisors or researchers (e.g. Bond and Bunce, 2001; Bourbonnais et al., 2006a, Bourbonnais et al., 2006b; Bourbonnais et al., 2010; Mikkelsen and Saksvik, 1999; Nielsen et al., 2006; Orth-Gomer et al., 1994; Park et al., 2004; Saksvik et al., 2002).

In our OHS case studies, the follow-up by the external advisors after the implementation of changes was noted as contributing to solving residual problems and to ensure the compatibility with the workplace context and coherence with prevention criteria. This non-linear process is stressed by Kristensen (2005, p. 207): "In real life situations there may be many feedback loops going from the exposure level back to the intervention and from health back to the previous two levels." This takes us to the implemented changes.

Proposed and implemented changes
In the multiple case study reported here, changes consisted of creating or modifying existing workplace structures and preventative activities (contextual issues)

and of changes that modified working conditions (and the psychosocial work environment), including workers' training, that are intended to act on health and safety (or mental health) intermediate (e.g. exposure to risk factors) and final outcomes (e.g. psychological distress, accidents and illnesses).

In the field of mental health at work intervention, the need to describe the actual changes that are implemented is also clearly stated. For example, Gilbert-Ouimet et al. (2010 and forthcoming) present a structured method for documenting the implemented changes in an intervention aimed at reducing adverse psychosocial work factors and related health problems. The study recorded "all the organizational changes introduced into the workplace with the explicit goal (or the plausible consequence) of improving the employees' situation with regard to one of the four adverse psychosocial work factors" which are high psychological demands, low decision latitude, low social support (Johnson and Hall, 1988; Karasek and Theorell, 1990) and high effort/low reward (Siegrist, 1996).

In the field of individual stress management interventions, it can be difficult to document what is defined here as "implemented changes", as individuals subjects have to "translate" the intervention content into either their behaviour or in the working conditions they control. Looking at this type of intervention, Murta et al. (2007, p. 250) defined the intervention content as the "nature of intervention", including themes discussed (e.g. conceptualization of stress), techniques used (e.g. assertiveness training), or actions implemented (e.g. committees to discuss problems at work and implement solutions). This definition partly corresponds to the one of "Activities" used here. Murta et al. (2007, p. 250) also use the concepts of dose delivered (i.e. amount of intended unit/components of the intervention provided by the intervention *provider*) and dose received (i.e. "the extent to which participants use materials, resources, or techniques recommended by the program, or the audience's engagement with the program"). "Dose delivered" would correspond here to the intervention's activities and proposed changes, and "Dose received" to the changes actually made by participants in their working conditions. As stressed by Kristensen (2005, p. 207):

> In laboratory research "intervention" and "change of exposure" is the same, since the researcher is by definition in control of the exposure. This is usually not the case in field studies, which makes it necessary to study this issue separately. (. . .) The next question is whether or not the changed exposure had the intended effect on health and other study outcomes.

Organizational-level intervention research in both fields may face a similar issue: the changes to implement are not set a priori, but defined along the intervention process, by workplace actors, in context. Tools are thus requested to register the actual changes that are implemented and evaluate whether employees have been exposed to them (Biron et al., 2011). In the multiple case study reported here, it was decided to note every single proposal for change that was formulated by any of the actors. The fate of each proposal was thus followed until the end of the intervention.

To face the methodological issue related to the fact that changes are not "pre-defined" but designed by the workplace actors to fit the workplace specific context and problems, some intervention studies on work stress or occupational health reported the use of tools like "action plans" or lists of planned activities or solutions (e.g. Bourbonnais et al., 2006a, 2010; Bond and Bunce, 2001; Gilbert-Ouimet et al., 2010 and forthcoming; Mikkelsen et al., 1999; Nielsen et al., 2006; Orth-Gomer et al., 1994; Park et al., 2004; Saksvik et al., 2002). Some of these studies provide examples of what is covered by the action plans (Bond and Bunce, 2001; Nielsen et al.; 2006; Park et al., 2004), sometimes with tables detailing the changes, distinguishing between proposed changes and implemented changes (e.g. Bourbonnais et al., 2006; Bourbonnais et al., 2010; Nielsen et al., 2006), or categorizing changes in order to measure workers' exposure to them (Biron et al., 2011).

The need to classify changes in categories is also observed in both fields. This enables synthesis and comparisons between workplaces in relation to their context. It also makes possible a comparison between intervention and control group, or between groups, both according to context and process factors. In the multiple case study exposed here, change proposals were classified into categories: Training, layout and equipment, system or procedure, work organization.[3] The fate of each of the change proposal was noted in a table and a proportion of those completely implemented or still being implemented at the end of the study could be calculated.

The nature of the implemented changes
One of the interests of qualitative studies is that they enable a precise and systematic description of the changes to the work situation (Bellemare et al., 2001; Montreuil et al., 2004). It makes it possible to examine how the initial objectives are translated into means, what are the more frequent, or most easily implemented types of changes. In our case studies, the characteristics influencing the fate of the change proposals were mainly attributed to:

- the nature of the proposal, or the object which it concerns;
- the capacities required to implement the changes;
- the stakes for each collective actor around each change proposal, and the relations between these actors around these stakes, i.e. their relative political capacity to influence a decision.

As for the nature of the change proposals, the analysis of the OHS intervention cases showed that proposals consisting in workers' training were the more frequently implemented. However, proposals concerning the conditions for the training itself (e.g. presence of supervisors, access to information and equipment) were less frequently implemented. After training, proposals respecting layout and equipment were the most frequently implemented. When they were not, barriers were related to financial resources or technical constraints that went beyond the local employer's capacities. Proposals consisting in the implementation of systems or process (e.g. a confined spaces entry procedure, which necessitates identifying each confined space and its specific risks, defining a procedure, providing equipment, ensuring the regular calibration of gas detectors, AND training workers) were less likely to be implemented fully. Unlike technological changes, their implementation takes more time and requires considerable investments (mainly in time), and collaboration from multiple actors and possibly changes to work organization. All these characteristics of the change proposal appeared to influence their fate.

Intervention studies on mental health have looked into the nature of change proposals in relation to outcomes. In their systematic review of the job-stress intervention evaluation literature, LaMontagne et al. (2007) categorized prevention programs as primary, secondary, or tertiary prevention levels. They showed that the more comprehensive programs which included all three levels of preventive activities (i.e. referred to as high system approach) were more frequent than in previous reviews.

Some studies go further in the analysis of the fate of categories of changes in a specific intervention. For example, in their intervention study, Brisson et al. (2006) used logbooks to systematically document the changes, in addition to focus groups with employees and follow-up meetings with managers. This allowed them to record the types of changes carried out and the number of workers affected by them.

Gilbert-Ouimet et al. (2010 and forthcoming) analyzed the logbooks registering changes implemented during the intervention on the psychosocial work environment described by Brisson et al. (2006). Their analysis illustrates how managers translated the intervention priorities identified by a priori risk assessment into changes, and attempts to determine whether these changes corresponded with the set objectives. It also allowed highlighting that the intervention group was the only one where changes targeting the workload were introduced.

The capacities required by the proposed/implemented changes
In our OHS case studies, changes most easily implemented by the employer appeared to be those that required the lowest level of capacities, and that were more of a one-off than recurring nature. Implemented changes tended to be of the same level of capacities as those that had been implemented previously in the workplace.[4] The paradox of this observation based on OHS interventions parallels the one discussed by Nielsen et al. (2006, p. 282) in the context of an intervention aiming to improve health, where a certain level of healthiness or readiness of the organization creates better conditions for intervention. These authors also explained the limited result of an intervention by a related phenomenon, i.e. that "the organizational structure was highly developed and thus the scope for further development was limited" (Nielsen et al., 2006, p. 280). The question is thus, on our point of view, how to overcome these obstacles.

The relationship between the level of capacities present in a workplace or resources allocated to an intervention and the type of changes implemented has also been observed in the field of mental health at work interventions where there is a marked preference for interventions focused on reducing the consequences of stress on individuals instead of on preventing them by modifying the working conditions (Cartwright et al., 1996; Kompier and Cooper, 1999).

The stakes of proposed/implemented changes
The stakes each change represents for actors is another important characteristic. For senior management as for line management, the association between OHS problems and actual or anticipated dysfunction in production is an important factor. This is in line with Nichols (1997) observation that the frequency and severity of production problems (e.g. interruptions, quality problems) are factors explaining the propensity of an employer to act with regard to OHS matters. Change proposals may also represent an issue for the employer regarding relations with workers and union. Another important stake concerns the resources needed to implement the change proposal. Resource allocation is influenced both by their absolute level and their one-off or recurring nature.

The judgment workers posed on the change proposals appeared to be influenced by the following factors:

- Their coherence and compatibility with the current "rules" that structured their work activity, in particular respecting time constraints (work should not be longer to complete, as time pressure was already important in most cases).
- The possibility to preserve margins of manoeuvre or decision latitude about their work activity.
- The "cost" of the proposed changes, in particular in relation to workload
- The coherence with quality criteria, with the ethic of doing good work.
- The coherence with the identity dimension of work (e.g. some professions define themselves as protecting others and saving lives, and preventive measures that would interfere with it may be resisted).
- The perceived preventive efficiency of the change proposals, in comparison with existing strategies.
- The presence of other issues like wages, qualification or employment.

Similar stakes influence the fate of change proposals in the field of mental health intervention. With regard to managers, Nytrø et al. (2000, p. 217) point out that:

> Managers generally have a preference for supporting individual-level interventions, such as employee assistance programmes and stress management training, and are much less likely to support organizational changes aimed at reducing the organizational sources of stress (Murphy, 1989; Reynolds, and Briner, 1994). They avoid issues concerning power, autonomy and work organization that are critical factors in understanding the development of occupational illness.
>
> (Bohle, and Quinlan, 2000)

Murphy and Sauter (2004, p. 82) propose that this may be explained by the fact that stress management interventions are perceived as less risky choices, that do not entail "what may be costly and disruptive changes in production practices".

Workers' judgment regarding the proposed changes has been addressed in different ways. Murta et al. (2007, p. 250) included participants' attitudes toward the quality of the intervention implemented amongst the process variables that have to be examined. Nytrø et al. (2000) also stress the importance of taking into account "unresolved anxieties, passive sabotage and non-intended subversion" (p. 219). The concept of "judgment of relevance" used by Trudel and Montreuil (1999) is in line with the proposal by Nytrø et al. (2000, p. 222) to assess "whether the proposed intervention(s) is regarded as suitable for the identified problem (e.g. reducing acute responses) or future challenges (e.g. promoting enhanced well-being)". In the same way, Biron et al. (2009, p. 455) include as determinant factors the "perceived significance" of changes for workplace actors and the "employee's support in reducing risks at the implementation of action plan phase". It seems that one important factor in this judgment is what Biron et al. (2009, p. 455) call the "attribution of responsibility for stress prevention". In the same way, "employees might, for example, question why they should have to strengthen their capacity to cope with the stress created by their work when they see that little is being done to redesign jobs" (Nytrø et al., 2000, p. 217).

Nytrø et al. (2000, p. 220) indicate that changes introduced too rapidly or in conflict with the organization culture may be perceived as an extension of management control. The cost of changes to workers is mentioned by Semmer (2006, p. 521) who stresses the importance of understanding compensatory and side effects, i.e. other changes that would occur in relation to those coming from the intervention, but alter the outcome, for example an increase in workload after a work group is given more autonomy. Finally, job security is also amongst the issues for workers. Nytrø et al. (2000, p. 220) recall that tendencies to inertia are particularly present if employees feel their job security is threatened. Overall, issues for workers seem quite similar in both OHS and mental health fields of intervention.

Conclusions

Despite the number of papers reporting intervention studies, it is hard to add to the existing knowledge, to compare between interventions and draw a global picture of their effectiveness, without a better description of the "how" of interventions. Qualitative description and analysis can provide this type of information. We hope this chapter shows that qualitative studies of process and context and actual changes can be completed in a structured way, and provide descriptions that go far beyond narrative accounts, with the use of well defined concepts, articulated in a model grounded both in theory and empirical observation. We exposed a model based on the analysis of OHS interventions. From selected papers reporting on mental health at work interventions, we then looked at the description of the intervention context, process and the implemented changes, and at what is said about the influence of these factors on interventions' outcomes. We pointed to many similarities between OHS and mental health at work interventions. We hope this chapter encourages the use of common models for systematic and structured descriptions of interventions and their context, to enable comparisons and thus a better understanding of levers and obstacles to OHS and mental health at work intervention's effectiveness, both at micro and meso level (in the workplace) and at macro level (public policy).

Notes

1 Detailed information about the methodology and the results of this study may be found in Baril-Gingras (2003) and Baril-Gingras, Bellemare and Brun (2004, 2006a, 2006b, 2007). This research was funded by the Québec Institut de recherche Robert-Sauvé en santé et sécurité du travail and by a doctoral fellowship from the Social Sciences and Humanities Research Council (GBG).
2 The literature search and its analysis were completed with the help of the L'Alliance de recherche universités-communautés (ARUC) sur les innovations, le travail et l'emploi, the Chaire en gestion de la santé et de la sécurité du travail and the Groupe de recherche sur l'organization et la santé au travail. The authors thank Julie Ross, Alexandre Côté, Françoise Proust and Jonathan Mercier for their help.
3 Personal protective equipment were not relevant in each case so this category was not included in the comparison table.
4 Space limits do not allow us to expand on how these conclusions could be drawn, nor to discuss how they relate to other intervention studies in OHS, but the interested reader can refer to Baril-Gingras (2003) and Baril-Gingras et al. (2004) for more information.

References

Bambra, C. (2011). *Work, worklessness, and the political economy of health.* Oxford: Oxford University Press.

Bambra, C., Egan, M., Thomas, S., Petticrew, M., & Whitehead, M. (2007). The psychosocial and health effects of workplace reorganization. 2. A systematic review of task restructuring interventions. *Journal of Epidemiology & Community Health, 61*(12), 1028–1037.

Baril-Gingras, G. (2003). *La production de transformations visant la prévention lors d'interventions de conseil externe en santé et sécurité du travail: un modèle fondé sur l'analyze d'interventions de conseillers d'associations sectorielles paritaires, dans le contexte du régime québécois.* Doctoral dissertation (Ph. D.), Université Laval, Québec.

Baril-Gingras, G., Bellemare, M., & Brun, J.-P. (2004). *Intervention externe en santé et en sécurité du travail: un modèle pour comprendre la production de transformations à partir de l'analyze d'interventions d'associations sectorielles paritaires.* Études et recherches, Rapport R-367. Montréal, Canada: Institut de recherche Robert Sauvé en santé et sécurité du travail.

Baril-Gingras, G., Bellemare, M., & Brun, J.-P. (2006a). The contribution of qualitative analyzes of occupational health and safety interventions: An example through a study of external advisory interventions. *Safety Science, 44*(10), 851–874.

Baril-Gingras, G., Bellemare, M., & Brun, J.-P. (2006b). Interventions externes en santé et en sécurité du travail: influence du contexte de l'établissement dans l'implantation de mesures préventives. *Relations industrielles / Industrial Relations, 61*(1), 9–43.

Baril-Gingras, G., Bellemare, M., & Brun, J.-P. (2007). Conditions et processus menant à des changements à la suite d'interventions en santé et en sécurité du travail: l'exemple d'activités de formation. *PISTES, 9*(1).

Baril-Gingras, G., Bellemare, M., Poulin, P., & Ross, J. (2010). *Conditions et processus de changement lors d'interventions externes en SST: Élaboration d'outils pour les praticiens*, Rapport de recherche (pp. 70). Montréal, Canada: Institut de recherche Robert Sauvé en santé et sécurité du travail.

Benach, J., Muntaner, C., Santana, V., chairs (2007). *Employment Conditions and Health Inequalities: Final Report of the WHO Commission on Social Determinants of Health (CSDH)*. Employment Conditions Knowledge Network (EMCONET).

Bellemare, M., Montreuil, S., Marier, M., Prévost, J., & Allard, D. (2001). L'amélioration des situations de travail par l'ergonomie participative et la formation. *Relations industrielles / Industrial Relations*, 56(3), 470–490.

Biron, C., Cooper, C. L., & Bond, F. W. (2009). Mediators and moderators of organizational interventions to prevent occupational stress. In S. Cartwright & C. L. Cooper (Eds.), *Oxford Handbook of Organizational Wellbeing*. Oxford: Oxford University Press.

Biron, C., Gatrell, C. & Cooper, C. L. (2010). Autopsy of a failure: Evaluating process and contextual issues in an organizational-level work stress intervention. *International Journal of Stress Management*, 17(2), 135–158.

Biron, C., Ivers, H., Brun, J-P., & Cooper, C. L. (2011, mai). *The more the merrier? A dose-response analysis of an observational study of organizational-level stress interventions*. Paper presented at the Work, Stress, & Health Conference, Orlando, Fl.

Bohle, P., & Quinlan, M. (2000). *Managing Occupational Health and Safety: A Multidisciplinary Perspective*. Melbourne: Macmillan.

Bond, F. W., & Bunce, D. (2001). Job control mediates change in a work reorganization: intervention for stress reduction. *Journal of Occupational Health Psychology*, 6(4), 290–302.

Bourbonnais, R., Brisson, C., Vinet, A., Vezina, M., & Lower, A. (2006a). Development and implementation of a participative intervention to improve the psychosocial work environment and mental health in an acute care hospital. *Occupational & Environmental Medicine*, 63(5), 326–334.

Bourbonnais, R., Brisson, C., Vinet, A., Vézina, M., Abdous, B., & Gaudet, M. (2006b). Effectiveness of a participative intervention on psychosocial work factors to prevent mental health problems in a hospital setting. *Occupational & Environmental Medicine*, 63(5), 335–342.

Bourbonnais, R., Brisson, C., & Vézina, M. (2010). Mental health long term effects of an intervention on psychosocial work factors among health care professionals in a hospital setting. *Occupational & Environmental Medicine*, published online 28 September 2010.

Brisson, C., Cantin, V., Larocque, B., Vézina, M., Vinet, A., & Trudel, L. (2006). Intervention Research on Work Organization Factors and Health: Research Design and Preliminary Results on Mental Health. *Canadian Journal of Community Mental Health*, 25(2), 241–259.

Brun, J.-P., Biron, C. & Ivers, H. (2008). *Strategic Approach to Preventing Occupational Stress*. Studies and Research Projects / Report R-577, Montréal, Canada: IRSST.

Cartwright, S., Cooper, C. L., & Murphy, L. R. (1995). Diagnosing a healthy organization: A proactive approach to stress in the workplace. In L. R. Murphy, J. J. J. Hurrell, S. L. Sauter, G. P. Keita (Eds.), *Job stress interventions* (pp. 217–233). Washington, DC, US: American Psychological Association.

Cohen, A., & Colligan, M. J. (1998). *Assessing occupational safety and health training, A literature review* (pp. 149). Cincinnati: National Institute for Occupational Safety and Health, Center for Disease Control and Prevention.

Cole, D., Rivilis, I., Van Eerd, D., Cullent, K., Irvin, E., & Kramer, D. (2005). *Effectiveness of participatory ergonomic interventions: A systematic review*. Toronto: Institute for Work and Health.

Cox, T., Griffiths, A. J., Barlowe, C. A., Randall, R. J., Thomson, L. E., & Rial-González, E. (2000). Organisational Interventions for Work Stress: A Risk Management Approach (pp. 193). Sudbury: HSE Books.

Cox, T., Karanika, M., Griffiths, A., & Houdmont, J. (2007). Evaluating organizational-level work stress interventions: Beyond traditional methods. *Work & Stress, 21*(4), 348–362.

Dawson, S., Willman, P., Bamford, M., & Clinton, A. (1988). *Safety at work: The limits of self-regulation*. Cambridge: Cambridge University Press.

Dejours, C. (1993). Travail et usure mentale. In Bayard Éditions (Ed.), *De la psychopathologie du travail à la psychodynamique du travail*. Paris: Bayard Éditions.

Denis, D., St-Vincent, M., Imbeau, D., Jette, C., & Nastasia, I. (2008). Intervention practices in musculoskeletal disorder prevention: a critical literature review. *Applied Ergonomics, 39*(1), 1–14.

Denis, D., St-Vincent, M., Jetté, C., Nastasia, I., & Imbeau, D. (2005). *Les pratiques d'intervention portant sur la prévention des troubles musculo-squelettiques: un bilan critique de la littérature* (p. 79). Montréal: Institut de recherche Robert-Sauvé en santé et en sécurité du travail.

Dorman, P. (2000). If safety pays, why don't employers invest in it? In K. Frick, P. L. Jensen, M. Quinlan, and T. Wilthagen (Eds.), *Systematic occupational health and safety management, perspectives on an international development* (pp. 351–365). New York: Pergamon.

Egan, M., Bambra, C., Petticrew, M., & Whitehead, M. (2009). Reviewing evidence on complex social interventions: appraising implementation in systematic reviews of the health effects of organizational-level workplace interventions. *Journal of Epidemiology & Community Health, 63*(1), 4–11.

Egan, M., Bambra, C., Thomas, S., Petticrew, M., Whitehead, M., & Thomson, H. (2007). The psychosocial and health effects of workplace reorganization. 1. A systematic review of organizational-level interventions that aim to increase employee control. *Journal of Epidemiology & Community Health, 61*(11), 945–954.

Eisenhardt, K. M. (1989). Building theories from case study research. *Academy of Management Review, 14*(4), 532–550.

Fairris, D. (1997). *Shopfloor Matters, Labor-management relations in twentieth-century American manufacturing*, London: Routledge, p. 234.

Fairris, D. (1998). Institutional Change in Shopfloor Governance and The Trajectory of Postwar Injury Rates in U.S. Manufacturing, 1946–1970. *Industrial and Labor Relations Review, 51*(2), 187–203.

Garrety, K., & Badham, R. (1999). Trajectories, social worlds, and boundary objects: a framework for analysing the politics of technology. *Human Factors and Ergonomics in Manufacturing, 9*(3), 277–290.

Gilbert-Ouimet, M., Baril-Gingras, G., Brisson, C., Cantin, C., Leroux, I., & Vézina, M. (2010). *The content of a work environment preventive intervention*. Paper presented at the Scientific Committee on Work Organization and Psychosocial Factors (WOPS) of the International Commission on Occupational Health (ICOH), Amsterdam.

Gilbert-Ouimet, M., Baril-Gingras, G., Brisson, C., Cantin, C., Leroux, I., & Vézina, M. (Forthcoming). The content of a work environment preventive intervention.

Goldenhar, L. M., LaMontagne, A. D., Katz, T., Heaney, C., & Landsbergis, P. (2001). The intervention research process in occupational safety and health: an overview from the National Occupational Research Agenda Intervention Effectiveness Research team. *Journal of Occupational and Environmental Medicine, 43*(7), 616–622.

Goldenhar, L. M., & Schulte, P. A. (1994). Intervention research in occupational health and safety. *Journal of Occupational Medicine, 36*(7), 763–775.

Graham, J., & Sakow, D. M. (1990). Labor market segmentation and job-related risk: Differences in risk and compensation between primary and secondary labor markets. *American Journal of Economics and Sociology, 49*(3), 305–323.

Green, L. H., & Kreuter, M. W. (1991). *Health promotion planning – an educational and environmental approach* (2nd Ed.). Mountain View: Mayfield.

Guastello, S. J. (1993). Do we really know how well our occupational accident programs work? *Safety Science, 16,* 445–463.

Guérin, F., Laville, A., Daniellou, F., Duraffourg, J., & Kerguelen, A. (2006). *Comprendre le travail pour le transformer: la pratique de l'ergonomie.* Collection Outils et Méthodes. Lyon: ANACT, p. 318.

Heacock, H., Koehoorn, M., & Tan, J. (1997). Applying epidemiological principles to ergonomics: a checklist for incorporating sound design and interpretation of studies. *Applied Ergonomics, 28*(3), 165–172.

Hirschman, A. O. (1972). *Exit, voice and loyalty.* Cambridge: Harvard University Press. French translation by C. Besseyrias, Face au déclin des entreprises et des institutions. Paris: Édtions ouvrières.

Huberman, A. M., & Miles, M. (1991). *Analyze des données qualitatives: Recueil de nouvelles méthodes.* Bruxelles & Montréal: ERPI & DeBoeck-Wesmael.

Hurrell, J. J., Jr, & Murphy, L. R. (1996). Occupational stress intervention. *American Journal of Industrial Medicine, 29,* 338–341.

Johnson, J. V., & Hall, E. M. (1988). Job strain, work place social support, and cardiovascular disease: a cross-sectional study of a random sample of the Swedish working population. *American Journal of Public Health, 78*(10), 1336–1342.

Karasek, R. A., & Theorell, T. (1990). *Healthy work: stress, productivity and the reconstruction of working life.* New York: Basic Books.

Karsh, B. T., Moro, F. B. P., & Smith, M. J. (2001). The efficacy of workplace ergonomic interventions to control musculoskeletal disorders: a critical analysis of the peer-reviewed literature. *Theoretical Issues in Ergononomics Science, 2*(1), 23–96.

Kompier, M., & Cooper, C. L. (1999). Introduction: Improving work, health and productivity through stress prevention. In M. Kompier & C. Cooper (Eds.), *Preventing Stress, Improving Productivity: European Case Studies in the Workplace,* pp. 1–8. London, New York: Routledge.

Kristensen, T. S. (2005). Intervention studies in occupational epidemiology. *Occupational & Environmental Medicine, 62*(3), 205–210.

Lamonde, F., & Montreuil, S. (1995). Le travail humain, l'ergonomie et les relations industrielles. *Relations industrielles / Industrial Relations, 50*(4), 695–740.

LaMontagne, A. D., Keegel, T., Louie, A. M., Ostry, A., & Landsbergis, P. A. (2007). A systematic review of the job-stress intervention evaluation literature, 1990–2005. *International Journal of Occupational and Environmental Health, 13*(3), 268–280.

LaMontagne, A. D., Youngstrom, R. A., Lewiton, M., Stoddard, A. M., Perry, M. J., Klar, J. M. (2003). An exposure prevention rating method for intervention needs assessment and effectiveness evaluation. *Applied Occupational & Environmental Hygiene, 18*(7), 523–534.

Levenstein, C. (1996). Policy implications of intervention research: research on the social context for intervention. *American Journal of Industrial Medicine, 29*(4), 358–361.

Mikkelsen, A., Saksvik, P. Ø, & Landsbergis, P. (2000). The impact of a participatory organizational intervention on job stress in community health care institutions. *Work & Stress*, *14*(2), 156–170.

Mikkelsen, A., & Saksvik, P. Ø. (1999). Impact of a participatory organizational intervention on job characteristics and job stress. *International Journal of Health Services*, 29(4), 871–893.

Montreuil, S., Bellemare, M., Prévost, J., Marier, M., & Allard, D. (2004). L'implication des acteurs dans l'implantation de projets d'amélioration des situations de travail en ergonomie participative: des constats différenciés dans deux usines. *PISTES*, *6*(2), 11.

Murphy, L. R. (1988). Workplace interventions for stress reduction. In C. L. Cooper & R. Payne (Eds.). *Causes, coping and consequences of stress at work*. Chichester: Wiley, pp. 301–339.

Murphy, L. R., & Sauter, S. L., (2004). Work organization interventions: state of knowledge and future directions. *Soz.- Präventivmed*, *49*, 79–86.

Murta, S. G., Sanderson, K., & Oldenburg, B. (2007). Process evaluation in occupational stress management programs: a systematic review. *American Journal of Health Promotion*, *21*(4), 248–254.

Needleman, C., & Needleman, M. L. (1996). Qualitative methods for intervention research. *American Journal of Industrial Medicine*, *29*, 329–337.

Nichols, T. (1997). *The Sociology of Industrial Injury*. London: Mansell.

Nielsen, K., Fredslund, H., Christensen, K., & Albertsen, K. (2006). Success or failure? Interpreting and understanding the impact of interventions in four similar worksites. *Work & Stress*, *20*(3), 272–287.

Nielsen, K., Randall, R., & Albertsen, K. (2007). Participants, appraisals of process issues and the effects of stress management interventions. *Journal of Organizational Behavior*, *28*, 793–810.

Nytrø, K., Saksvik, P. O., Mikkelsen, A., Bohle, P., & Quinlan, M. (2000). An appraisal of key factors in the implementation of occupational stress interventions. *Work & Stress*, *14*(3), 213–225.

Orth-Gomer, K., Eriksson, I., Moser, V., Theorell, T., & Fredlund, P. (1994). Lipid lowering through work stress reduction. *International Journal of Behavioral Medicine*, *1*(3), 204–214.

OTA (Office of Technology Assessment, U.S. Congress). (1985). *Preventing illness and injury in the workplace*, pp. 173–185. OTA-H-256. Washington, DC, USA.

Park, K., Schaffer, B., Griffin-Blake, C., Dejoy, D., Wilson, M., & Vanderberg, R. (2004). Effectiveness of a Healthy Work organization Intervention: Ethnic Group Differences. *Journal of Occupational & Environmental Medicine*, *7*(46), 623–634.

Polanyi, M. F., Cole, D. C., Ferrier, S. E., Facey, M., & The Worksite Upper Extremity Research Group (2005). Paddling upstream: a contextual analysis of implementation of a workplace ergonomic policy at a large newspaper. *Applied Ergonomics*, *36*(2), 231–239.

Quinlan, M., Mayhew, C., & Bohle, P. (2001) The global expansion of precarious employment, work disorganisation and occupational health: a review of recent research. *International Journal of Health Services*. 31(2), 335–414.

Randall, R., Nielsen, K., & Tvedt, S. D. (2009). The development of five scales to measure employees' appraisals of organizational-level stress management interventions. *Work & Stress*, *23*(1), 1–23.

Reynolds, S., & Briner, R. B. (1994). Stress management at work: With whom, for whom and to what ends? *British Journal of Guidance and Counselling*, *22*(75–89).

Robson, L. S., Shannon, H. S., Goldenhar, L. M., & Hale, A. R. (2001). *Guide to evaluating the effectiveness of strategies for preventing work injuries: How to show whether a safety intervention really works* (pp. 121): National Institute for Occupational Safety and Health.

Saksvik, P. Ø., Nytrø, K., Dahl-Jørgensen, C., & Mikkelsen, A. (2002). A process evaluation of individual and organizational occupational stress and health interventions. *Work & Stress, 16*(1), 37–57.

Saksvik, P. Ø., Tvedt, S. D., Nytrø, K., Andersen, G. R., Andersen, T. K., Buvik, M. P., Torvath, H. (2007). Developing criteria for healthy organizational change. *Work & Stress, 21*(3), 243–263.

Semmer, N. K. (2006). Job stress interventions and the organization of work. *Scandinavian Journal of Work, Environment & Health, 32*(6), 515–527.

Shannon, H. S., Robson, L. S., & Guastello, S. J. (1999). Methodological criteria for evaluating occupational safety intervention research. *Safety Science, 31*, 161–179.

Siegrist, J. (1996). Adverse health effects of high-effort/low-reward conditions. *Journal of Occupational Health Psychology, 1*(1), 27–41.

Simard, M., & Marchand, A. (1997). *La participation des travailleurs à la prévention des accidents du travail: formes, efficacité et déterminants*, Rapport R-154, Montréal, Canada: Institut de recherche en santé et en sécurité du travail.

Teiger, C., & Montreuil, S. (1995). Les principaux fondements et apports de l'analyze ergonomique du travail en formation. *Éducation permanente, 124*, 13–28.

Theberge, N., Granzow, K., Cole, D., Laing, A., & The Ergonomic Intervention Evaluation Research Group (2006). Negotiating participation: Understanding the "how" in an ergonomic change team. *Applied Ergonomics, 37*, 239–248.

Trudel, L., & Montreuil, S. (1999). Understanding the transfer of knowledge and skills from training to preventive action using ergonomic work analysis with 11 female VDT users. *Work, 13*(3), 171–183.

Zwerling, C., Daltroy, L. H., Fine, L. J., Johnston, J. J., Melius, J., & Silverstein, B. A. (1997). Design and conduct of occupational injury intervention studies: A review of evaluation strategies. *American Journal of Industrial Medicine, 32*, 164–179.

9 What works, for whom, in which context?
Researching organizational interventions on stress and well-being using realistic evaluation principles

Caroline Biron

Introduction

This chapter aims to use realistic evaluation (RE) principles to analyze organizational stress interventions. Although there is a clear call for more attention to be paid to the intervention process and its context, not much guidance is offered on *what* methods may be appropriate and *how* they can be used to answer this call. Instead of only focusing on the effects of an intervention program, RE is an approach proposed by Pawson and Tilley (1997) which attempts to identify mechanisms and contextual characteristics in relation to outcomes. In this chapter, I first explore the reasons why, so far, stress interventions have produced inconsistent results. As Pawson and Tilley (1994) underline, there are several well-recognized failures of experimental evaluation of social programs, and the field of stress intervention is no exception. Nevertheless, there are some indications that organizational-level interventions can produce positive effects (Semmer, 2011). However, it remains unclear how they might be produced, by which mechanisms, in which context, and for the benefit of whom. This *black box problem* (where all the emphasis is placed on describing outcomes) presents a real barrier to progress in the prevention of psychosocial risks and occupational stress. In this chapter, I explore how the principles of RE can be used to evaluate organizational stress interventions. Before applying these principles to analyze a case of implementation failure in a UK private company, I discuss the paradigmatic wars characterizing this field and their implications on the production of knowledge on effective organizational interventions for stress and well-being.

According to Robson (1993), evaluation research has two main purposes: to provide evidence regarding the merit and worth of social interventions, and to improve practice itself to respond to the changing contexts, for the betterment of society. With large sums of money being invested in social programs, it is crucial that policy makers and practitioners are able to draw useful evidence as a result of program evaluation. The results of such evaluation are, however, too often disappointing. Evaluations and meta-evaluations of social programs rarely confirm with clarity or certainty that X caused Y because of Z.

This is also the case in the field of stress at work. For example, Biron, Cooper, and Bond (2009) reviewed the literature on organizational-level interventions to reduce stress and improve well-being. Although several reviews have been published during the last decade or so, knowledge is still scarce regarding the effectiveness or organizational interventions. Biron, Cooper, and Bond (2009) report that most reviews still focus on the effects of individual-level interventions. Overall, the literature focusing on organizational-level interventions is somewhat promising, although inconsistent and mainly using a black box approach with no focus on the process and the reasons why outcomes were or were not produced. Several reviews (Briner and Reynolds, 1999; Graveling et al., 2008; Parkes and Sparkes, 1998; Richardson and Rothstein, 2008; Van der Klink et al., 2001) have concluded that there is insufficient evidence and more research is needed to draw strong conclusion about the effectiveness of organizational stress interventions. The results obtained by these systematic reviews and meta-analyses are also counter-intuitive, since, as pointed out by Semmer (2006), intervening on the conditions of work which are known to be deleterious to health should, theoretically, be an effective strategy to reduce these health problems. On the one hand, very few organizational intervention studies meet the strict methodological criteria required in order to be included in systematic reviews. Yet, on the other hand, occupational stress remains a costly problem for organizations, and although several studies have shown promising effects of organizational level interventions (Bambra et al., 2007; Bond and Bunce, 2001; Bourbonnais et al., 2006; Egan et al., 2007), there is still little for policy-makers to build on based on the knowledge produced by systematic reviews. Moreover, these reviews provide no information on the change process of the interventions, the mechanisms by which interventions produce certain outcomes, or the contextual elements influencing their effectiveness. Understanding these processes can increase the effectiveness of interventions (Nielsen et al., 2010).

Paradigmatic wars

Before discussing RE and its applicability for the study of organizational-level interventions, this section takes a step back to review the methodological and epistemological background behind the scarcity of knowledge on organizational interventions. The point here is not to make a case for less rigour in intervention research. As argued by Cox et al. (2007), the methods used to determine the effectiveness of organizational-level interventions, are not always "fit for purpose" (Cox et al., 2007) or, in other words, the correct approach to obtaining data of appropriate quality (Thompson and Fearn, 1996) judged against the purpose of obtaining these data. The choice of method should be determined by it being fit for purpose instead of by a paradigmatic orthodoxy. The following section reviews the paradigmatic debates and how a broader and more encompassing methodological framework is a necessary condition to understand how and why interventions are effective at reducing exposure to psychosocial risks and improving health and well-being.

Methodological dichotomization

Traditionally, according to a positivist perspective, the ideal way of demonstrating a causal effect is by using a randomized controlled trial (RCT). An RCT implies randomization of participants in a group receiving a treatment, an intervention, or a manipulation of an independent variable by the researcher. Another group of participants, comparable in all respect to the intervention group, receives a placebo, or no treatment, or the independent variable is not manipulated by the researcher. The outcomes of interest are assessed before and after the "treatment" (i.e. the intervention) is administered. Because the control and treatment group are comparable prior to the treatment, if differences are found, it can be assumed with confidence that they are due to the treatment. This research design is considered by positivists as the "golden standard". Heaney and Goetzel (1997) critically reviewed evaluation studies of the health-related effects (i.e. health risk modification and reduction in worker absenteeism) of multi-component worksite health promotion programs. Their results suggest that randomized control trials (RCTs) have less probability of finding positive effects of a treatment (i.e. and intervention) compared to non-randomized with comparison groups. Both RCT and non-randomized comparison group designs yield a lower probability of positive effects compared to studies with no comparison group at all. This suggests that studies using a less restrictive design are positively biased. However, according to RE, it could also imply that researchers using more traditional methods neglect the mechanisms and contextual aspects of interventions.

Methodologically, empirical research amounts to "discovering regularities, invariant phenomena, and the relationships among them" (Hugues and Sharrock, 1997, p. 47). Although suited for the study of natural sciences such as physics, there are problems with quasi-experimental designs and true experiment (RCTs) because they work better in laboratory settings where the researchers have more control over the conditions. In real-life organizational settings, where there are several unavoidable constraints or uncontrollable events, RCTs are more challenging. As Kristensen (2008) highlights, RCTs are well suited for well-defined, controllable and "simple" interventions, but this is far from being the case in organizational interventions. In fact, Kristensen (2008), Griffiths (1999), and Cox et al. (2007) all agree that trying to force organizational interventions into a strict natural science methodological paradigm is misleading and counterproductive. Griffiths (1999) points out that:

> Highlighting the difficulties and limitations of thoroughly evaluated 'outcome' research is not an argument for its abandonment [...], rather, the acknowledgement of these challenges represents a plea to researchers, journal editors, practitioners, and policy makers alike for more realistic expectations, more appropriate criticism, more in-depth interpretations, and a greater awareness of alternative (but complementary) approaches.
>
> (p. 590)

Opening the door to evidence without throwing away the methodological rigour

There have been calls for more attention to the process by which interventions are developed and implemented, as well as to their context. Using mixed methods can provide access to new and interesting research questions and augment the explanatory power of research. Pawson (2008) even goes as far as claiming that "method mix is the new methodological Holy Grail" (p. 121), although I tend to think more of mixed methods as an interesting opportunity to revisit the possibilities and limits of quantitative and qualitative methods. In work psychology, and in particular in stress intervention research, there has been a strict use of quantitative methods, albeit with a few exceptions (e.g. Biron, Gatrell, and Cooper, 2010; Nielsen, Randall, and Albertsen, 2007; Randall, Cox, and Griffiths, 2007). As discussed in the introductory chapter of this volume, mixed methods and a broader conceptual framework ought to be used to enrich our knowledge of how and why interventions produce certain effects.

Organizational-level interventions should not be evaluated using an approach that used to be considered as the ivory tower of epistemic knowledge, but instead should be more reflective in order to address the ongoing challenges in this field. As argued by Cox et al. (2007), there is a need to use a broader framework and a more eclectic mix of methods in order to study real-life situations over which we have little control.

Realistic evaluation

RE principles are not disconnected from organizational change theories proposing frameworks that specifically consider non-linear and non-universal models of change. These theories tend to use a contingency framework for analyzing change, considering the organization's characteristics and environment. This is parallel to Pawson's and Tilley's (1997) notion that the effectiveness of social programs (change) depends on their triggering mechanisms and the context in which they are triggered. Contingency theorists are not attempting to find a "one-size fits all" or a universal recipe for organizational change. Instead, they "aim to develop useful generalizations about appropriate strategies and structures under different typical conditions" (Dawson, 1994, p. 18). In this sense, RE can be considered as contingency theory, and can offer a broader framework to study organizational-level interventions.

According to RE principles, evaluation research should not consider only the effects of interventions on a specific range of outcomes in order to replicate these results. Researchers should also aim to uncover the *mechanisms* triggered in certain *contexts* which produce the intervention outcomes. Inconsistencies in the evaluation of social programs are due to the very low probabilities of finding regularities between one type of program and one set of outcomes. Indeed, the complexity of organizational reality renders it difficult to find consistent results using traditional scientific methods given that they change rapidly and rarely offer the necessary conditions required under the natural science paradigm (Griffiths,

1999; Lipsey and Cordray, 2000; Randall, Griffiths, and Cox, 2005; Rossi, Lipsey, and Freeman, 2000). Pawson (2002) argues that "it is not 'programs' that work: rather it is the underlying reasons or resources that they offer subjects that generate change" (p. 342). Researchers should attempt to understand "what works, for whom and under what circumstances" (Pawson and Tilley, 1997, p. 342). The realists see the causal potential of an intervention program in its underlying mechanisms, namely the theory about how program resources may influence stakeholders' actions. The context determines whether or not these mechanisms will be triggered, and produce certain outcomes. Context in realistic terms also refers to the context of the intervention, of the organization in which the intervention takes place, as well as the extra-organizational context.

The aim of RE is not to discover a universal regularity, or in other words, an intervention program which produces the intended outcomes without exception, but instead to examine how causation in the social world should be construed. Explanation takes the form of conceiving and testing some underlying mechanism, which are triggered by certain contextual characteristics. Once triggered, this mechanism generates some outcome pattern, some regularity. In Giddens' (1987) terms, it equates to making propositions about the interplay between structure and agency, or in other words, about how certain mechanisms are triggered in particular contexts.

Pawson (2006) describes realism as a methodological orientation, or broad logic of inquiry which is grounded in the philosophy of science and the social sciences. It is mainly considered as post-positivist perspective, meaning that it aims to link empiricism and constructivism. The post-positivistic paradigm frees researchers from the philosophical assumptions of logical positivism, according to which quantitative methods are inherently superior and qualitative research offers a weaker set of methods (Fitzgerald and Matthews Rasheed, 1998). For example, Greene and Caracelli (1997) use a realist perspective to examine effective combinations of complementary methodologies. In terms of methods, RE does not favour the qualitative or quantitative nor does it search for universal laws. Rather, it is open to a range of types of evidence, drawing from the whole repertoire of research and evaluation approaches. It learns from real-world phenomena, including program failure such as the one presented in this chapter. There is no magic bullet or a one-size-fits-all in social programs. Pawson and Tilley (1997) critically confront the methodologies in evaluation research by using examples from crime prevention studies. They argue that experimental evaluation and its epistemological assumptions about causation are inappropriate for studying social programs. They make a case by illustrating how experimental evaluation works with a logic which prioritizes certain sets of observational categories. This logic overlooks "the real engine for change in social programs which is the process of differently resourced subjects making choices about the opportunities provided" (p. 46).

Mechanisms of transformation

The focus on mechanisms could be abstractly taken to imply that programs work only if people choose to make them work. Programs offer a range of monetary, skill, material or other types of resources. However, whether people choose to use these resources in the way they were intended will affect whether the program succeeds of fails, in part or in its totality. This is what Pawson and Tilley (1997) describe as a "generative" approach to causation, as opposed to a "successionist" one. The generative view of causality explains changes by considering both internal and external features of that which is changed. One event can trigger another only under certain circumstances and the evaluator has to understand these in order to link an intervention with its outcomes. In contrast, the successionist view of causality considers causation as unobservable, and we can therefore make inferences only based on observable data. This successionist view follows the work of Hume (1739), according to whom the only way to distinguish causal and a spurious relationship is to use observational data in an established controlled sequence, like the RCT. This view, according to RE, masks the actual engine of change, which is an internal feature of the person, case, thing, or system, or whatever is the subject of transformation. Successionists view causation as external, in that we do not observe the causal forces at work, or in other words, we cannot witness the increase in well-being after a stress initiative has been implemented. Instead, we recognize that such change has occurred by measuring it rigorously and systematically, so we can infer that the change observed is brought about by the intervention (an external event) and not by something else. The generative view of causation suggests that we also consider the internal features of what is being changed. For example, the gunpowder explodes when ignited (a cause-effect relationship), but not if it is wet, not compacted, or has the wrong chemical properties. The generative causation approach seeks to understand causation as acting both internally and externally. One event triggers another, but only in the right circumstances, and it is these underlying levels (internal and external causes) which need to be investigated.

Context

An example given by Pawson and Tilley (1997) to illustrate the inappropriateness of the natural science paradigm and its associated experimental methodologies concerns the lack of attention paid to differences in the social context. They illustrate the role of context by describing a study by Bennett (1991) on the effectiveness of a police-initiated fear-reducing strategy. The study was methodologically very complex and Pawson and Tilley use it as an example because of its excellence. The program was rolled out in two intervention areas in Britain, London, and Birmingham, with three and two control areas respectively. These sites were characterized by high levels of fear of crime based on high victimization rates, and signs of visual disorder such as broken windows, graffiti, and litter.

Evaluation of the program showed that it had been implemented. Police presence in these neighbourhoods had increased and there was a marked increase in the percentage of residents who knew a police officer by name or sight in both intervention areas, compared to the control areas. However, crime rates remained the same, fear of victimization in the intervention groups was similar to those in the control groups, perceptions and satisfaction with the area improved in Birmingham, whereas sense of community improved in London. This very rigorous design did not produce much useable information for policy makers since it remained unknown whether, where, or how community policing could be taken forward. Indeed, the considerable differences between the two intervention areas were unexplained.

According to the realistic approach, the range of activities that were implemented (leafleting, counselling, problem identification, monitoring actions, security advice, etc.) were quite demanding on the schedules of the police teams, and the nature of police presence should have been part of the focus of this investigation. Instead, the program was studied as a series of mechanical steps to uncover whether performance targets were reached. This approach does not explain how the patrols might lead to a reduction in crime rates, in victimization, in fear of crime, and in improved perceptions of the community: "Social programs are the product of the volition, skilled action and negotiation by human agents and are not reducible to the facticity of a given event" (p. 50). The researchers could have investigated, for example, how police presence can influence the residents' perceptions about crime.

Although realists seek to choose an intervention on the basis that it has reasonable chances of being replicated successfully in another organization or context, the positive outcome patterns are not achieved by applying and repeating a magical formula or an "intervention recipe": "We know that there are no universal panaceas and no magic bullets in the world of social and public programs" (Pawson, 2006, p. 22). To identify causal connections, there is a need to understand outcome *patterns* instead of searching for outcome *regularities*. This implies going beyond outcome evaluation, and paying more attention to process to understand how programs work. Any intervention is inserted into a context with certain constraints. Some contextual characteristics may be more favourable than others to the intervention program. Programs are inserted in local, organizational, economical, political institutions, which influence the stakeholders, their choices and resources. Both successful and unsuccessful programs should be used as guidelines to inform future research, since lessons can be learnt from both.

Context, mechanisms, outcomes

To increase understanding of how interventions work, there is a need to accrue evidence linking the intervention mechanisms (M) with their contexts (C) and outcome (O) patterns. Accumulation of evidence implies traversing between a general theory (abstract C+M=O configurations) and empirical case studies (focused configurations). Accumulation of evidence in RE leads to the production of a middle-range theory. The idea here is not to collect evidence in the experimentalist way by attempting to find universal laws (this program always leads to this outcome), but instead to build up a range of testable families of configurations. In Pawson and Tilley's words, the accumulation of evidence proceeds as follows: "what develops is not a 'master theory' or a mass catalogue of inconsistent results but a typology of

170 *Challenges and methodological issues*

Table 9.1 Examples of elements of context, mechanisms and outcomes for potential avenues for research using realistic evaluation theory

Context (of the program, the organization, or wider/external context)

Communication about the program
Resources (Financial/human/expertise/skills)
Motives (e.g., legal/economical/political)
Appointment of coordinating body (e.g. steering body sufficiently powerful/composed of people who have influence on well-being-related issues)
Social climate of leaning from failure/ability to detect obstacles early
Role and importance of intervention in relation to other ongoing changes/projects

Mechanisms (program theory, characteristics of the intervention and its stakeholders)

Stakeholders' support and commitment (at local and organizational levels)
Based on needs assessment (is an intervention needed/is the need felt by stakeholders?)
Based on risk assessment which is theory-sound (identify groups most at risk)
Comprehensive (i.e., including primary, secondary and tertiary levels)
Clearly defined interventions with measurable and specific goals
Feasibility
Designed to be participative (multi-level) with clearly defined roles
Stakeholders' willingness to participate/readiness to change/expectations

Examples of outcomes

Possible outcomes as a result of the intervention

Individual outcomes (physical and mental health, well-being, job satisfaction)
Organizational outcomes (absenteeism, presenteeism, turnover)

Elements of each phase of the risk management cycle considered as an outcome

Preparation phase (who is involved, participants' buy-in, characteristics of commitment type, how the business case is built)
Risk assessment phase (resources invested, resources planned for the implementation phase, type of risk assessment methods)
Translation phase (consensus on actions to implemented, resources to invest, implementation timeline, alignment of actions retained with organizational culture)
Implementation phase (reaching participants, delivery of intended interventions, fidelity of interventions implemented vs planned, impact of interventions on social relations)
Evaluation of effects of the intervention on individual and organizational outcomes, evaluation of the process (interventions delivered as planned and received produce the desired effects / side effects on which subgroup of participants)

Note. Based on: Biron, Cooper, and Bond (2009); Brun, Biron, and Ivers (2008); Cox et al. (2000); Giga, Faragher, and Cooper (2003); Goldenhar et al. (2001); LaMontagne et al. (2007); Linnan and Steckler (2002); Noblet and Lamontagne (2009); Nytrø et al. (2000); Tvedt, Saksvik, and Nytrø (2009).

broadly based configurations" (p. 126). Elements of these configurations are proposed in Table 9.1. These configurations could be tested in further intervention studies, or even using meta-analytic procedures. These are modest beginnings, but they do illustrate how attention paid to the context and mechanisms such as the characteristics of the stakeholders and of the intervention can help making sense of outcome patterns. They can help to understand why an intervention failed or succeeded when implemented and how it can be used as formative evaluation to improve a program, and can be linked with the results of summative evaluation in order to facilitate interpretation and strengthen the conclusions.

Applying realistic evaluation to study organizational stress interventions

Understanding the mechanisms and context of an intervention is important since one of the main challenges in occupational health psychology is to transform the extensive and rapidly growing body of knowledge on stress and health into successful interventions. Studying interventions that modify aspects of the organization is difficult. Indeed, there are several aspects which are not under the control of the researchers, such as the organizational context, the volition of the stakeholders, and the way in which they receive, perceive, and make use of the intervention. As Roen et al. (2006) argue, methods for appraising the methodological quality and explanatory potential of evidence should be extended to include information on the implementation process. They suggest different criteria than the ones traditionally used in systematic reviews in order to account for the implementation process. Murta, Sanderson, and Oldenburg (2007) in the field of stress management interventions and Egan et al. (2009) also conclude that there is insufficient consideration of the implementation process in complex intervention studies.

In the next section, I describe a case study in order to illustrate how RE principles can be used to evaluation organizational stress interventions. After describing the program theory (i.e. how the intervention was supposed to work), RE principles are used to uncover configurations of context, mechanisms, and outcomes.

Case study

The case study used here to illustrate how RE can be useful for evaluating organizational stress interventions is described in greater depth in Biron, Gatrell, and Cooper (2010). The study took place in a large private utility organization in the UK where a department of 205 workers was selected to participate in the research. An organization-wide intervention was to be implemented, supported by external consultants who provided a stress risk assessment (SRA) tool to be used (confidentially) by all managers in the organization with their respective teams. According to the program theory, for interventions to be successfully implemented, the following steps had to be completed:

1 Managers attend a meeting where they receive training on occupational stress and how to use the SRA tool, and they subsequently invite their team to complete it.
2 Each employee receives the electronic invitation from their manager to complete the SRA tool.

3   After completion, each employee receives an individualized confidential stress profile.
4   Once 60 per cent of the team completes the SRA tool, the manager receives a group report, which points out the main sources of stress in the team, and compares the risk level associated with each stressor with the rest of the company as a benchmark.
5   The manager arranges a group meeting to discuss any particular stressors that are potential risks to health. After discussion, an action plan is agreed and implemented.
6   The manager then informs the Occupational Health staff which intervention has been implemented, and the tool can be used to monitor changes over time.

At the beginning of the study, utilization of the SRA (which is intended to bring about the development and implementation of interventions) was known to be poor in the entire company (only 15 per cent of managers used the SRA according to the organization's records). Since the difficulties associated with the implementation of interventions following a risk assessment are arguably typical, a department from this company was chosen to investigate in depth the reasons why the utilization of the SRA was low, and evaluate if employees in teams where the SRA tool was used would report positive changes (in which case, the marketing of the tool could be enhanced and perhaps the use of the SRA increased). Table 9.2 summarizes the underlying assumptions behind each of these steps, or in other words, the program theory for this intervention to be successfully implemented. The research design to evaluate this intervention involved two waves of questionnaires. The first questionnaires were sent before and the second questionnaires were despatched nine months after the initial training was delivered to managers (Step 1). After six months, interviews with managers in the first training group (out of two groups who were to be trained at six-month intervals) revealed that none of them had implemented interventions and most had not used the SRA tool, for a variety of reasons. Some had left the company, did not believe the tool was necessary, had been moved to another job, or had changes in the composition of their team. This highlights the necessity of using mixed-methods research designs: had it not been for the qualitative data collected after six months, the effects of a phantom program would be measured, drawing conclusions about the effects of an intervention which was not implemented.

In the following sections, I use elements of program theory to illustrate possible CMO configurations by comparing this case with other intervention studies and theories. Without aiming to be clear-cut on the context (C), mechanisms (M), and outcomes (O) configurations emerging from the program theory described in Table 9.2, I attempt to make a few suggestions of potential emerging avenues of testable configurations. This presentation of configurations is not an exhaustive list of all contextual issues, mechanisms, and outcomes, but is a humble stone on the path towards applying both a broader framework and more eclectic methods to evaluate organizational stress interventions.

Table 9.2 Description of the program theory of an organizational intervention in a private company

Intervention: Managers use a team-based stress risk assessment tool	Program theory
Stress risk assessments in a large UK utilities company ($N = 10\,000$)	All managers will conduct a team-based stress risk assessment which will serve as basis for group discussions and development of an action plan
Participating department, $n = 205$	Underlying assumptions: • HR staff are trained to teach the workshop and managers attend
Anticipated impact:	• Managers understand and see benefits in using the tool
Managers implementing tailored interventions to decrease exposure to stressors	• Occupational health department's data is up to date so that all managers receive the SRA tool • The team has more than six members (otherwise the tool cannot be used due to risk of confidentiality breach) • Employees are committed to completing the tool and believe their workplace will be improved as a result of the project • Team composition (including the manager) remains stable, so the diagnostic is valid and interventions suited to the needs of team • Manager remembers / chooses to close the questionnaire phase and prints group report • Manager understands and interprets the results in a meaningful way • Level of trust between employees and manager is high and group discussions are constructive not accusative • Consensus is reached concerning priorities and how to improve • Manager has the human and technical skills, resources and will to implement interventions • Employees support and are involved in the process • Communications are made on accomplishments/changes • Interventions are delivered with fidelity, and are reaching their target

Getting beyond the risk assessment phase

The intervention was developed with resources planned only for the risk assessment phase (C1), out of legal requirement (C2), and within a highly changing context (C3). Indeed, the interviews with managers revealed that several of them were either no longer supervising a team, were supervising a different one, or had seen the membership in their team change to an extent where the risk assessment was no longer considered valid. Based on the program theory, mechanisms through which the intervention was expected to work include the managers taking ownership (M1) of the risk assessment tool and prioritizing it as a means for improvements in their team, managers acknowledging at least some level of responsibility for stress prevention (M2), and getting support from some internal or external source in order to translate the results of the risk assessment phase into concrete action plans (M3). These mechanisms through which the intervention could have progressed beyond the initial phase of the risk management cycle were not triggered. This led to a low rate of implementation of the risk assessment tool throughout the company including the participating department (O1), no interventions getting implemented in the department (O2), and approaching significance but clear tendency for decreased commitment levels in employees whose manager used the risk assessment tool given that no subsequent change was implemented (O3) (see Biron, Gatrell, and Cooper, 2010, for more details).

Several possible C-M-O configurations can be explored and could either be tested more specifically in future studies, or contrasted with existing ones in order to develop the evidence. For example, the context where the intervention is motivated by a legal requirement instead of arising from a more intrinsic need for change led in this case to low ownership from managers to implement a stress intervention. This is, in turn, associated with low levels of implementation, which was found to be associated with decreased commitment among employees. Obviously, this is not the fate of all change "nudged" by legislation, as shown by numerous accomplishments and improvements in health and safety in organizations since the introduction of relevant legislation and guidance. It seems more plausible to think that a legal requirement might be sufficient to persuade certain stakeholders, such as the top management and occupational health staffs, to take measures regarding the risks of occupational stress. In that sense, M1 (managers' ownership) is triggered by C2 (legal requirement), but only for top managers and not for managers at all hierarchical levels. Indeed, several line managers in the participating department felt that the only ways of persuading people to use the risk assessment tool was to have it included in their job description, or to ask their line manager to inscribe it in their operations along with a level of priority and a clear deadline. It thus seems more plausible to think the amalgamation of families of contextual issues failing to trigger families of mechanisms led to a certain range of outcomes, which are in this case considered as negative and unintended from a research perspective. In Pawson and Tilley's terms, this could be considered as one typology of broadly based configurations.

The role of external resources

Another element emerges from the comparison between the program theory and the way this program was actually rolled out in this company. Indeed, in this case the role of external support appears to be strikingly important and a trigger for moving the intervention unit beyond the risk assessment phase. External here refers to resources outside the participating unit, not necessarily outside the organization. In this case, the consultant only provided a risk assessment tool, namely a stress questionnaire, which was processed electronically. Figure 9.1 illustrates the role of external resources on involvement and participation of stakeholders. In this case, no steering committee (C) either at the local or corporate level was mandated to support the process, and the role of consultants did not include coaching or any form of implementation support (M), which can in turn be associated with the low level of involvement and ownership of the project by line managers.

The role of consultants in managing psychosocial risks has often been cited as a helpful resource, and they can be seen as agents of change. The topic of change agents is well covered in the organizational change literature, although it has been thinly covered in the field of organizational stress interventions. Bennis (1966) for example describes the variety of roles that change-agents can play, as counsellors, teachers, trainers, and line managers. Yet their role involves some "diagnosis of organizational health which pivots on interpersonal and group relationships and the implications of these changes on structures and tasks" (Bennis, 1966, p. 114). The role of external consultants as change agents in stress intervention initiatives has not been researched at as much depth as in the organizational change literature, yet it has received its part of criticism. On one hand, as pointed out by Noblet and LaMontagne (2009), it is problematic when there is an excessive use of external help and adoption of pre-packaged or one-size-fits-all solutions. The plethora and accessibility of these strategies marketed by consultants can appeal to organizations as a quick-fix solution, but their appropriateness for the specific organisational context is often questionable. Another issue with external resources mentioned by Noblet and LaMontagne can arise when the sole responsibility for developing the intervention is externally resourced. Heavy reliance on external resources leads to a loss of momentum once the outside resource ceases to be involved. As the present case illustrates, the development of interventions needs to be "owned" by the ones in charge of implementing them.

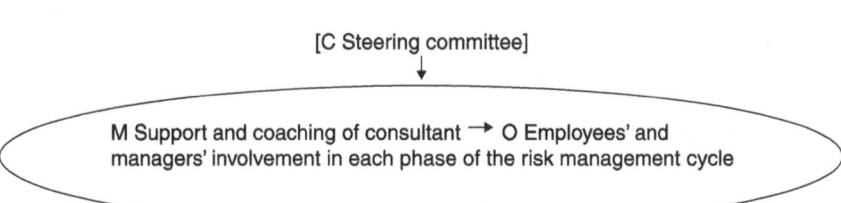

Figure 9.1 Realistic configuration illustrating the role of external resources on involvement and participation of stakeholders.

The role played by the consultant in this case was independent from the implementation process. However, as described by Brun, Biron, and Ivers (2008), the consultant can often play several roles in the project, such as resolving tensions, support the members agreeing on an action plan, discuss participants' reservations about the project, and give managers training on how to conduct a face-to-face meeting on stress-related issues with their employees. In the present case, survey data at follow-up showed that 32 per cent of managers felt uncomfortable meeting their team to discuss stress-related issues. The effects of external coaching on managers could be used as a mechanism to help them progress through the phases of the risk management cycle. The following section aims to explore with more depth the role of external resources and how it is associated with the fact a unit volunteers to participate in an intervention.

Volunteering effect

The volunteering effect is not something a realist would try to control for, but can be seen as a mechanism *through* which a program works. Instead of attempting to control for this self-selection effect, realists would consider it as a relevant research topic, and explore differences between volunteers and non-volunteers. In a traditional experiment, the researcher attempts to eliminate this self-selection, which may pose a threat to the internal validity of the study (participants in the intervention group might improve only because they volunteered, due to higher motivation level), by allocating participants randomly to either the intervention or the control group. Realists would instead verify what the makeup of people who volunteer for an intervention project is instead of attempting to maintain the equivalence between the two groups. "Choice is the very condition of social and individual change and not some sort of practical hindrance to understanding that change" (Pawson and Tilley, 1997, p. 37). In other words, choices are constrained in the sense that interventions offer people opportunities, but it belongs to them to decide whether to participate in the program, to retain information when given training, to apply whatever they have learned in their daily practices, and to maintain acquired skills over time.

The managers in the present case had to attend a training session on using the tool. Yet, this session is a learning process and some individuals are more likely to have the appropriate characteristics to allow them to benefit from the experience. In clinical trials, human volition is considered a contaminator. In RE, because participants can miss, ignore, forget, find boring, challenge, and dispute programs, interventions are considered as active. This is in line with the study by Nielsen et al. (2007) which showed that participants actively evaluate an intervention and that this evaluation influences the benefits that they get from it. This has profound implications for research methodology since the traditional experimental design, with all its other priorities, tends to suppress this type of data. We should instead appreciate that participants join a program with a wide range of expectations and attitudes, from being open-minded and receptive, highly motivated and committed, to being cynical or disengaged. This compendium of attitudes reinforces the idea of conceiving programs as offering a whole range of opportunities, and we should focus more on how groups of participants use these opportunities.

In realistic terms, I consider obtaining employee buy-in on an intervention project as a first target outcome (O). Employee participation is a key factor for the initiation and success of an intervention (Cox et al., 2000; Cox, Randall, and Griffiths, 2002; Kompier, Cooper, and Geurts, 2000). Knowing how important the role of the manager is going to be in this project, I consider top management and line managers as mechanisms (M) by which employees are willing to participate.

Brun et al. (2008) discuss how employees were initially sceptical because they hardly believed any significant change would be achieved and sustained following the intervention. Subsequent focus groups agreed to participate because the director was there, face-to-face, assuring them that she would be considering their suggestions seriously and responding to them as extensively as possible. Brun et al. (2008) describe an intervention context characterized by a project which was well resourced and supported by a consultant (C), a research team (C), the H&S department (C), the H&R department (C), supported by the union (C), and was part of the priorities of an organization's permanent steering committee on psychological health (C), a committee funded by the top management, thus reflecting their preoccupation for this topic (C). Figure 9.2 illustrates a possible realistic C-M-O configuration.

The volunteering effect is an important effect throughout all the phases of the risk management cycle. As Sugarman (2001) underlines in an analysis of learning organizations : "They were volunteers, not under orders to do this; and in presenting it to their followers, they sought volunteers who wanted to become engaged in the initiative. The emphasis was on 'growing' support, not on 'driving' a program forward" (p. 75).

The volunteering effect has various implications in terms of who we target for stress interventions and the optimal effects we can expect as a result. In the context of criminal offenders, Pawson and Tilley (1997) describe the "Andrews principle" (Andrews, Bonta, and Hoge, 1990) which stipulates that programs are more effective when targeted at people who represent a high risk, in this case of recommitting an offence after being released from prison. Andrew's principle is fully in accordance with the approaches usually adopted in terms of stress management. On the

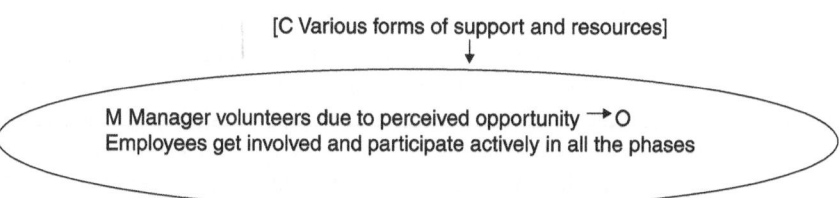

Figure 9.2 Realistic configuration illustrating the role of support and resources on manager's involvement.

one hand, delivering stress interventions to workers with low risks of stress would be like delivering smoking cessation to non-smokers (see Semmer, 2011). In this sense, the Health & Safety Executive's Stress Management Standards in Great Britain (see Mackay et al., this volume) are to be used to target high-risks areas and investigate which stressor(s) constitute high risks to health. However, managers who volunteer to take part in extensive projects might not be the ones who are at most risk nor the ones with the poorest leadership skills. Targeting managers who do not volunteer is certainly more challenging and, based on the Andrews principle, more desirable. Yet, if this leads to an implementation failure, it can raise employee cynicism and affect them negatively (Reichers, Wanous, and Austin, 1997).

Roles of researchers

Parkes and Sparkes (1998) reviewed organizational-level stress interventions studies and recommend that researchers involved in evaluating interventions are independent of those involved in identifying stress problems, and devising and implementing solutions. Their recommendation arises from participative action research, where the researcher both evaluates the intervention and facilitates the changes associated with it. Parkes and Sparkes (1998) consider the risk management cycle as an extended process which makes it impossible for the researcher to remain neutral and to be seen as independent at the moment of conducting the evaluation of the effects.

In the present case, I was "allowed" to conduct my research in parallel to their intervention project, but was never considered a part of it. My role was remote from the action on the field, since I was expected to evaluate an existing program, and was never expected to comment on its structure or its implementation. The role of researchers can be considered in terms of the context it provides. Indeed, as discussed in Brun et al. (2008), a research project can sometimes provide a specific set of opportunities to the participating units. For example, Brun et al. (2008) describe how a research project provided the pilot participating units with a specific budget, the implication of an external consultant to get through each phase of the risk management cycle, attention from the top management, and visibility within the larger organization. As such, the research played a certain role in triggering or at least influencing mechanisms such as the manager and the unit volunteering to participate, and the allocation of a budget to support the implementation of interventions, which then led to positive outcomes in terms of reduction of exposure to psychosocial constraints, decrease in psychological distress, and increase in job satisfaction.

In the present case, the context of the research was considered an opportunity to uncover whether the initiative had any benefits. However, it did not trigger a higher volunteering effect from the managers and employees of the participating unit. The intervention was neither circumscribed clearly within a research project, nor was it allocated a specific budget with resources. Instead, it was made part of the daily business, but as such, it was compulsory yet not enforced, not part of the job description, and managers were not evaluated on it. The motivation for using the tool, from the manager's perspective, was thus extrinsic. The initiative would have had a better chance of getting implemented had the managers been intrinsically motivated (Ryan and Deci, 2000).

Generalizability vs. sustainability

The top-down approach used implies that the managers at the top might have more at stake since they approved the project and its application company-wide, whereas the managers at lower hierarchical levels do not share this specific concern. Notwithstanding the fact that the program was inherently flawed and based on underlying assumptions difficult to meet (see Table 9.2), the scope of the project has important implications. On one hand, a small-scale, pilot project might have more chances of producing positive results (Brun et al., 2008) because of the nature, structure, context and resources of the project. On the other hand, the chances of a pilot project becoming an organization-wide practice are small.

As Noblet and LaMontagne (2009) suggest, it is good practice to implement highly visible and quick to implement interventions in order to maintain motivation, but implementing irreversible and permanent solutions ensures the initiatives will not be swept away with the next wave of organizational changes. A permanent decision committee implemented as a result of the intervention can be considered as an outcome of an intervention and ensure the sustainability of the solutions. The issue of finding the right balance between a sustainable approach within each unit, and yet one which is more widely applicable is still problematic. There is a need to further our understanding of:

1 the necessary "dose" of intervention required to produce what is considered as acceptable results,
2 whether some combination or components of interventions can be considered as optimal,
3 where to focus our efforts, on the high-risk units and/or those most likely to succeed and be considered as models to imitate by other units,
4 what type of organizational context is most favourable to trigger the volunteer effect and progression from one phase of the risk management cycle to the other.

Conclusions

With stress being one of the leading causes of work disability, there is a need to improve our approaches to evaluate complex interventions in order to go beyond the conclusion often reached by systematic reviews, and to address the fact that there is insufficient evidence on the effects of organizational-level interventions. This conclusion is based on the few studies carefully selected for meta-analyses, which are poorly placed to answer questions regarding *how* and *why* interventions produce their results.

Although paradigm wars will continue and it is not the purpose of this chapter to provide a definitive answer to these debates, this discussion shows how a combination of methods and approaches can be used, and how mixed methods can be fruitful for the study of this topic typically researched without methodological reflexivity. The use of RE theory as a framework and its application to the study of organizational-level stress interventions and occupational health and safety interventions demonstrates the relevance of process and contextual issues for understanding the effects of interventions (Pedersen, Nielsen and Kines, 2012). There is a need for methodological reflexivity in order to start cumulating knowledge in this field. The propositions made in this chapter can be seen as tentative conceptions of potential avenues for future research in directions where, eventually, we might find less inconsistencies and more useful answers to the question "do organizational-level stress interventions work?".

By considering these families of contextual factors that trigger specific mechanisms and lead to a range of outcomes, we can start to provide some answers regarding the reasons why intervention programs work, what are the necessary or favourable conditions for them to be implemented beyond the risk assessment, and whether any of their particular components are more likely to have an effect on outcomes. The accumulation of evidence here is not meant to be immutable or to provide laws of interventions programs: RE is a contingent theory. As such, it is expected that the effects of interventions will always be affected by micro and macro social forces (see LaMontagne, Noblet, and Landbergis, in this volume). Progress can, however, be made by further developing and testing such middle-range theories. Using a realistic framework helps to generate hypotheses about potential pathways where contexts trigger mechanisms which then lead to certain outcomes. Without these clearly defined theories, we are unable to capture and understand how interventions work. These hypotheses can then be used to guide and focus our research agendas and the further elaboration of stress interventions in organizations.

References

Andrews, D., Bonta, J., & Hoge, R. (1990). Classification for effective rehabilitation. *Criminal Justice and Behavior, 17*, 19–52.

Bambra, C., Egan, M., Thomas, S., Petticrew, M., & Whitehead, M. (2007). The psychosocial and health effects of workplace restructuring interventions reorganization. 2. A systematic review of task. *Journal of Epidemiology and Community Health, 61*, 1028–1037. doi: 10.1136/jech.2006.054999

Bennett, T. (1991). The effectiveness of a police-initiatied fear-reducing strategy. *British Journal of Criminology, 36*, 567–573.

Bennis, W. G. (1966). *Changing organizations: Essays on the development and evolution of human organization.* New York: McGraw-Hill.

Biron, C., Cooper, C. L., & Bond, F. W. (2009). Mediators and moderators of organizational interventions to prevent occupational stress. In S. Cartwright & C. L. Cooper (Eds.), *Oxford Handbook of Organizational Well-being* (pp. 441–465). Oxford: Oxford University Press.

Biron, C., Gatrell, C., & Cooper, C. L. (2010). Autopsy of a failure: Evaluating process and contextual issues in an organizational-level work stress intervention. *International Journal of Stress Management, 17*(2), 135–158.

Bond, F. W., & Bunce, D. (2001). Job control mediates change in a work reorganization intervention for stress reduction. *Journal of Occupational Health Psychology*, 6(4), 290–302.

Bourbonnais, R., Brisson, C., Vinet, A., Vézina, M., Abdous, B., & Gaudet, B. (2006). Effectiveness of a participative intervention on psychosocial work factors to prevent mental health problems in a hospital setting. *Journal of Occupational & Environmental Medicine*, 63, 335–342.

Briner, R. B., & Reynolds, S. (1999). The costs, benefits, and limitations of organizational level stress interventions. *Journal of Organizational Behavior*, 20, 647–664.

Brun, J.-P., Biron, C., & Ivers, H. (2008). Strategic Approach to Preventing Occupational Stress (R-577). Québec, Canada: Institut de recherche Robert-Sauvé en santé et en sécurité du travail.

Cox, T., Griffiths, A. J., Barlowe, C. A., Randall, R. J., Thomson, L. E., & Rial-González, E. (2000). *Organisational Interventions for Work Stress: A Risk Management Approach* (p. 193). Sudbury: HSE Books.

Cox, T., Karanika, M., Griffiths, A., & Houdmont, J. (2007). Evaluating organizational-level work stress interventions: Beyond traditional methods. *Work & Stress*, 21(4), 348–362.

Cox, T., Randall, R., & Griffiths, A. (2002). *Interventions to control stress at work in hospital staff* (p. 160). Nottingham: The Institute of Work, Health and Organizations, University of Nottingham.

Dawson, P. (1994). *Organizational change: A processual approach*. London: Paul Chapman/Sage Publications.

Egan, M., Bambra, C., Thomas, S., Petticrew, M., Whitehead, M., & Thomson, H. (2007). The psychosocial and health effects of workplace reorganization. 1. A systematic review of organisational-level interventions that aim to increase employee control. *Journal of Epidemiology and Community Health* 61, 945–954. doi: 10.1136/jech.2006.054965

Egan, M., Bambra, C., Petticrew, M., & Whitehead, M. (2009). Reviewing evidence on complex social interventions: appraising implementation in systematic reviews of the health effects of organisational-level workplace interventions. *Journal of Epidemiology and Community Health*, 63(1), 4–11. doi: 10.1136/jech.2007.071233

Fitzgerald, J., & Matthews Rasheed, J. (1998). Salvaging an evaluation from the swampy lowland. *Evaluation and Programming Planning*, 21, 199–209.

Giddens, A. (1987). *La constitution de la société: Éléments de la théorie de la structuration*. Paris, France: Presses universitaires de France.

Giga, S., Faragher, B., & Cooper, C. L. (2003). Identification of good practice in stress prevention/management. In J. Jordan, E. Gurr, G. Tinline, S. Giga, B. Faragher, & C. L. Cooper (Eds.), *Beacons of excellence in stress prevention* (Vol. HSE Research Report 133, pp. 1–45). Sudbury, England: HSE Books.

Goldenhar, L. M., LaMontagne, A. D., Heaney, C., & Landsbergis, P. (2001). The intervention research process in occupational safety and health: An overview from NORA Intervention Effectiveness Research Team. *Journal of Occupational and Environmental Medicine*, 43(7), 616–622.

Graveling, R., Crawford, J. O., Cowie, H., Amati, C., & Vohra, S. (2008). Workplace interventions that are effective for promoting mental well-being: Synopsis of the evidence of effectiveness and cost-effectiveness. From National Institute for Health and Clinical Excellence (NICE). See: www.nice.org.uk/nicemedia/pdf/WorkplaceMental-HealthSynopsisOfEvidence.pdf.

Greene, J. C., & Caracelli, V. J. (1997). Defining and describing the paradigm issue in mixed-method evaluation. In J. C. Greene & V. J. Caracelli (Eds.), *New Directions for Evaluation* (Vol. 74, pp. 5–18). San Francisco: Jossey-Bass.

Griffiths, A. (1999). Organizational interventions: Facing the limits of the natural science paradigm. *Scandinavian Journal of Work and Environment Health, 25*(6), 589–596.

Heaney, C. A., & Goetzel, R. Z. (1997). A review of health-related outcomes of multi-component worksite health promotion programs. *American Journal Of Health Promotion, 11*(4), 290–307.

Hugues, J., & Sharrock, W. (1997). *The philosophy of social research*. Harlow: Longman Group UK Ltd.

Hume, D. (1739). *A treatise of human nature*. London: John Noon.

Kompier, M. A. J., Cooper, C. L., & Geurts, S. A. E. (2000). A multiple case study approach to work stress prevention in Europe. *European Journal of Work & Organizational Psychology, 9*(3), 371.

Kristensen, R. T. (2008). Psychosocial intervention studies – Opportunities and challenges. Quebec: Third ICOH International Conference on Psychosocial Factors at Work.

Linnan, L., & Steckler, A. (2002). Process evaluation for public health interventions and research: An overview. In A. Steckler & L. Linnan (Eds.), *Process evaluation for public health interventions and research.* (pp. 1–21). San Fransisco: Jossey-Bass Publishers.

Lipsey, M. W., & Cordray, D. S. (2000). Evaluation methods for social intervention. *Annual Review of Psychology, 51*, 345–375.

Murta, S. G., Sanderson, K., & Oldenburg, B. (2007). Process Evaluation in Occupational Stress Management Programs: A Systematic Review. *American Journal of Health Promotion, 21*(4), 248–254.

Nielsen, K., Randall, R., & Albertsen, K. (2007). Participants, appraisals of process issues and the effects of stress management interventions. *Journal of Organizational Behavior, 28*, 793–810.

Nielsen, K., Taris, T. W., & Cox, T. (2010). The future of organizational interventions: Addressing the challenges of today's organizations. *Work & Stress, 24*(3), 219–233.

Noblet, A., & LaMontagne, A. D. (2009). The challenges of developing, implementing, and evaluating interventions. In S. Cartwright & C. L. Cooper (Eds.), *Oxford Handbook of Organizational Well-being*. Oxford: Oxford University Press.

Parkes, K. L., & Sparkes, T. J. (1998). Organizational interventions to reduce work stress: Are they effective? A review of the literature *HSE books* (p. 52). Sudbury: Health and Safety Executive.

Pawson, R. (2002). Evidence-based Policy: The Promise of 'Realist Synthesis'. *Evaluation, 8*(3), 340–358. doi: 10.1177/135638902401462448

Pawson, R. (2006). *Evidence-based policy – A realist perspective*. London: Sage Publications Ltd.

Pawson, R. (2008). Method mix, technical hex, and theory fix. In M. M. Bergman (Ed.), *Advances in mixed methods research* (pp. 120–137). London: Sage.

Pawson, R., & Tilley, N. (1997). *Realistic evaluation*. London: Sage Publications.

Pawson, R., & Tilley, N. (1994). What Works in Evaluation Research? *British Journal of Criminology, 34*(3), 291–306.

Pedersen, L. M., Nielsen, K. J., & Kines, P. (2012). Realistic evaluation as a new way to design and evaluate occupational safety interventions. *Safety Science, 50*, 48–54.

Randall, R., Cox, T., & Griffiths, A. (2007). Participants' accounts of a stress management intervention. *Human Relations, 60*(8), 1181.

Randall, R., Griffiths, A., & Cox, T. (2005). Evaluating organizational stress-management interventions using adapted study designs. *European Journal of Work and Organization Psychology*, *14*(1), 23–41.

Reichers, A. E., Wanous, J. P., & Austin, J. T. (1997). Understanding and managing cynicism about organizational change. *Academy of Management Executive*, *11*(1), 48.

Richardson, K. M., & Rothstein, H. R. (2008). Effects of occupational stress management intervention programs: a meta-analysis. *Journal of Occupational Health Psychology*, *13*(1), 69–93.

Robson, C. (1993). *Real World Research*. Oxford UK: Blackwell Publishers.

Roen, K., Arai, L., Roberts, H., & Popay, J. (2006). Extending systematic reviews to include evidence on implementation: Methodological work on a review of community-based initiatives to prevent injuries. *Social Science & Medicine*, *63*(4), 1060.

Rossi, P. H., Lipsey, M. W., & Freeman, H. E. (2000). *Evaluation – a systematic approach* (7th ed.). Thousand Oaks, CA: Sage Publications Inc.

Ryan, R. M., & Deci, E. L. (2000). Self-Determination Theory and the Facilitation of Intrinsic Motivation, Social Development, and Well-Being. *American Psychologist*, *55*(1), 68–78.

Semmer, N. (2011). Job stress interventions and organization of work. In J. C. Quick & L. E. Terrick (Eds.), *Handbook of Occupational Health Psychology* (2nd ed., pp. 299–318). Washington, DC: APA

Semmer, N. K. (2006). Job stress interventions and the organization of work. *Scandinavian Journal of Work and Environmental Health*, *32*(6, special issue), 515–527.

Sugarman, B. (2001). A Learning-Based Approach to Organizational Change: Some Results and Guidelines. *Organizational Dynamics*, *30*(1), 62–76.

Thompson, M., & Fearn, T. (1996). What exactly is fitness for purpose in analytical measurement? *Analyst*, *121*, 275–278.

Van der Klink, J. J. L., Blonk, R. W. B., Schene, A. H., & Van Dijk, J. H. (2001). The benefits of interventions for work-related stress. *Journal of Public Health*, *91*(2), 270–276.

Part 2
Addressing process and context in practice

10 Evaluation of an intervention to prevent mental health problems among correctional officers

Renée Bourbonnais, Nathalie Jauvin, Julie Dussault, and Michel Vézina

Introduction

The objective of this chapter is to document the development, implementation and effectiveness phases of a participative intervention project aimed at reducing adverse psychosocial factors and improving mental health of correctional officers (COs), and to pinpoint factors facilitating or hindering this intervention process. Qualitative and quantitative methods were used. Intervention groups composed of various workplace representatives (e.g. correctional officers, management, local unions) identified changes needed to improve the psychosocial work environment. In conclusion, interventions may act in preventing further increases in the prevalence of adverse psychosocial work factors. The three phases of this research address a critical need identified in the literature: that of identifying the factors that facilitate or hinder the intervention process.

Statement of the problem

Like many workplaces, correctional institutions are currently facing higher operating costs, budgetary restraint and occupational shortage, all of which have an impact on staff (Griffin et al., 2010; Lambert and Paoline, 2010). New challenges have also emerged in recent years in detention facilities. In particular, the inmate population has grown and diversified (Liebling, 2006). A rise in imprisonment has increased the prevailing emotional burden (Liebling, 2006). Improved prison conditions, through greater recognition of inmates' rights and the development and diversification of social reintegration programs, have also had an impact on correctional officers and has increased their qualitative and quantitative workload (Cheng et al., 2004, Lambert, Lambert, and Ito, 2004; Owen, 2006). They have also led to an increase in formal procedures to be followed, growing bureaucratization of the system, and a decrease in the decisional latitude of correctional agents in daily negotiations inherent in relations with inmates (Béthoux, 2000). Correctional officers also face work-related constraints owing to the very nature of their jobs. The prison work environment is renowned for being more stressful than many other environments (Armstrong and Griffin, 2004). Safety risks are ever present, as is latent unpredictability (Jauvin et al., 2006). Work in a correctional environment is often effectively difficult, dangerous and belittled by the general population (Finn, 1998; Griffin et al., 2010). This social non-recognition towards correctional officers is frequently coupled with a lack of consideration by prison administration (Finn, 1998).

The effects of working in a prison environment on the health of correctional officers are generally acknowledged as being very intense. Stress affects a great number of officers, often quite severely (Finn, 1998). Recourse to sick leave and turnover offer a glimpse of the severity of the situation (Camp and Lambert, 2006; Finn, 1998; Lambert and Paoline, 2010). Some effects have repercussions on both physical and psychological health (Golberg, 1994; Tartaglini and Safran, 1997). More correctional officers than employees in other groups of professionals suffer from anxiety, high blood pressure, psychosomatic problems and the consumption of psychotropic medication (Bourbonnais et al., 2005; Johnson et al., 2005; Lavigne and Bourbonnais, 2010). The few studies conducted on the subject have highlighted the need for special attention on burnout among correctional officers (Griffin et al., 2010).

Although individual factors should not be ignored, available research has identified organizational factors in the workplace as being more highly correlated with burnout among prison staff (Cheng et al., 2004; Griffin and Hepburn, 2005). A review of 43 investigations conducted in nine European and American countries showed a high incidence of work constraints and burnout among correctional officers (Schaufeli and Peeters, 2000). Among other things, one study linked negative interaction between colleagues to the stress experienced by prison staff (Moon and Maxwell, 2004). Overwork (Triplett, Mullings, and Scarborough, 1996), low social support, and conflicts with colleagues (Bourbonnais et al., 2005) and superiors (Armstrong and Griffin, 2004; Auerbach, Wuick, and Pegg, 2003; Bourbonnais et al., 2005) were also associated with psychological distress.

Epidemiological studies support the association between stress-related health and exposure to psychosocial risks in the workplace (Bourbonnais et al., 2005; Johnson et al., 2005). In the last decades, two models of job strain have dominated empirical research on psychosocial work factors: the demand-control-support model (DCS) (Johnson and Hall, 1988; Karasek and Theorell, 1990) and the effort-reward imbalance model (ERI) (Siegrist et al., 1990). Through the two main components of demands and decision latitude or control, Karasek and Theorell have managed to highlight the possible combinations between job requirements and workers' autonomy (Karasek and Theorell, 1990). According to this model, mental and physiological strain occurs at work when high demands are accompanied by low control. Social support from colleagues and supervisors mediates the association between high job strain and mental health problems (isostrain hypothesis). Siegrist has developed a complementary model based on the imbalance between effort made at work and the rewards obtained. Balance lies in the possibility of having access to legitimate benefits expected given the effort put in at work. According to this model, work situations that demand high effort and offer little reward could have adverse effects on physical and emotional health (Siegrist et al., 1990). Many empirical studies support the effect of these constraints on physical health (cardiovascular diseases) and mental health (depression and burnout) (Stansfeld and Candy, 2006; van der Doef and Maes, 1998, 1999). Used with great success in many countries and widely disseminated in the scientific community, these models have become a permanent reference in the field of occupational mental health.

Supported by these theories (Johnson and Hall, 1988; Karasek and Theorell, 1990; Siegrist et al., 1990), one study reported that mental health problems experienced by prison staff were strongly associated with extreme demands and a lack of control and influence over work (Ghaddar, Mateo, and Sanchez, 2008). Combined exposure to these two constraints was also identified as an important factor in the emergence of job-related mental health problems among the population under study. Another study reported a connection between very weak participation in decision-making in the workplace and burnout among correctional staff (Lambert, Hogan, and Jiang, 2010).

Although many studies have documented the effects of adverse psychosocial work factors on the incidence and the prevalence of mental health problems, only a few studies have evaluated the impact of interventions aimed at reducing adverse psychosocial work factors and their health effects, and available studies have important methodological limitations (Kompier et al., 2000; Kompier and Kristensen, 2001; Semmer, 2006; van der Hek and Plomp, 1997). These include a lack of a theoretical model for the intervention, lack of a control group, lack of *a priori* evaluation of risks, and lack of prospective designs (Kompier and Kristensen, 2001), an insufficient follow-up period following the implementation of the intervention, and insufficient qualitative documentation of the specific context and the implementation of the intervention (Dahl-Jørgensen and Saksvik, 2005). Moreover, despite correctional officers' poor health, little knowledge has emerged to-date on the efficacy of different approaches used to help mitigate stress at the source (Keinan and Malach-Pines, 2007). Nevertheless, studies that have measured improvements in psychosocial work factors, have reported significant decreases, between 9 per cent and 55 per cent, in symptoms associated with mental health and sick leave (Bond and Bunce, 2001; Karasek, 1985). Moreover, recent literature reviews on organizational level and task restructuring interventions show that they generally have a positive effect on workers' health, particularly when they are aimed at increasing their level of control (Bond and Bunce, 2001; Egan et al., 2007) and even more so when this is associated with a reduced work load (Bambra et al., 2007; Bambra et al., 2009).

Several authors (Aust et al., 2010; Kristensen, 2005; Semmer, 2006) highlight the importance of studying the entire intervention process, from planning to evaluation, in addition to implementation. Documentation of the entire process serves to better identify success factors of the intervention and circumstances that are conducive or not to its success. Among models available for intervention process assessment, the model proposed by Goldenhar and her colleagues (Goldenhar et al., 2001) is of particular interest. It consists of a three-phase process for conducting occupational health and safety intervention research: development, implementation and effectiveness of the intervention. Few studies integrate these phases.

The research presented in this chapter originated from a request made by representatives of the provincial Correctional Officers' Union (COU) and of the provincial Public Security Department (PSD) in Québec, Canada. Faced with growing absenteeism and employment injury encountered by correctional officers, the CPU and the PSD asked the research team to examine the situation in order to understand and help remedy it. A joint project *steering committee* was established in December 1999, with representatives of management, union and human resources at the provincial level of the PSD, representatives of local correctional facilities (management and union), and three members of the research team (18 members in total). In 2000, a survey on correctional officers' working conditions and health was carried out in all correctional facilities in the province of Québec (Bourbonnais et al., 2005) This survey showed that the COs were more exposed to adverse psychosocial working conditions and reported more health problems than a comparable sample of workers in Québec. High psychosocial demands combined with low decision latitude, and an imbalance between expended efforts and perceived rewards, low social support at work, and conflict with colleagues and superiors were directly and significantly associated with psychological distress.

Following this first survey, 54 individual and nine group interviews were conducted with 132 COs between 2001 and 2003 to understand the roots of health complaints, using a phenomenological approach based on workers' accounts of their experience (Jauvin et al., 2008; Vézina et al., 2006). Analyses showed that a large part of the distress experienced by correctional officers in Quebec was related to the changes affecting prison establishments. Feelings of distress were rooted in:

i   the job and the organization (organizational dimension),
ii social relations (relational dimension), and
iii identity (individual dimension).

The presence of stress and imbalance between these three dimensions accounted for an increased risk of suffering among workers, whereas balance, on the contrary, was associated with job satisfaction and well-being. This comprehensive study served to develop a first account of possible intervention paths to address psychosocial constraints in the workplace. Such paths would supersede individual and interpersonal interventions to integrate interventions at the organizational level. Supported by this evidence, in 2004 we began the last phase of this study among correctional officers in Quebec, which consisted of a preventative intervention on psychosocial constraints in the workplace.

The objective of this chapter is to document the three phases (development, implementation, and effectiveness) of the intervention process aimed at reducing four theory-grounded and empirically supported adverse psychosocial work factors (high psychological demands, low decision latitude, low social support, and low reward), and their mental health effects among correctional officers. This chapter also aims to document the factors facilitating or hindering the intervention process.

Methods

The intervention population included all correctional officers in Quebec with permanent tenure (full and part time) and casual employees working an average of more than 24 hours per week for over three months. The officers worked throughout the province in 18 prison institutions under provincial jurisdiction. Half of the inmates in the establishments had been sentenced to less than two years and the other half were awaiting trial. Often, the latter were potentially more dangerous because many would eventually be redirected to a federal penitentiary upon their sentencing to two or more years of prison.

The methodological approach for this intervention study was based on the research framework proposed by Goldenhar et al. (2001; see Figure 10.1). The development phase aims to answer questions related to the changes needed and the best ways to bring them about, the barriers preventing these changes from happening and the theories that might apply in the specific intervention context. The implementation phase is concerned with the means put in place in order to produce changes to the work environment. It describes the nature and intensity of changes implemented, what difficulties and facilitators were encountered. Lastly, the effectiveness phase aims to demonstrate whether the intervention was successful in reducing the prevalence of adverse work factors and improve mental health of COs.

Our interpretation of the results presented in this chapter is based on the entire process encompassing the three distinct phases of the intervention research. Qualitative methods were used to ensure follow-up of the development and

Phases	1. Development	2. Implementation	3. intervention effectiveness
	What is the importance of constraints, their causes, and their effects?	Type of changes	To what extent does the intervention reduce: • Adverse work organization factors • Psychological distress?
	What changes are needed? What are the best ways to bring about changes?	Quality of the implementation How many workers are affected by changes?	
Quantitative	Prior risk evaluation (questionnaire) M0		Quasi-experimental design with control group; post intervention measure M1 at 12 months; follow-up intervention measure M2 at 36 months
Qualitative	Involvement of intervention centres		
	• Participative Intervention (IG) • Observation	• Follow up of changes with IG in the three intervention centres and with the CG • Analysis of implementation and appropriation • Follow up of changes with key informers and observation	• Follow up of changes with key informers

Figure 10.1 Phases and methods of the research. Adapted from Goldenhar et al. (2001).

implementation of the interventions in detention centres: observation (made possible within the framework of the participative approach to the entire project), follow-up of the interventions in three intervention centres and exchange with key informers within the intervention and control centres. Likewise, quantitative methods (pre- and post-intervention questionnaires at 12 and 36 months) were used to evaluate the evolution of a variable set (described below).

The *development phase* started in 2004 and aimed to identify concrete targets for the preventive interventions using both qualitative and quantitative methods. First, a *prior risk evaluation* was performed between May and October 2004, in all correctional facilities of the province of Québec, using a quantitative approach. A self-administered questionnaire was used to determine the prevalence of adverse psychosocial work factors and health problems among COs compared to an appropriate reference population. It was also used to identify correctional facilities which were the most susceptible to gain from an intervention, i.e. those who had a high prevalence of adverse psychosocial work factors and/or mental health problems compared to other facilities and to the reference population. The questionnaires were administered to all COs with at least three months tenure in the 18 correctional facilities (N=1881). The reference group was composed of all workers who had participated in the Quebec Health Survey (QHS) in 1998 (Institut de la statistique du Québec, 2000). The QHS was conducted in a stratified random sample of all Quebeckers appearing in the records of the Quebec Insurance Board (which covers more than 95 per cent of the population). The weighted sample is representative of non-institutionalized Quebeckers at the time of the survey (Institut de la statistique du Québec, 2000). A subgroup was selected from this sample with characteristics similar to the COs under study (holding a paid job, a college or university degree (same educational range as the COs)). The reference population thus comprised 9475 workers (5102 male and 4373 female) holding a range of jobs in a range of industrial sectors. Primary comparison data were available, consisting of a self-administered questionnaire with the same instruments as used in our study.

Validated instruments based on the DCS and ERI models and measuring correctional officers' psychosocial work environment and state of health were used. Demands, decision latitude, and social support were evaluated using 26 items from the Job Content Questionnaire (JCQ) (Karasek, 1985; Karasek and Theorell, 1990). The sound psychometric qualities of the JCQ have been demonstrated (Karasek, 1985; Karasek and Theorell, 1990). Demands included the amount of work, intellectual requirements, and time constraints of the job, whereas decision latitude covered the use and development of skills and control at work, work autonomy, and participation in decision-making. Social support at work covered socio-emotional support or esteem, which is of a socio-psychological or interpersonal nature; instrumental support, which measures extra resources or assistance with work tasks; and a negative level of support, hostility or conflict. Reward at work was measured using 11 items from Siegrist's original instrument, for which factorial validity and internal consistency have been documented (Niedhammer et al., 2000; Siegrist, 2001). It included three dimensions:

1 socio-emotional reward measured the respect and esteem received at work from superiors and colleagues,
2 organizational reward measured promotion prospects and job security, and
3 monetary reward, which is related to salary earned.

The effort dimension of Siegrist's ERI model was substituted in this study by the demands dimension. In our study, internal consistency based on Cronbach's coefficient alpha was 0.78 for job decision latitude, 0.78 for demands, 0.86 and 0.78 for social support from supervisors and colleagues, and 0.81 for reward. In addition, two types of interpersonal violence (or any type of vertical or horizontal violence among members of a work organization (individuals or groups of individuals excluding violence from inmates) were measured: psychological harassment and intimidation. Respondents were asked whether they had been exposed to these indicators (one item each) in the past 12 months.

Psychological distress was measured using an abridged version (14 items) of a validated instrument, the Psychiatric Symptom Index (PSI) (Bellerose et al., 1995; Ilfeld, 1976). Work-related burnout was measured using questions from the Copenhagen Burnout Inventory (Kristensen, 1999), for which strong support for its validity has been provided (Kristensen, 1999; Kristensen and Borritz, 1999). Sleeping problems were measured using five questions from the Nottingham Health Profile (NHP) (Hunt et al., 1980). In our data, the Cronbach's alpha coefficients were 0.91 for the total score of PSI, 0.86 for work-related burnout, and 0.78 for sleeping problems. Perceived general health status was measured by one question from the Short-Form Health Survey (SF-36) which assesses an individual's perceived general health compared to a person of the same age (McHorney, Ware, and Raczek, 1993) This question was validated in French by the Québec Health Survey (QHS) (Bellerose et al., 1995). Three questions on the use of health- or social services during the two weeks preceding the questionnaire were likewise taken from the QHS (Bellerose et al., 1995).

Next, *three correctional centres* were targeted for the intervention on the basis of a particularly high prevalence of problems associated with work organization: one large (consisting of 315 COs), one medium (81 COs) and one small facility (49 COs). The development phase in each intervention centre was initiated with a formal agreement to participate in the research. A control group included all other correctional facilities in the province (n=15).

An *intervention group* (IG) was implemented in each of the three intervention centres according to the principles of German health circles which have shown their effectiveness in the prevention of stress at work in Europe (European Agency for Safety & Health at Work, 2002). Health circles principles include:

1 operating in small groups;
2 group members of different hierarchical levels;
3 regularly-scheduled work meetings, preferably eight to ten;
4 meetings led by an external moderator (here, the researcher); and

5   individual knowledge of team members used as input for finding solutions to adverse psychosocial factors.

The ultimate objective is to identify and eliminate problems at their source (primary prevention). Each IG was composed of correctional officers (COs), supervisors, and two researchers. Their aim was to identify adverse psychosocial work factors on which interventions should focus and seek possible solutions. IG members were also responsible for disseminating information from the meetings to their colleagues and for providing the IG with feedback (comments and reactions from the COs). A *coordination group* (CG) was also formed. Its role was to facilitate generalization of local changes between intervention centres and to make recommendations to the *steering committee* (SC) for provincial generalization. The CG included members from the three intervention groups (management and union members). Figure 10.2 presents the structure of the intervention project. Researchers were present in each IG, the CG and the SC.

The *implementation phase* consisted in a continuous monitoring of organizational interventions (changes) throughout the research period:

1   follow-up of plans of actions by IG in each intervention centre;
2   follow-up of changes that happened independently of the research in the intervention and three paired control groups;
3   identifying special contexts enabling natural experiments (e.g. new laws or governmental programs, union negotiations, new director, etc.).

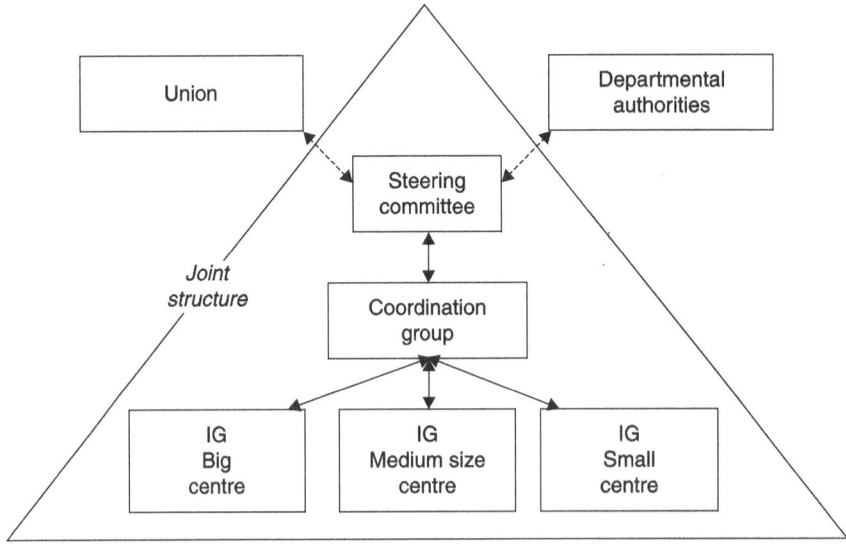

Figure 10.2 Structure of the intervention project.

Certain changes were also made by the Coordination Group (CG) on a provincial basis. Throughout the process, local and national authorities remained responsible for the implementation of all organizational changes proposed. Follow-up of the implementation was completed through interviews with key informers in the three intervention centres. The key informers were identified by members of the national union executive (for correctional officer representatives) and by correctional services management (for local management representatives) based on predefined criteria (e.g. knowledge of the milieu, credibility). The semi-structured interviews with informants lasted about 1.5 hours.

The *effectiveness phase* aimed to assess the short- and long-term effects of the intervention. It used a quasi-experimental design with the three intervention correctional facilities, 15 control facilities and pre-post intervention measurement. The research included all correctional officers. Self-administered questionnaires on COs' exposure to psychosocial job factors and health status were completed pre-intervention (at baseline, M0), post intervention one year later (M1), and at follow-up three years after the intervention (M2). The pre and post-intervention assessment included the same instruments as described above.

Analyses

Qualitative analyses

The inductive-type analysis sought to better describe the development and implementation of interventions, to determine more clearly the contextual aspects of the intervention, and to provide an in-depth understanding of the qualitative data (individuals' accounts and researchers' observations conducted in the context of a participative approach). This qualitative phase of the analysis also served to explore more particularly the issue of factors conducive and not conducive to the implementation of such interventions.

We carried out a vertical and horizontal thematic analysis of the data (Blanchet and Gotman, 1992), and then proceeded to interpret the data first for each intervention centre and then for the intervention approach overall. A summary was drafted for each of the interviews summarizing their content as faithfully as possible and outlining the researchers' reflections. A cross-sectional review of these summaries provided a more global overview of the qualitative data. Likewise, a summary of each interview was prepared by the researchers and sent to the respective interviewee for validation. Once the corrections were made, a synthesis report was prepared grouping together all organizational changes implemented by the establishment.

At different stages in the analytical process, the preliminary findings were discussed with members of different committees, correctional officers and employer representatives who were asked to compare it to their own experiences as key players in the correctional milieu. This served to compare and verify the results of the preliminary analysis against the perceptions of workers from the correctional environment.

Statistical analyses

The prevalence of psychosocial constraints in the work environment and health problems were measured (Rothman and Greenland, 1998). A total score was calculated for each of the measures. Psychological distress was deemed high when the score was in the upper quintile of the score distribution observed among workers from the population in Quebec (>28.57 for women and >23.81 for men) (Bellerose et al., 1995). Demand was deemed high when the score was equal to or greater than the median observed among workers from the population of Quebec (≥ 9) and decisional latitude was deemed weak when the score was equal to or lower than the median of this same population (≤ 72) (Larocque, Brisson, and Blanchette, 1998). This method is the one proposed by Karasek for a dichotomized analysis (Karasek, 1985). The effort/reward imbalance was determined by the ratio of demands to reward, and the measure adjusted by means of a correction factor (serving to eliminate the difference between the number of questions in the numerator in relation to the denominator), as recommended by Siegrist (2001). Data for QHS 1998 were weighed according to the method recommended by Santé-Québec (Daveluy et al., 2000) in order to obtain a representative sample of the non-institutionalized population in Quebec at the time the survey was conducted.

The prevalence of psychosocial constraints and health problems were analysed with a generalized linear model with a binomial distribution and an identity link. Generalized estimating equation (GEE) were used to take into account the correlation between measures taken on the same individual at different times (Everitt and Fidell, 2001; Tabachnik and Fidell, 1996). The model included a group effect (intervention or control), a time effect (M0, M1 and M2), and a group by time interaction effect. Differences in prevalence between M1 and M0 and between M2 and M0 were assessed within each group by means of contrasts.

Results

The development phase

Commitment of the organization: There was a collaborative agreement (at local and provincial level) on a research intervention project (2004–2009) defining the involvement of management and union by: appointing representatives to the intervention group; releasing staff from work duties to attend meetings and facilitate the follow-up; supporting the implementation of action plans proposed by the intervention group (IG).

The individual and collective interviews with workers revealed many problems that had to be addressed. These included:

1 high workload: overcrowded centres, mentally ill and violent offenders (many had committed major crimes but were awaiting trial in a provincial prison);

2   lack of authority (credibility), recruits insufficiently trained, lack of support from management (a lot of turnover, supervisor not often nor very present in the workplace), lack of consultation (militaristic top down structure), negative social image, both inside and outside the prison;
3   involvement in social rehabilitation was often denied despite the work being officially orientated towards it because of increasing security workloads, conflicting demands between security and rehabilitation, and lack of guidelines and of supervisor support.

The prior risk evaluation showed a higher prevalence of adverse psychosocial factors and mental health problems at baseline (M0) for Québec's correctional officers in comparison with the reference population (36.6 per cent vs 16.7 per cent for job strain, 84.8 per cent vs 18.1 per cent for intimidation, and 40.2 per cent vs 22.5 per cent for distress among males; and 44.7 per cent vs 22.1 per cent for job strain, 82.4 per cent vs 21.2 per cent for intimidation and 37.5 per cent vs 22.5 per cent for distress among females). Furthermore, a comparison of 2000 and 2004 data for COs showed that the prevalence of these problems were similar and therefore unlikely to be remedied without any interventions (Bourbonnais et al., 2007). These results supported the need for a participative intervention aimed at reducing the four theory-grounded and empirically supported adverse psychosocial work factors (high demands, low decision latitude, low social support, and low reward), and interpersonal violence (intermediate outcomes), in order to reduce their effects on mental health (ultimate outcomes).

The *intervention group* (IG) in each intervention correctional facility identified adverse psychosocial work factors using the theoretical model of the research, the DCS model and the ERI model, which were presented by the researchers to the members of the IG. Meetings took place regularly at each of the three IGs. The number of meetings, their duration and the periods during which they were held varied between groups depending on their circumstances (specific problems, concurrent projects, etc.). Globally, an average of ten meetings were held over an average 29-month period. During these meetings, two researchers supported the IGs in identifying problems and changes needed to reduce identified negative health outcomes. At the first meeting, the IGs examined the results of the questionnaire (M0) to establish links with the theoretical models, and provided alternative explanations for the identified high psychosocial constraints experienced by correctional officers.

The IGs then suggested potential solutions to these constraints, determined intervention feasibility, priorities and time schedules, and appointed a person responsible for the follow-up for each solution. Some of the interventions identified were: establishing clear goals and rules, and granting more freedom to apply these rules and in decision-making; improving new recruits' induction and training process, and implementing a mentoring system; creating opportunities for exchange; allocating time for regular meetings during shifts; promoting greater transparency around decision-making and the allocation of promotions, training opportunities or different mandates; developing a better communication system both vertically between management and COs and horizontally between COs; restoring the public image of the profession; encouraging increased presence of superiors in the workplace; devising better methods of evaluation; and offering more formal recognition.

A final report was presented to the steering committee and the management of each intervention centre. The report contained recommendations for solutions (action plans). Several solutions had already been applied to several of the constraints identified during the work of the IG in each intervention centre. Solutions focused on conditions that could be solved easily and managed locally, including time for meetings, transmission of information on the evening and night shifts. Other solutions concerned organizational constraints that required the support and consent of top management of the PSD and were brought to the steering committee. Many of these recommendations were applicable to all three intervention centres.

Implementation phase

An intervention was defined as objective changes undertaken by management to reduce adverse psychosocial work factors. This included solutions proposed by the IG and other changes introduced in the workplace with the explicit goal (or the clear consequence) of improving any one of the adverse psychosocial factors identified. The project implementation was spread over the whole research period; some projects were implemented within months and others within years. In each IG, a review meeting was held several months after the withdrawal of the researcher team for an update on projects implemented and the management of activities. In total over 30 intervention projects were developed during an average 29-month period. Each project required the involvement of IG members and the participation of other external resources, managerial staff and colleagues who took an active part in the implementation of certain activities. These projects can be grouped in three categories: adopting more participative means of doing things, adjusting work methods, and improving interpersonal relations and individual well-being (see Table 10.1). Each of these categories presented below includes examples of activities developed by the three local IGs.

Provincial projects were also implemented. These included: providing all COs with an official badge and number; instituting an awards ceremony to deliver the badges; a new training program for recruits (this project started in one intervention centre before it was adopted by the coordination group as a provincial project); and establishing associations with a humanitarian organization (one promoting breakfast in elementary schools) to foster COs' better self-image and a good public image of the profession in the general population. According to participants' accounts and qualitative data collected during the research process, the interventions implemented locally and developed in the course of work completed by IGs were chosen on the basis of the problems that were specifically encountered in each of the centres, the variables targeted by the study, and their feasibility.

Table 10.1 Changes introduced in intervention facilities

	Underlying principles	Examples of activities developed	Examples of targeted/achieved effects
Adopt more participative means of doing things that allow recognition of the importance of each individual	Encourage greater transparency.	Democratizing training and mandates offered through tenders (or offers) to all.	• Decreased perceptions of favouritism and injustice • Valorization and recognition of those who achieve mandates • Development of a consultative practice • Increased involvement of COs in the decision-making process
	Improve communication processes.	Structuring communications to reduce role difficulties and perceived danger through the creation of a procedure to systematically disseminate particular events and/or a crisis.	• Heightened transparency and exactitude of facts with respect to a situation experienced • Decreased propagation of rumours • Improved work methods • Improved dissemination of inter-sectorial information • Heightened feelings of belonging • Increased support from management
	Create spaces for exchange, places to talk, discuss things.	Holding daily meetings in lodging units (first on the day shift and then the evening shift).	• Improved exchange of information between COs and superiors • Strengthened work group and standardization of procedures • Decrease in daily stressors among work groups • Decrease in the stress caused by lack of communication and information • Improved support among colleagues and from immediate superior

(*Continued overleaf*)

Table 10.1 Continued

Underlying principles	Examples of activities developed	Examples of targeted/achieved effects
Identify and implement formal methods of recognition.	Increasing participation in decision making, encouraging worker participation in various advisory committees.	• Improved workers decision latitude • Increased feelings of belonging • Improvements in the exchange of information • Improved work methods
Adjust work methods to establish guidelines on ways of doing things Create tools to better establish work guidelines.	Creation of a memo (paper and computerized) to remind workers of the routine of tasks or procedures to follow at different work stations.	• Knowledge transfer • Work facilitated for newcomers and those occupying different positions on an alternating basis • Decreased workload for permanent employees • Reduction in tension between senior and new employees • Improved work methods • Greater efficiency in the execution of tasks • Improved safety
Favour a more regular presence of supervisory staff in the workplace, on the "floor."	Increase in the amount of time spent by supervisors on the floor thanks to the sharing of certain special mandates with the CO.	• Bringing better acknowledgement of COs work and more delegation of task increasing decision latitude • Improvements in the dissemination of inter-sectorial information • Improved exchange of information between COs and superiors • Decrease in daily stressors among work groups • Improved support among colleagues and the immediate superior • Improved ability of the superior to properly evaluate correctional officers

	Improve the welcome and follow-up offered to recruits.	Pairing new officers with more experienced officers during initial training; to achieve this, creating new tools to support and to assess the work.	<ul><li>Improved content of basic training for newcomers (more in touch with the reality of daily life in the field)</li><li>Improved training tools</li><li>Improved knowledge transfer between senior employees and recruits</li><li>Identification of strong and weak points among recruits to improve their potential job performance (possible adjustments during initial training)</li><li>Appreciation and recognition of those involved in the training of recruits</li><li>Improved support offered to recruits</li><li>Improved climate between workers, creation of social relations, improved exchange</li><li>Increase confidence of correctional officers in recruits</li><li>Better potential retention of new staff</li></ul>
Devise methods to encourage the development of healthy interpersonal relations and individual well-being	Improve the integration process of recruits.	Implementing a mentorship program.	<ul><li>Recognition of the expertise of more experienced officers</li><li>Improved support offered to newcomers</li><li>Improved knowledge transfer between senior employees and recruits</li><li>Improved climate between workers, creation of social relations, improved exchange</li><li>Appreciation and recognition of those involved in mentorship</li><li>Increased efficiency in the execution of tasks</li></ul>

(*Continued overleaf*)

Table 10.1 Continued

Underlying principles	Examples of activities developed	Examples of targeted/achieved effects
Set up activities to reunite the groups.	Creating a peace officers' week (various activities such as honorary meals, exhibits of artwork by officers, open door events, etc.).	• Greater feelings of safety with respect to newcomers • Better potential retention of new staff • Improved ability for teamwork based on a shared goal going beyond work • Greater solidarity among participating officers • Appreciation of the officer's role • Greater feelings of individual recognition • Ability to view or get to know colleagues "differently" • Improved team spirit
Create a mutual aid system by peers.	Implementing a voluntary aid system among peers, first-line interveners, in potential crisis situations experienced by a peer.	• Implementation of early, preventive action plans to mitigate personal crises faced by some officers (more rapid control over crisis situations) • Development of specific knowledge through specialized training for volunteers • Development and strengthening of quality interpersonal relations between employees • Appreciation and recognition of those involved in mentorship • Increased support between colleagues
Make organizational values known in terms of healthy interpersonal relations and clamp down on those who fail to comply with established norms.	Adopting a code of ethics and values.	• Development of healthier interpersonal relations • Decrease in mobbing/psychological violence in the workplace • Increased feeling of pride in the trade • Increased feelings of belonging

The effectiveness phase

This participative intervention produced significant results one year later (measured at M1; see Table 10.2). In the intervention group, although demands and job strain had increased, three positive changes were noted. Specifically, two constraints and one health problem decreased significantly: harassment from supervisors, intimidation by supervisor, and medical consultation. There were no positive changes between the pre- and post-intervention assessment in the control group. There was, however, a significant deterioration in most of the psychosocial factors measured in the study and most health indicators: demands; job strain; rewards; effort-reward imbalance; social support from supervisor; harassment from supervisor; intimidation by supervisor; psychological distress; sleep problems; burnout; mental health consultation; and general health.

At the 2nd post intervention measure (M2), three years after the start of the intervention, there was an increase in demands and job strain in both intervention and control centres but also an improvement in rewards for both groups, and an increase in decision latitude in control centres. However, there were significant

Table 10.2 Prevalence (PR in % respondents) of psychosocial factors at work and health, comparison between pre (M0 in 2004), post (M1 in 2007) and follow-up intervention (M2 in 2009) assessment in the intervention and the control groups

	Intervention group			Control group		
	PR M0	PR M1	PR M2	PR M0	PR M1	PR M2
Psychosocial factors						
Demands +	52	59*	61**	43	56***	51***
Decision latitude –	91	91	88	90	89	86*
Job strain	44	53**	54*	36	49***	41*
Supervisor Support –	63	58	48***	53	60***	55
Colleague Support –	67	70	56**	65	65	66
Reward –	63	66	50***	55	65***	49**
Effort-reward imbalance	65	70	60	55	69***	58
Harassment colleagues	33	32	30	37	37	32*
Harassment supervisors	37	29**	31!	31	35*	28
Intimidation colleagues	35	34	31	35	35	34
Intimidation supervisors	47	34***	37**	32	40***	32
Health						
Psychological distress	40	41	38	38	46***	37
Sleeping problems	37	35	36	38	43*	36
Work burnout	38	41	38	37	49***	38
Mental health consultation	5	6	3	5	6*	5
Medical consultation	50	43*	49	48	47	51
General Health	8	8	8	8	11*	8

Note. Intervention group M0(N=361), M1(N=323), M2(N=229). Control group M0(n=1138), M1(n=957), M2(n=722); ***$p \leq 0.001$; **$p \leq 0.01$; *$p \leq 0.05$; $p \leq 0.10$.

Statistical analyses used a generalized linear model with identity link and binomial distribution, with GEE, to test within each group.

204 *Addressing process and context in practice*

improvements only in intervention centres for social relations with supervisors (low social support, harassment, and intimidation), and support among colleagues. In the control group, there was only one significant improvement in harassment from colleagues. There were no significant changes in health indicators in either intervention or control centres. However, there was a tendency toward improvement in the intervention group, in four out of six health indicators and none deteriorated, while in the control group two health indicators showed a tendency to improve and two deteriorated.

Factors facilitating or impeding the approach

The analyses allowed us to document a range of factors that intermittently impeded or facilitated the implementation of interventions in correctional establishments in Quebec, both locally and at the provincial level. They combined all factors of the developmental and implementation approach and were based on interviews with key informers, our own observations, and information collected throughout the research from the various project committees (intervention group, coordination and steering committees). Globally, a set of context and process variables were identified. Results are presented in Table 10.3.

Table 10.3 Facilitating and hindering factors of the intervention process

Facilitating factors

Context variables	Support from management	• Support from top management and involvement of all levels of management at the local correctional facility and the regional and provincial levels where policies are adopted. • Financial support, where and when required, to pursue the research process between research grants and also for the implementation of some of the proposed changes.
	Openness to work in partnership	• Changes to the established method of doing things (hierarchical, authoritarian) translated by openness to developing interventions with partners using an egalitarian approach (participative, democratic).
	Leadership at different levels	• Participation in different aspects of the intervention project of affirmative leadership reflecting, among other things, prioritization of the project by management (government department, correctional facilities) but also by local and provincial trade union authorities. Among the IGs, the presence of members having acknowledged leadership capabilities, benefiting from strong credibility and the capacity for communication and listening.
Process variables	Sound theoretical and methodological background and rigorous approach	• Insuring a choice of targets and solutions and providing validated measures for psychosocial factors and mental health indicators.

		• Translation of work irritants into higher-order theoretical concepts thereby increasing the level of understanding concerning the impact of psychological work conditions on mental health which is a necessary condition to the appropriation of the intervention by the workers.
	Prior risk evaluation	• Use of recognized tools in the careful evaluation of existing risks present in the environment where an intervention is to take place. • Prioritization of interventions to implement based on the results of this evaluation.
	Employee participation in the entire process	• Employee participation in the identification of problems and solutions since they are the experts in what is going on in their work. • Granting of time to participate in the GI activities.
	Open, transparent communication	• Mandatory implementation of methods of communication to inform one and all of the process. • Feedback to participants in IG, to and from all workers and local management. • Creation of a communication toll (Info-RIPOST) aimed to announce the different phases and main results of the research and to maintain the interest of COs for the process.
	Rigorous implementation and follow up of changes	• Respect of commitments made and rigorous implementation of required changes among targeted groups of workers, changes that were proposed and obtained consensus as to their relevance and feasibility. • Appropriation of the process and of changes by the organization in order to insure perennity of the process of identification of constraints and solutions.

Hindering factors

Context variables	Mistrust within the organization	• A kind of transfer from the mistrust between the prisoners, and between COs and prisoners, has an effect on COs between themselves and opposite recruits and also opposite management. • Owing to past experience, some workers, often the more senior employees, have become disenchanted and sceptical of the real potential for change in the wake of new approaches, irrespective of their nature.
	Adoption, during the project, of new policies, new ways of doing things	• New departmental policy implemented by senior management (since the research process takes a lot of time) that can possibly interfere in some projects implemented or planned in establishments.
	Increased stress in work relations	• Emergence of a critical period in work relations (difficult negotiations, trade union advertising campaigns that partially foil work undertaken locally and provincially on the COs image and trade recognition). • During more stressful periods, difficulty for some players involved in the research to adjudicate between action taken under the aegis of the intervention project and traditional management-union negotiations.

(*Continued overleaf*)

206   *Addressing process and context in practice*

Table 10.3 Continued

	Hindering factors (cont.)	
	Lack of financial and human resources	• Long duration of the approach implies an important investment for all partners in time and resources. • Financial imperatives often remained a very difficult obstacle to overcome.
	Deterioration of working conditions (workload, turnover, stressing events . . .)	• Workload already significant and increasing and not much possibility to reduce at least the quantity of work. • Presence of a high turnover rate among COs and management, making communication, information transfer, understanding work underway and adhesion to the project itself more difficult. • The occurrence of difficult events that affect workers and their job conditions: suicide of a colleague, a prisoner riot . . .
Process variables	Unwieldy research process	• Bureaucracy and competition for funding (university research). • The formalities and length of the research process overshadow more immediate and pressing needs within the milieu.
	Changes in committee composition	• A lot of mobility at the local and provincial levels in management, trade union and staff representatives in the IG sometimes made the dissemination of information and the maintenance of commitment difficult. Some employees retired, others moved from one facility to another.
	Lack of control of participants in the process	• Given the existence of a centralized provincial network and a firmly established hierarchal structure, some changes potentially applying to the entire network, needed to be discussed at the provincial level. • In these specific cases, participants at the origin of proposed interventions had to withdraw partially or totally from the approach (issue handled thereafter at a provincial level).
	Unexpected results	• Results that sometimes fail to provide the anticipated or expected outcome or take more time than expected.
	Use of results on other tribunes for political reasons	• Emergence of political pressure (departmental, administrative and trade union authorities) to influence work underway or discredit the process when it concerned issues where there was disagreement. • Use of results by the authorities (departmental, administrative and trade union) to exercise political pressure on other tribunes; dissatisfaction with other partners in the project not in agreement with this parallel use of results.
	Difficult appropriation of the process by the organization	• The workplace becomes dependent of external expertise and in certain cases, has difficulty ensuring continuation of project offshoots once the team of researchers is withdrawn.

Discussion

These results suggest that interventions may act in preventing further increases in the prevalence of adverse psychosocial work factors and mental health problems. Post intervention there were three beneficial changes in the intervention centres. Although we could have expected more positive results, solutions to identified problems at the local level were not yet all implemented when this first post intervention measure was taken and demands had increased substantially since baseline. Moreover, in the control group, there was a significant deterioration of most of the psychosocial risk factors and the health indicators measured.

At follow-up, although demands increased in both intervention and control groups, rewards also increased in both groups and this was probably associated with positive interventions targeting the correctional officers' public image and their sense of value. These measures were first implemented in the intervention centres and then, through the steering committee, in all the detention centres. In the control centres, there appeared to be a stabilization of constraints between post-intervention and follow-up; this may also be partly explained by the roll-out of changes implemented in the intervention centres to all detention centres. However, only the intervention centres experienced a significant improvement of social relations with supervisors (increased social support, and reduced harassment and intimidation). The improvement in social relations in the intervention centres, with supervisors and colleagues, was undoubtedly one of the spinoffs of the intervention study brought about by a better understanding of roles and improved communication. In fact, health indicators did not improve significantly in the intervention group at follow-up, but four out of six of these health outcomes improved (although not significantly) and none deteriorated. A deterioration might have been expected since the high expectations of a positive effect were not met and demands and job strain increased at the same time. A protective effect of better social relations in the intervention centres is consistent with Léveillé, Blais, and Hess (2003) who argue that employee management methods that are rooted in support, provide autonomy and utilize staff skills, can have a positive impact on correctional officers' stress and burnout levels.

The increase in demands in all detention facilities could not be prevented by the intervention research. In addition to the constraints mentioned in the literature reviewed, which are known to exist among correctional officers in Quebec (Griffin et al., 2010; Lambert and Paoline, 2010; Liebling, 2006), this increase in demands was mainly due to overcrowded centres with higher numbers of mentally ill and violent offenders, and many prisoners awaiting trial who had committed crimes more serious than the ones committed by inmates of provincial prisons compared to federal prisons. This increase in demands and the lack of improvement in control may explain the smaller than expected impact on health of the intervention research. Indeed, a recent systematic review of the effects of organizational-level and task restructuring interventions on health and well-being reported that increased control had a positive effect on health while decreased demands and increased support had a less consistent effect (Egan et al., 2007).

Lower demands and higher control generally resulted in improved health (Bambra et al., 2007).

The decrease in harassment from colleagues in both intervention and control groups (although statistically significant only in the control group), might be partly explained by the introduction of new legislation on harassment in the workplace (Government of Quebec, 2004). Since these new legislative provisions, several training and awareness-raising activities have been implemented in workplaces, which might have an incidence on the results of this study.

The literature also supports our overview of factors facilitating or impeding the intervention approach. Several context and process variables emerging from our analyses have also been identified by other researchers in similar contexts. Note, for instance, the singular importance of managerial support throughout the approach identified as a condition for success in organizational interventions (Semmer, 2006). Too often, evaluative studies are carried out without significant support from corporate management, thereby diminishing the scope of organizational changes implemented (Kristensen, 2005). In addition, employee involvement throughout the approach was also highlighted in this study. The interventions most likely to prove efficient are those that adopt a participative approach in the analysis and resolution of problems (DeJoy et al., 2010; Semmer, 2006). Interventions conducted in the absence of real employee involvement in the process, however, tend to lead to decreased employee support for the changes (Dahl-Jørgensen and Saksvik, 2005). Rigour in the implementation and follow-up of changes is another aspect also documented by others. The degree of implementation of planned interventions also greatly influences the effect of the intervention (Kobayashi et al., 2008). Lack of structure in the intervention process, lack of clarity in the definition of the roles and responsibilities of players involved in the intervention have been identified as obstacles to the success of organizational interventions (Aust et al., 2010). It is important to highlight that an organizational intervention necessarily creates expectations among employees and that if these expectations are not addressed to a sufficient degree, they can have deleterious effects on the quality of interpersonal relations between staff and management (Aust et al., 2010). Prudence and rigour are, therefore, mandatory.

We were also able to observe throughout the entire evaluative approach that the scope of an intervention can exceed the mere implementation of specific projects at the local and provincial levels. Thus, over and above the organizational changes implemented, we noted that the research/intervention approach itself contributed to certain changes in the study environments. Based on observations and participants' accounts gathered during the study, the researchers noticed that the research project instilled new dynamics within groups supporting the intervention, and the coordination group and the follow-up committee tasked, among other things, with tackling problems related to disrupted social relations. Indeed, participants viewed working in partnership and participation as the cornerstones of the intervention process overseen by the research team. At all the decision groups (the steering committee, the coordination committee and the intervention group), researchers, correction officers, and management were united around a single goal: improving relations between detention centre staff. The correctional milieu is generally administered in an autocratic rather than bureaucratic mode, in a spirit of confrontation and negotiation. These stakeholders set aside their traditional ways of doing things to adopt new perspectives of a shared problem. The approach proposed was based entirely on partnership, transparency and full participation. Mere participation in the intervention could give workers and their supervisors an opportunity to develop a shared understanding of the situation and improve communication (Kobayashi and Kawakami, 2008). This, incidentally, is the principle underlying quality circles: involving employees in the decision-making process, increasing their control over the adjustment of demand and the resources available to them, and improving communication between them and with their superiors (Aust and Ducki, 2004).

As indicated by the participants – and as observed in the field within the different project committees – this change of mentality did not take place overnight. It required a great deal of openness among the participants and the will to achieve something together, not necessarily easy in such an environment where, in the past, projects tended to generate more expectations than results and contributed to increasing confusion and resistance among officers regarding any type of intervention. Thus, recruiting participants to the project was crucial to the approach. Despite participants' good will, "reminders" were required when discussions digressed from the goals pursued by the groups or when the discussions diverged from the joint participative framework established at the onset. It was not simple to maintain the idea that all effort should be directed at a same goal, that of improving interpersonal relations between correctional staff members.

One of the challenges often raised by the intervention participants was, as stated by a number of them, sharing these participative values with all other employees and management in the establishments. Within the intervention groups, it was not always easy to introduce this new method of operation that was often at variance with the old, more autocratic way of doing things. Managerial staff turnover in the intervention groups made the task even more difficult. It should be noted that the spread of this innovative and participative mode depended quite often on the management philosophy of the directors in post. Interviews clearly showed that some were more inclined to deal with new ideas, but others less so. It should be noted that this challenge is great in an environment as structured as correctional services based on military tradition.

The arrival of the team of researchers and the establishment of intervention groups generated a certain level of enthusiasm in the three intervention centres and at the provincial scale. Thus, other completely independent interventions were implemented or are planned in the centres based on the same principles as those established during the research project. This interest bears witness to a certain empowerment in the centres and the dynamic effect of the intervention project.

Finally, one last limitation to this specific type of intervention in work organizations in constant movement is that researchers have no control over action taken outside the study framework. Thus, in addition to interventions undertaken within participant groups in the three intervention centres and provincial interventions that had an impact on all centres, other activities were initiated in control centres, but also, in some cases, in the intervention centres. These initiatives may have had an impact on the evaluation of the process and the effects of the intervention, making it difficult to distinguish precisely the results of initiatives implemented in the study. Likewise, in addition to interventions carried out at the centres, other events over which the researchers had no control might have contributed favourably or unfavourably to action undertaken. For example, the suicide of an employee in the workplace during the project could have modified the desired impact of certain interventions. The same applies to decisions, for example, by authorities that potentially and on occasion may mitigate efforts directed at implementing certain projects. Another example is the major increase in the number of inmates and the complex nature of the inmates that eluded the research process and had an impact on the overtime to be worked and possible effects on health indicators measured in the study.

Conclusions

Optimizing the psychosocial work environment creates winning conditions for workers' health and well-being. Even if the adverse psychosocial environment conditions and solutions identified in this study are specific to the correctional sector, the intervention process used (participative problem-solving) appears highly exportable to other organizations. Therefore, recommendations will potentially be useful for other sectors, organizations, and workers exposed to psychosocial risk factors at work. The methodological strengths of this research are worth mentioning because while avoiding the limits of prior research, they favour the generalization of the process outside of the correctional sector. The intervention research was based on sound theoretical models insuring a choice of targets and solutions based on psychosocial work factors known to have an impact on workers' health. It used a participative process which involves increased dialogue between workers' and employers' representatives and a better understanding and acceptation of their respective roles and responsibilities. A quasi-experimental design was used including a comparable control group. Indeed, in this case of a lesser than expected impact of the intervention in the intervention group because general conditions in the PSD deteriorated during the study period (budget cuts, higher number of inmates, etc.), the control group nevertheless presented a worse overall condition. The pre-intervention assessment, in addition to supplying the baseline information to assess the effectiveness of the intervention, allowed the identification of at-risk groups of workers and provided evidence for the need for the intervention. A post and a follow-up intervention assessment permitted it to go beyond a potential Hawthorne effect and also gave an opportunity to measure the longevity of changes. The high number of subjects in the research provided analytical power. Finally, the use of validated instruments insured the validity of the results and conclusions. Most importantly, the documentation and report of the three phases of the research address a critical need identified in the reviewed literature.

Acknowledgements

This research was funded by the Fonds québécois de recherche sur la société et la culture (FQRSC), the Social Science and Humanities Research Council of Canada, the Institut de recherche en santé et sécurité du travail du Québec (IRSST), the Public Security Department (PSD) and the Syndicat des agents en services correctionnels du Québec (SPASCQ). We wish to acknowledge the contribution of the members of the intervention groups, the advisory committee, the coordination committee, and all the correctional officers and their supervisors. Our thanks go also to all those who have contributed to the study at a specific time: Martin April, Julie Gagnon, Éric Lavigne, and finally to Michel Gaudet, Stéphanie Camden and Myrto Mondor for the statistical programming and analyses.

References

Armstrong, G. S., & Griffin, M. L. (2004). Does the job matter? Comparing correlates of stress among treatment and correctional staff in prisons. *Journal of Criminal Justice, 32*, 577–592.

Auerbach, S., Wuick, B., & Pegg, P. (2003). General job stress and job-specific stress in juvenile correctional officers. *Journal of Criminal Justice, 31*, 25–36.

Aust, B., & Ducki, A. (2004). Comprehensive health promotion interventions at the workplace: Experiences with health circles in Germany. *Journal of Occupational Health Psychology, 9*(3), 258–270.

Aust, B., Rugulies, R., Flinken, A., & Jensen, C. (2010). When workplace interventions lead to negative effects: Learning from failures. *Scandinavian Journal of Public Health, 38*(Suppl 3), 106–119.

Bambra, C., Egan, M., Thomas, S., Petticrew, M., & Whitehead, M. (2007). The psychosocial and health effects of workplace reorganization 2: A systematic review of task restructuring interventions. *Journal of Epidemiology & Community Health, 61*(12), 1028–1037.

Bambra, C., Gibson, M., Sowden, A. J., Wright, K., Whitehead, M., & Petticrew, M. (2009). Working for health? Evidence from systematic reviews on the effects on health and health inequalities of organisational changes to the psychosocial work environment. *Preventive Medicine, 48*(5), 454–461.

Bellerose, C., Lavallée, C., Chénard, L., & Levasseur, M. (1995). *Et la santé, ça va en 1992–93? Rapport de l'enquête sociale et de santé 1992–1993.* Montréal: Ministère de la Santé et des Services sociaux, Gouvernement du Québec.

Béthoux, É. (2000). La prison: recherches actuelles en sociologie. (Note critique.) *Terrains & Travaux, 1*, 71–89.

Blanchet, A., & Gotman, A. (1992). *L'enquête et ses méthodes: l'entretien.* Paris: Nathan.

Bond, F. W., & Bunce, D. (2001). Job control mediates change in a work reorganization intervention for stress reduction. *Journal of Occupational Health Psychology, 6*(4), 290–302.

Bourbonnais, R., Jauvin, N., Dussault, J., & Vézina, M. (2007). Psychosocial work environment, interpersonal violence at work and mental health among correctional officers. *International Journal of Law and Psychiatry, 30*, 355–368.

Bourbonnais, R., Malenfant, R., Vézina, M., Jauvin, N., & Brisson, I. (2005). Les caractéristiques du travail et la santé des agents en services de détention. *Revue d'Épidémiologie et de Santé Publique, 53*(2), 127–142.

Camp, S. D., & Lambert, E. (2006). The influence of organizational incentives on absenteeism: Sick-leave use among correctional workers. *Criminal Justice Policy Review*, *17*, 144–172.

Cheng, E. M., Daly, J., Hancock, K. M., Bidewell, J. W., Johnson, A., Lambert, V. A., Lambert, C. E. (2004). Workplace stressors, ways of coping and demographic characteristics as predictors of physical and mental health of Japanese hospital nurses. *International Journal of Nursing Studies*, *41*, 85–97.

Commission des normes du travail. See article 81.18 of Loi sur les normes du travail et les suivants, 1 June 2004.

Dahl-Jørgensen, C., & Saksvik, P. O. (2005). The impact of two organizational interventions on the health of service sector workers. *International Journal of Health Services*, *35*(3), 529–549.

Daveluy, C., Pica, L., Audet, N., Courtemanche, R., Lapointe, Coté, L., & Beaulne, J. (2000). *Enquête sociale et de santé 1998. Social and health survey 1998*. Québec: Institut de la statistique du Québec.

DeJoy, D. J., Wilson, M. G., Vandenberg, R. J., McGrath-Higgins, A. L., & Briffin-Blake, C. S. (2010). Assessing the impact of healthy work organization intervention. *Journal of Occupational and Organizational Psychology*, *83*, 139–165.

Egan, M., Bambra, C., Thomas, S., Petticrew, M., Whitehead, M., & Thomson, H. (2007). The psychosocial and health effects of workplace reorganization.1: A systematic review of organisational-level interventions that aim to increase employee control. *Journal of Epidemiology & Community Health*, *61*(11), 945–954.

European Agency for Safety and Health at Work. (2002). *How to tackle psychosocial issues and reduce work-related stress*. Luxembourg: Office for Official Publications of the European Communities.

Everitt, B., & Fidell, L. (2001). *Analysing medical data using S-Plus*. New York: Springer-Verlag.

Finn, P. (1998). Correctional officer stress: A cause for concern and additional help. *Federal Probation*, *62*(2).

Ghaddar, A., Mateo, I., & Sanchez, P. (2008). Occupational stress and mental health among correctional officers: A cross-sectional study. *Journal of Occupational Health*, *50*(1), 92–98.

Golberg, P. (1994). Santé et conditions de travail des personnels de l'administration pénitentiaire. *Droit et société*, *28*, 649–654.

Goldenhar, L. M., LaMontagne, A. D., Katz, T., Heaney, C., & Landsbergis, P. (2001). The intervention research process in occupational safety and health: an overview from the National Occupational Research Agenda Intervention Effectiveness Research team. *Journal of Occupational and Environmental Medicine*, *43*(7), 616–622.

Government of Quebec (2004). 'Psychological harassment at work – Commission des normes du travail du Québec" '. Retrieved November 10, 2011 from: www.cnt.gouv.qc.ca/en/in-case-of/psychological-harassment-at-work/index.html

Griffin, M., & Hepburn, J. R. (2005). Side bets and reciprocity as determinants of organizational commitment among correctional officers. *Journal of Criminal Justice*, *33*(6), 611–625.

Griffin, M. L., Hogan, N. L., Lambert, E. G., Tucker-Gail, K. A., & Baker, D. N. (2010). Job involvement, job stress, job satisfaction, and organizational commitment and the burnout of correctional staff. *Criminal justice and behavior*, *37*(2), 239–255.

Hunt, S., McKenna, S. P., McEwen, J., Backette, E. M., Williams, J., & Papp, E. (1980). A quantitative approach to perceived health status: a validation study. *Journal of Epidemiology and Community Health*, *34*(4), 281–286.

Ilfeld, F. W. (1976). Further validation of a psychiatric symptom index in a normal population. *Psychological Reports*, *39*, 1215.

Institut de la statistique du Québec. (2000). *Enquête sociale et de santé 1998*. Québec: Gouvernement du Québec.

Jauvin, N., Vézina, M., Bourbonnais, R., & Dussault, J. (2006). Violence interpersonnelle en milieu de travail: Une analyse du phénomène en milieu correctionnel québécois. *PISTES*, *8*(3).

Jauvin, N., Vézina, M., Bourbonnais, R., & Dussault, J. (2008). Au cœur de l'administration de la peine: Les surveillants et leur propre souffrance. Conference << Le pénal aujourd'hui: pérennité ou mutation >> www.erudit.org/livre/#Actes.

Johnson, J. V., & Hall, E. M. (1988). Job strain, workplace social support, and cardiovascular disease: A cross-sectional study of a random sample of the Swedish working population. *American Journal of Public Health*, *78*(10), 1336–1342.

Johnson, S., Cooper, C., Cartwright, S., Donald, I., Taylor, P., Millet, C. (2005). The experience of work related stress across occupations. *Journal of Managerial Psychology*, *20*(2), 178–187.

Karasek, R., & Theorell, T. (1990). *Healthy work: Stress, productivity and the reconstruction of working life*. New York: Basic Books.

Karasek, R. (1985). *Job content questionnaire and user's guide*. Los Angeles, CA: Department of Industrial and System Engineering, University of Southern California.

Keinan, G., & Malach-Pines, A. (2007). Stress and burnout among prison personnel: Sources, outcomes, and intervention strategies. *Criminal Justice and Behavior*, *34*, 380–398.

Kobayashi, Y., Kaneyoshi, A., Yokota, A., & Kawakami, N. (2008). Effects of a worker participatory program for improving work environments on job stressors and mental health among workers: A controlled trial. *Journal of Occupational Health*, *50*(6), 455–470.

Kompier, M., & Kristensen, T. S. (2001). Organizational work stress interventions in a theoretical, methodological and practical context. In A. Véga (Ed.), *Stress in the workplace: Past, present and future* (164–190). London & Philadelphia: Whurr Publishers.

Kompier, M., Aust, B., van den Berg, A. M., & Siegrist, J. (2000). Stress prevention in bus drivers: Evaluation of 13 natural experiments. *Journal of Occupational Health Psychology*, *5*(1), 11–31.

Kristensen, T. S. (1999) Challenges for research and prevention in relation to work and cardiovascular diseases. *Scandinavian Journal of Work Environment & Health* 25(6, special issue), 550–557.

Kristensen, T. S. (2005). Intervention studies in occupational epidemiology. *Occupational and Environmental Medicine*, *62*(3), 205–210.

Kristensen, T. S., & Borritz, M. (1999). *The Copenhagen Burnout Inventory (CBI): A new questionnaire for measuring burnout*. Copenhagen, Denmark: National Institute of Occupational Health.

Lambert, E., & Paoline, E. A. (2010). Take this job and shove it: An exploratory study of turnover intent among jail staff. *Journal of Criminal Justice*, *38*(2), 139–148.

Lambert, E., Hogan, N. L., & Jiang, S. (2010). A preliminary examination of the relationship between organisational structure and emotional burnout among correctional staff. *The Howard Journal of Criminal Justice*, *49*(2), 125–146.

Lambert, V. A., Lambert, C. E., & Ito, M. (2004). Workplace stressors, ways of coping and demographic characteristics as predictors of physical and mental health of Japanese hospital nurses. *International Journal of Nursing Studies*, *41*, 85–97.

Larocque, B., Brisson, C., & Blanchette, C. (1998). Cohérence interne, validité factorielle et validité discriminante de la traduction française des échelles de demande psychologique et de latitude décisionnelle du Job Content Questionnaire de Karasek. *Revue d'Épidémiologie et de Santé Publique*, *96*, 371–381.

Lavigne, É., & Bourbonnais, R. (2010). Psychosocial work environment, interpersonal violence at work and psychotropic drug use among correctional officers. *International Journal of Law and psychiatry*, *33*(2), 122–129.

Léveillé, C., Blais, M., & Hess, U. (2003, March). *A Test of the Professional Autonomy Model in Correctional Facilities*. Paper presented at the Challenges in a Changing Workplace – The Fifth Interdisciplinary Conference on Occupational Stress and Health, Toronto.

Liebling, A. (2006). Prisons in transition. *International Journal of Law and psychiatry*, *29*, 422–430.

McHorney, C. A., Ware, J. E., & Raczek, A. E. (1993). The MOS 36-item short-form health survey (SF-36). II: Psychometric and clinical tests of validity in measuring physical and mental health constructs. *Medical Care*, *31*(3), 247–263.

Moon, B., & Maxwell, S. R. (2004). The sources and consequences of corrections officers' stress: A South Korean example. *Journal of Criminal Justice*, *32*, 359–370.

Niedhammer, I., Siegrist, J., Landre, M. F., Goldberg, M., & Leclerc, A. (2000). Étude des qualités psychométriques de la version française du modèle du Déséquilibre Efforts/ Récompenses. *Revue d'Épidémiologie et de Santé Publique*, *48*, 419–437.

Owen, S. S. (2006). Occupational stress among correctional supervisors. *Prison Journal*, *86*(2), 164–181.

Rothman, K. J., & Greenland, S. (1998). *Modern epidemiology* (2nd ed.). Philadelphia, PA: Lippincott Williams & Wilkins.

Schaufeli, W. B., & Peeters, M. C. W. (2000). Job stress and burnout among correctional officers: a literature review. *International Journal of Stress Management*, *7*(1), 19–49.

Semmer, N. K. (2006). Job stress interventions and the organization of work. *Scandinavian Journal of Work, Environment & Health*, *32*(6), 515–527.

Siegrist, J. (2001). The model of effort-reward imbalance. Retrieved March 23, 2001.

Siegrist, J., Peter, R., Junge, A., Cremer, P., & Seidel, D. (1990). Low status control, high effort at work and ischemic heart disease: Prospective evidence from blue-collar men. *Social Science & Medicine*, *31*(10), 1127–1134.

Stansfeld, S., & Candy, B. (2006). Psychosocial work environment and mental health: A meta-analytic review. *Scandinavian Journal of Work, Environment & Health*, *32*(6), 443–462.

Tabachnik, B. G., & Fidell, L. (1996). *Using multivariate statistics*. Pearson Education.

Tartaglini, A. J., & Safran, D. A. (1997). A topography of psychiatric disorders among correction officers. *Journal of Occupational and Environmental Medicine*, *39*, 569–573.

Triplett, R., Mullings, J. L., & Scarborough, K. (1996). Work related stress and coping among correctional officers: Implications from organizational literature. *Journal of Criminal Justice*, *24*, 291–308.

van der Doef, M., & Maes, S. (1998). The job demand-control (-support) model and physical health outcomes: A review of the strain and buffer hypotheses. *Psychology and Health*, *13*, 909–936.

van der Doef, M., & Maes, S. (1999). The job demand-control (-support) model and psychological well-being: A review of 20 years of empirical research. *Work & Stress*, *13*(2), 87–114.

van der Hek, H., & Plomp, H. N. (1997). Occupational stress management programmes: A practical overview of published effect studies. *Occupational Medicine, 47*(3), 133–141.

Vézina, M., Jauvin, N., Dussault, J., & Bourbonnais, R. (2006). L'éclairage de la psychodynamique du travail pour comprendre la souffrance des agents de la paix en services correctionnels. In Institut de psychodynamique du travail du Québec (Ed.) *Espace de réflexion, espace d'action en santé mentale au travail: Enquêtes en psychodynamique du travail au Québec* (143–167), Québec, Canada: Les Presses de l'Université Laval.

11 The vital role of line managers in managing psychosocial risks

Rachel Lewis, Joanna Yarker, and Emma Donaldson-Feilder

Introduction

This chapter provides an introduction into the role of line managers in managing risks to employee psychosocial health. It looks at how line managers (or immediate supervisors, used interchangeably) can influence the success of interventions for psychosocial issues through their impact on process and contextual issues, as well as the direct role of line managers in causing or reducing psychosocial risks. It also considers interventions that can reduce psychosocial risks by developing appropriate line manager behaviours and some key process and contextual factors that influence the success of such interventions. It aims to provide a balance between academic research and practical guidance to help professionals engage managers in managing psychosocial risks. In addition to an academic review of the role that managers play in managing psychosocial risks, the authors draw from a range of consultancy experiences and from a five year research programme working with managers to prevent and reduce stress. The programme of research was sponsored by the Health & Safety Executive (HSE), the Chartered Institute of Personnel Development (CIPD), Investors in People (IiP), and a consortium of participating organizations; it involved over 30 UK organizations and 300 managers. It identified not only what manager behaviours are important in this context, but also ways of helping managers change their behaviour and the process issues that can help and hinder them in doing so.

Why managers are vital to managing psychosocial risk

It is widely recognized (albeit often implicitly) that managers are in a position to affect most, if not all, aspects of work design (Offerman and Hellman, 1996). Managers typically 'hold the key' to work re-design initiatives and organizational development/change initiatives more generally (e.g. Parker and Williams, 2001; Parker, Chmiel, and Wall, 1997; Saksvik et al., 2002). Not only may they be responsible for the implementation, or allowing and enabling the implementation, of interventions within their team, but their support of, and buy-in to, the intervention is likely to be important for its success (Nielsen and Randall, 2009). Without the support and approval of both line managers and senior managers, participation and support from employees will be harder to gain, and the consequence will be that the intervention will be more likely to fail (French and Bell, 1995).

Managers also have a key role to play in identifying pressure and stress in employees within everyday people management, and as part of team meetings, one-to-one meetings, and appraisals and performance reviews with employees. If stress is identified, managers are also increasingly likely to be involved in designing and implementing solutions to alleviate that stress such as risk assessments and staff surveys (Thomson, Rick, and Neathey, 2004).

Further, there is an increasing body of literature causally linking management behaviour with stress-related outcomes in employees. Tepper (2000), for instance found the supervisor–employee relationship to be the most commonly reported source of stress in the workplace. Similarly, Hogan, Curphy and Hogan (1994) found that for 60 to 75 per cent of employees, the 'worst or most stressful aspect of their job is their immediate supervisor' (p. 494). In the 2008 CIPD Absence Management report (CIPD, 2008), stress was found to be most common cause of long-term absence in non-manual workers (p. 21), and the top three causes of stress were workload, management style, and relationships at work (p. 26). Managers are able to impact upon stress-related outcomes in employees in a number of ways. The most direct, or obvious way is by their behaviour, with managers causing (or conversely preventing) stress in their employees by their behaviour towards them (e.g. Tepper, 2000; Hogan et al., 1994). Research on the impact of manager behaviour on employee strain and well-being is explored more fully below.

Managers' behaviour is also likely to impact upon the presence or absence of psychosocial hazards in the working environment of employees (e.g. van Dierendonck et al., 2004; Cherniss, 1995). For instance, a manager may be able to 'gate-keep' or prevent stressors, such as additional workload and deadlines, being passed on to employees. Further, the stressor–strain relationship may be affected by the way that a manager behaves, for example, supervisor support can reduce or exacerbate the impact of the work environment on employee outcomes such as absenteeism (e.g. Nielsen et al., 2006). As the line manager is the person directly responsible for much of the day-to-day communication with the employee, it is logical to assume that the manager can also influence the way the employee perceives their working environment, and consequently the impact of that environment upon the employee. Literature demonstrating both direct and indirect effects of manager support (e.g. Rooney and Gottlieb, 2007), transformational leadership (e.g. Nielsen et al., 2008), and leader-member exchange (e.g. Harris and Kacmar, 2005) would support this assumption. Therefore, it is important to consider both the direct impact of manager behaviour on employee outcomes, and also the role of the manager in affecting, or moderating, the relationship between psychosocial hazards and employee well-being.

Thus, line managers have a vital role to play in organizational psychosocial risk management not only from the perspective of the context they set for such interventions, the impact they have on the implementation process, and their role in interventions for which they are directly responsible; they also have a very direct role in affecting, or determining, the psychosocial environment and levels of well-being of their employees. This presents the possibility of intervening within organizations to reduce psychosocial risks by changing manager behaviour.

Management behaviours to reduce psychosocial risks

With the realisation that managers are vital to the effective management of psychosocial risks, there has been a growing interest in the specific manager behaviours that foster or reduce well-being. This research can be seen to fall into five clusters placing a focus on specific leadership styles or theories: supportive behaviours, task-relationship focused behaviours, transactional and transformational leadership behaviours, negative leadership behaviours, and other types of supervisory indices. Together, this large body of research not only provides a powerful argument for placing a focus on the line manager when considering psychosocial interventions, but provides a steer for the ways in which managers can make a difference.

Manager supportive behaviours

Much of the research investigating the link between manager behaviour and employee well-being has been focused upon the level of support provided by managers. There have been numerous studies documenting the positive consequences of manager support, relating higher levels of support to employee well-being and job satisfaction (e.g. Offerman and Hellman, 1996; Amick and Celantano, 1991; Baker, Israel, and Schurman, 1996; Moyle and Parkes, 1999), reductions in stress and burnout (e.g. Lee and Ashforth, 1996; Schaufeli and Enzmann, 1998), lower turnover intentions (e.g. Thomas and Ganster, 1995) and productivity and performance (e.g. Baruch-Feldman et al., 2002; Thomas and Ganster, 1995). The majority of research has been cross-sectional, and demonstrates strongly the positive relationship between manager support and employee outcomes. Of the few studies that have been longitudinal however, evidence is limited and inconclusive (Van Dierendonck et al., 2004). Although the majority of studies point to the direct effects of manager support on employee well-being, a small number have explored the moderating effect of manager support in the stressor–strain relationship (e.g. Moyle and Parkes, 1999; Dekker and Schaufeli, 1995; Stephens and Long, 2000), though with mixed results. Therefore, although there is strong support for the cross sectional relationship between manager support and employee outcomes, evidence for the long term or the buffering impact of manager support remains inconclusive.

Research has been criticized for the use of generic, out-dated and positively focused measures of manager support that may not capture the full range of manager behaviours (e.g. Yarker et al., 2007; Rooney and Gottlieb, 2007). Studies which have explored the types of behaviours that underpin manager support (Beehr, King and King, 1990; Fenalson and Beehr, 1994) have found that communication about positive aspects of the job and about topics unrelated to their job (such as people's interests outside of work) have been found to be particularly important components of support. The body of research, although clearly identifying that a lack of manager support is associated with negative employee outcomes, has only provided limited information on the specific supervisory behaviours that play a role in determining employee well-being outcomes.

Whilst manager behaviours that impact upon perceived support are clearly important, it is unlikely that all manager behaviours associated with employee outcomes are those that fall under the definition of manager support. Parallel to the support literature, there has been an increasing body of literature that draws from traditional leadership models. The majority of literature linking leadership and health refers to two theoretical models: that of behavioural models of leadership (or task and relationship focused behaviours), and the transformational and transactional leadership models.

Task and relationship focused behaviour

This model (Stogdill, 1950) identifies relationship-based (or consideration-based) leader behaviours, including behaviours such as supporting employees, showing respect for employees' ideas, increasing cohesiveness, developing and mentoring, looking out for employees' welfare, managing conflict, and team building (e.g. Arnold, Cooper and Robertson, 1995; Levy, 2003; Nyberg, Bernin, and Theorell, 2005; Seltzer and Numerof, 1988; Sosik and Godshalk, 2000). It contrasts these with task-based (or initiating structure-based) leader behaviours, which include behaviours such as planning and organizing, assigning people to tasks, communicating information, monitoring performance, defining and solving work-related problems, and clarifying roles and objectives.

Although the majority of studies on the behavioural approach have focused on outcomes of performance and productivity (e.g. Judge and Piccolo, 2004; Keller, 2006), a number of studies have investigated the relations between these two distinct types of supervisory behaviour and employee well-being (e.g. Duxbury et al., 1984; Seltzer and Numerof, 1988; Sheridan and Vredenburgh, 1978). Overall, this body of research (e.g. Duxbury et al., 1984; Landweerd and Boumans, 1994; Kuoppala et al., 2008) suggests that consideration/relationship behaviours have a positive impact on employee well-being but that the impact of leaders' initiating structure/task behaviours on employees' health may be more complex. To elaborate, research suggests that high levels of initiating structure behaviours can have a detrimental effect on employee well-being, but that this negative impact may be reduced if the same supervisors also exhibit a range of more consideration-based behaviours. When it comes to implementing organizational interventions for psychosocial issues, it seems likely that both structure/task behaviours and consideration behaviours would be needed. For example, a task/structure focus would be required to project manage and drive forward implementation, but consideration behaviours would be needed to support employee engagement and consultation through the implementation process.

Transformational and transactional leadership behaviours

Transformational leadership (Bass, 1985, 1998) involves generating enthusiasm for a 'vision', a high level of individualized consideration, creating opportunities for employees' development, setting high expectations for performance, and acting as a role model to gain the respect, admiration and trust of employees (e.g. Bass, 1999; Bass and Avolio, 1994; Rubin, Munz and Bommer, 2005). Unlike the original conceptualisation of transformational leadership by Burns (1978), Bass (1985, 1998) argued that it was not enough for leaders to be only transformational, but that they also needed to display transactional behaviours. Transactional leadership involves a more straightforward exchange between a leader and a direct report, whereby the employee is suitably rewarded for good performance (referred to as contingent reward behaviour). Therefore leaders who are more transactional than transformational are likely to explain to employees what is expected of them, and the likely outcomes of meeting those expectations, without necessarily explaining how they can personally develop and grow within the role and organization (Levy, 2003). Bass also specified a third category of leader behaviour, called laissez-faire leadership. This, or non-transactional leadership, is viewed as the least effective (and could even be damaging) as it is characterized by a passive leadership style, an avoidance of action, a lack of feedback and communication, and a general indifference to employee performance (Sosik and Godshalk, 2000).

A handful of studies have examined the positive effects of transformational leadership and negative effects of laissez-faire leadership on employee outcomes, such as retention (McDaniel and Wolf, 1992), empowerment and self-efficacy (e.g. Brossoit, 2001; Hetland, Sandal and Johnsen, 2007), meaningfulness (Arnold et al., 2007), optimism and happiness (Bono et al., 2007), conflict (e.g. Skogstad et al., 2007; Hauge, Skogstad, and Einarsen, 2007) and a variety of strain outcomes (e.g. Sosik and Godshalk, 2000; Alimo-Metcalfe and Alban-Metcalfe, 2001; Hetland et al., 2007; Nielsen et al., 2008; Kuoppala et al., 2008).

Evidence of the negative impact of laissez-faire leadership is provided largely by research into employee experiences of bullying. A study by Skogstad et al. (2007) found that laissez-faire leadership was positively correlated with role conflict and role ambiguity in employees, and was also related to increased numbers of employee conflicts. Further, through path modelling it was found that laissez-faire leadership was directly associated with employees' experience of bullying. In a related study, Hauge et al. (2007) found a link between laissez-faire leadership and bullying, and that, where immediate supervisors avoided intervening in and managing the stressful situation, bullying was more likely to occur.

Negative leadership behaviours

Relatively recent contributions to the leadership behaviour research suggest that supervisors can also perform behaviours that are negative, characterising these behaviours as bullying (Rayner and McIvor, 2006), undermining (Duffy, Ganster, and Pagon, 2002), 'health endangering' (Kile, 2000), tyrannical (Einarsen,

Aasland, and Skogstad, 2007), destructive (Einarsen, Aasland, and Skogstad, 2007), hostile (Schaubroek et al., 2007) and abusive (Tepper, 2000). Tepper (2000) argues that the majority of leadership research has focused upon constructive, effective and successful leadership rather than addressing negative leader behaviours and their effects. The growing body of research into negative leadership behaviour demonstrates that the behaviours involved, although negatively correlated with positive manager behaviours, appear to form separate factors in exploratory factor analyses, and demonstrate independent effects on well-being (Yagil, 2006). Further, research by Duffy et al. (2002), demonstrated that managers who combine both positive and negative behaviours produce more deleterious outcomes compared to negative supervision alone, perhaps due to the lack of consistency demonstrated.

Laissez-faire leadership is conceptualized as a passive leadership behaviour, where the leader effectively abdicates their responsibilities, whereas the negative leadership behaviour studies tend to focus upon active forms of leadership (where the leader actively belittles or abuses the employee) and don't explicitly define passive forms of leadership. That said, evidence is gathering (e.g. Hauge, Skogstad and Einarsen, 2007; Hetland, Sandal and Johnsen, 2007; Skogstad et al., 2007) of the deleterious effects of passive negative leader behaviour on employee outcomes.

Perhaps the construct with the strongest body of research to date is the concept of abusive supervision. Abusive supervision is defined as 'the sustained display of hostile verbal and non-verbal behaviours, excluding physical contact' (Tepper, 2000; p. 178) and may include behaviours such as using derogatory names, engaging in explosive outbursts, intimidating by use of threats and job loss, withholding needed information, aggressive eye contact, silent treatment and humiliating, ridiculing or belittling employees in front of others (Tepper, 2000). It involves prolonged emotional and psychological mistreatment (Harvey et al. 2007) rather than isolated instances of the behaviours. Initial research has suggested that abusive supervision affects an estimated 13.6 per cent (Harvey et al., 2007) or between 10 and 16 per cent (Tepper et al., 2004) of US workers. Although the incidents of abusive supervision are relatively infrequent, a small but growing body of research demonstrates links between abusive supervision and job dissatisfaction (e.g. Ashforth, 1994; Tepper, 2000; Tepper, 2007), intention to leave (e.g. Schat, Frone and Kelloway, 2006; Tepper, 2000), reduced organizational commitment (Ashforth, 1994; Tepper, 2000), employee deviance behaviours (Tepper et al., 2009) and a range of psychological outcomes such as anxiety (Harris and Kacmar, 2005; Tepper, 2000), depression (Tepper, 2000), burnout (Tepper, 2000; Yagil, 2006) and somatic health complaints (Duffy et al., 2002).

Consideration of negative manager behaviours presents a potentially fruitful line of research regarding organizational interventions, both in terms of the impact of negative manager behaviour on the effectiveness of organizational interventions to reduce psychosocial risk and in terms of conducting interventions to reduce negative manager behaviour. For example, where managers are made responsible for implementing interventions to reduce psychosocial risks within their team, this assumes that the managers are not a source of psychosocial risk themselves (Biron, Gatrell and Cooper, 2010); if these managers show negative manager behaviours, this may reduce both the degree to which the intervention is implemented and its ability to tackle the real problem (manager behaviour). If a key psychosocial risk within an organization is managers showing negative, bullying or abusive behaviours, then, in order to reduce this risk, it may be appropriate to implement organizational interventions to help managers behave differently: for example, using learning and development, selection, or performance appraisal to enhance managers' people management skills.

Other supervisory behaviour indices

Some occupational stress researchers have recently noted the limitations of simply adopting prominent leadership theories and measures (e.g. Gilbreath, 2004; Gilbreath and Benson, 2004; Nyberg, Bernin and Theorell, 2005; Offerman and Hellmann, 1996). As a result, a number of researchers have developed and/or employed other specific supervisor scales that perhaps more clearly reflect the wider research into work design and occupational health. Offerman and Hellmann (1996) explored the relationships between leadership and employee strain from the perspective of the managers, their bosses and their direct reports. The survey of management practices was used, which contained three factors of communication, leader control and delegation. Analyses revealed that high levels of delegation and communication and low levels of leader control predicted lower employee strain. The study also found that emotional support behaviour (approachability, team building, interest in growth and building trust) was related to lower levels of strain.

Gilbreath and Benson (2004) developed items for their supervisory behaviour scale via interviews with managers and employees in healthcare and retail organizations. The scale measured a range of supervisory behaviours relating to job control, communication, consideration, social support, group maintenance, organizing and looking out for employee well-being. Gilbreath and Benson, as predicted, found that these behaviours were significantly related to employees' mental health even after accounting for non-supervisory behaviour factors.

Van Dierendonck et al. (2004) administered a multidimensional leader behaviour scale and a measure of mental health to over 500 healthcare staff at four time points over a 14 month period. They found leadership behaviour and employee well-being were linked in a 'feedback' loop, in that more effective leader behaviour was related to higher employee well-being at one time point, and that higher levels of employee well-being led to more favourable perceptions of leader behaviour at another time point. This suggests that if it were possible to influence managers' behaviour (for example, through a learning and development intervention, see below) so that they were showing more effective leadership, this could potentially lead to a virtuous cycle of improvement in well-being and perceptions of leadership behaviour.

In summary, the research considered here leads to an overall conclusion that line managers play a vital role in managing psychosocial risks and that they do this through a wide array of positive and negative behaviours. This means there is great potential for organizational interventions that shift manager behaviour to be used as a mechanism for reducing psychosocial risks. Indeed, a recent review by Kelloway and Barling (2010) concludes that there is now sufficient evidence to allow 'the unambiguous conclusion that organizational leadership is related to, and predictive of, health and safety-relevant outcomes in employees' (p. 275) and recommends that leadership development interventions should be a main target in interventions for reducing psychosocial risks.

Introducing a behaviour-based competency approach

Although research has demonstrated numerous behaviours to be empirically linked to employee well-being and strain, a list of the management behaviours specific to the management of stress and well-being had not been developed. This presents challenges for both research and practice. First, the range of leadership measures used within the research mean that studies are tapping into different manager behaviours and therefore the line managers' role is not consistently measured. Second, while traditional leadership theories capture what it is to be a good leader and research clearly shows that good leadership is associated with a range of well-being indices, it is not clear whether these theories encompass all that is relevant to managing psychosocial issues – might there be something missing? Finally, without a clearly defined set of the management behaviours that are important for reducing psychosocial risk, it is difficult to design interventions to target the relevant behaviours. With many organizations using behaviour-based competency approaches to underpin a range of people management processes, there was an opportunity to use this approach to explore the role of management competencies for preventing and reducing stress at work. Further, taking a competency approach allows clear definition of relevant behaviours that can be the focus for interventions such as learning and development, selection and performance appraisal.

Introduction to the Management Competencies for Preventing and Reducing Stress research programme

The Management Competencies for Preventing and Reducing Stress (MCPARS) research programme (Yarker et al., 2007; Yarker, Donaldson-Feilder and Lewis, 2008; Donaldson-Feilder, Lewis and Yarker, 2009) was a three-phase programme sponsored by the HSE in Great Britain, CIPD, and a consortium of participating organizations. A behaviour-based competency approach was adopted for a number of reasons:

i   to put stress management into an organizationally-friendly language,
ii to provide clear guidelines and parameters to managers, and

iii   to enable the embedding of the behaviours into interventions and into existing organizational policies and practices.

Over the three phases, this research programme aimed to: identify the behaviours required by managers to prevent and reduce stress at work (Phase One); ensure the resultant framework was reliable, valid and, importantly, usable by managers, human resource (HR), health and safety (H&S) and occupational health (OH) professionals (Phase Two); and evaluate learning and development interventions to help line managers develop the relevant behaviours (Phase Three).

Phase One

The research in Phase One took a qualitative approach, using critical incident-based interviews to elicit the behaviours associated with both the effective and ineffective management of stress in employees. In total, 216 employees and 166 line managers were interviewed and completed written exercises, and 54 HR practitioners were involved in focus groups and also completed written exercises. Participants were drawn from the five sectors that had the highest prevalence of work-related stress in the UK at that time (Finance, Education, Healthcare, Local Government, and Central Government; see also chapter by Mackay et al., this volume). Interviews were transcribed and content analysis was used to extract themes and develop a coding framework. Behaviours were also elicited from the focus groups and completed written exercises. Content analysis was again used to fit this data into the coding framework. This allowed triangulation of the findings and a preliminary validation of the framework. The final framework included 19 competencies, all of which (with the exception of one) included both positive and negative behavioural indicators.

There were not found to be any significant differences in behaviours between the five sectors, suggesting that the framework would be relevant across a range of organizations. A mapping exercise was also conducted between the MCPARS framework and other general management frameworks (including both sector specific frameworks and the most popular leadership frameworks such as the Transformational Leadership Questionnaire (TLQ; Alimo-Metcalfe and Alban-Metcalfe, 2001), Multifactor Leadership Questionnaire (MLQ; Bass and Avolio, 1997) and Leader Behaviour Description Questionnaire (LBDQ; Stogdill, 1963)). In this comparative exercise it was found that all 19 competencies were included in at least one of the frameworks, but no framework covered all the competencies; therefore suggesting existing management and leadership frameworks may not tap into all the behaviours relevant to employee well-being.

Phase Two

The second phase of the research aimed to revise the framework using evidence from three sources (qualitative evidence, reliability analysis and exploratory factor analysis and literature review), design a stress management competency indicator tool and examine the usability of the framework.

In order to revise the current MCPARS framework and construct a stress management competency indicator tool, a combined quantitative and qualitative approach was taken. The Phase One data was re-probed to extract specific observable behaviours to use as questions for an indicator tool. These were then tested both qualitatively with stakeholders and experts (n = 21) and quantitatively with a snowball sample of employees (n = 292). Following reliability analyses, the resultant questionnaire was used as an upward feedback measure in 22 organizations. Participants included 152 managers and 656 direct reports. The direct report data was then subjected to exploratory factor analysis and a four factor solution was revealed. Workshops were held with stress experts (n = 38) to name the four factors and develop sub-clusters, resulting in three sub-clusters per competency (factor). Following final revisions, the questionnaire, or stress management competency indicator tool, consisted of 66 questions across four competencies. The final framework including each of the four competencies, the sub-competencies and descriptions of each sub-competency is shown in Table 11.1.

Table 11.1 Management competencies and sub-competencies

Competency	Sub-competency	Description of sub-competency
Respectful and responsible: Managing emotions and having integrity	Integrity Managing Emotions Considerate approach	Respectful and honest to employees Behaves consistently and calmly Thoughtful in managing others and delegating
Managing and communicating existing and future work	Proactive work management	Monitors and reviews existing work, allowing future prioritisation and planning
	Problem solving	Deals with problems promptly, rationally and responsibly
	Participative/empowering	Listens and consults with team, provides direction, autonomy and development opportunities to individuals
Reasoning/managing difficult situations	Managing conflict	Deals with conflicts fairly and promptly
	Use of organizational resources	Seeks advice when necessary from managers, HR and occupational health
	Taking responsibility for resolving issues	Supportive and responsible approach to issues
Managing the individual within the team	Personally accessible Sociable	Available to talk to personally Relaxed approach, such as socialising and using humour
	Empathetic engagement	Seeks to understand the individual in terms of their motivation, point of view and life outside work

For the usability testing, a qualitative approach was taken, including one-to-one interviews and workshops with 47 managers and six stakeholders from the same five sectors used in Phase One; and 38 stress experts (independent stress practitioners, HR, OH and H&S professionals). The interview and workshop data was transcribed and content analysis used to extract themes. The findings suggested that the framework was relevant and useful within organizations, in terms of both a management development tool and a stress management tool. It was suggested that the best format for the indicator tool was either as an upward feedback or a 360 degree feedback measure.

Intervention to help managers behave in ways that reduce psychosocial risks for their employees

Phase Three – learning and development intervention to help line managers change their behaviour

Although the first two phases of the MCPARS research identified those management behaviours important for the prevention and reduction of stress at work, there remained a need to establish whether line managers could be helped to increase their ability to prevent and reduce stress or, in other words, change their behaviour. The objectives of this final phase were therefore: to create a learning and development intervention to develop managers' competencies in managing stress in others (re-titled positive manager behaviour); to evaluate the effectiveness of the intervention; to refine the intervention as a result of data obtained; and to understand the process and contextual issues that impact on the effectiveness of this kind of intervention.

Participants were drawn from 16 UK organizations in the five sectors outlined above. The intervention group was made up of 58 managers and 209 employees; the control group 95 managers and 385 employees. In addition, two stakeholders from each organization participated in quarterly meetings to capture data on the process of implementing the intervention, and the barriers and facilitators in achieving manager behaviour change.

The intervention itself was made up of two elements:

- Upward feedback: Managers and their direct reports were asked to complete the stress management competency indicator tool (developed in Phase Two). Provided at least three direct reports responded to this questionnaire, the manager received a feedback report. This allowed the manager to compare their own responses with those of their direct reports for each question and each competency. This feedback was generated at two time points; time 1, which was approximately one week prior to attending the workshop; and time 2, which was three months after the workshop.
- Half-day workshop: The workshop aimed to introduce managers to the importance of positive manager behaviour, increase their self awareness of their own behaviour, and equip them with the tools to further enhance their

stress management competence. The workshop also provided the opportunity to explore their upward feedback further. The interactive design of the workshop combined individual reflection, case studies, vignettes, small group work, plenary discussions, debate and analysis.

In order to evaluate the interventions and explore process/contextual issues, along with the behaviour measurements at two time points, data was also gathered on manager reactions and learnings (at the end of the workshop and then three months later) and stakeholder reactions and learnings (evaluated through the quarterly discussions). The manager behaviour change was also compared to that of a control group. The control group was made up of managers and their direct reports who had received neither upward feedback nor workshop, but had self and direct report responses to the upward feedback questionnaire at two time points.

Manager behaviour change

The data was categorized into three groups, where managers' data was rated (by themselves and/or their direct reports) as effective, average or ineffective. The behaviour change data was analysed using paired sample t-tests to examine differences between the initial score (time 1) and the subsequent score (time 2) in each of the three groups across each of the four competencies. The intervention was found to have a different impact depending on the extent to which the manager already showed stress management competency (their effectiveness):

Ineffective managers: Managers who perceived themselves as ineffective at time 1 reported improvements in their behaviour whether in the intervention group or the control group. Direct report perceptions of these 'ineffective managers' suggest that their behaviour changed over time only if they were in the intervention group. They perceived managers in the control group to be largely unchanged over the three month period, whereas managers in the intervention group were seen to have significantly improved over time.

Average managers: Managers who perceived themselves as average at time 1 saw their behaviour as largely unchanged over time. Direct report perceptions again suggested that manager behaviour changed over time only where the manager was in the intervention group. Interestingly however, direct reports perceived managers in the control group to be largely unchanged over time, while direct reports perceived managers in the intervention group to have become significantly less effective over time.

Effective managers: Those managers who perceived themselves as effective at time one, and were in the control group perceived their behaviour to have become significantly less effective over time; those in the intervention group however saw their behaviour as largely unchanged. Direct reports viewed 'effective' managers, whether they had received the intervention or were in the control group, as becoming significantly less effective over time. These findings are summarized in Table 11.2, which has been adapted from Donaldson-Feilder, Lewis, and Yarker (2009).

Table 11.2 Changes in effectiveness for the three groups of managers

	Manager initially seen as 'ineffective'	Manager initially seen as 'average'	Manager initially seen as 'effective'
Manager self-perceptions of changes in management behaviour			
Control group	↑	–	↓
Intervention group	↑	–	–
Direct report perceptions of changes in their managers' management behaviour			
Control group	–	–	↓
Intervention group	↑	↓	↓

The results indicate therefore that the intervention was most effective in securing behaviour change over time for those managers who were initially perceived as 'ineffective' by their direct reports. The findings with regards to average and effective managers are interesting and there are a number of possible explanations: it could have been that completing the questionnaire the first time raised both manager and direct report awareness of the stress management behaviours and therefore focus upon them; both self and direct report expectations could have been raised from the process of responding; the absence of change may have felt frustrating to those direct reports who had invested in the completion of the questionnaire; and/or the intervention provided (feedback and workshop) might have been most suitable for 'ineffective' managers, whereas those who were already average or effective might have benefited more from a different approach (for example, a more advanced workshop or more specific intervention, perhaps coaching, tailored to their behavioural profile).

Manager perceptions of the intervention

Responses were received from 112 managers who attended the workshop (16 of which did not receive an upward feedback report). Whether or not they had received upward feedback, managers' initial responses suggested that the workshop was useful in enabling the exploration of positive manager behaviours and their own self awareness. Managers also felt that they had been equipped with the skills to go on and apply their learning in their roles. Receiving an upward feedback report was, however, found to be significant in terms of helping managers to understand their own behaviour: 13 per cent of managers who had not received feedback felt that their understanding was still 'poor' at the end of the workshop; compared to none that had received upward feedback.

Responses to a questionnaire asking about experiences of and learning during the intervention process were received from 40 managers three months after the workshop. The majority of managers reported that, in those three months, they had discussed the feedback with their manager and their direct reports. Thirty of these managers felt that they had been able to make changes as a result of the intervention. Of those that hadn't, explanations included team structure changes, a lack of necessity to change (if they had been found to be working effectively), a lack of time to consider development and confusion about what to focus on first.

Managers were also asked about how supportive their manager and the wider organization had been of the process. Although 72 per cent felt that their manager had been supportive and positive, only 56 per cent felt that the organization would be supportive of action taken, and only half felt that they had the resources needed to address the issues identified.

Organizational factors that can improve the success of the intervention and help managers show the relevant behaviours

During Phase Three, information was gathered from managers and stakeholders about the organizational factors that would affect success of interventions. Stakeholders were specifically asked to provide ideas on how the process could be improved to increase participation rates from managers and the user experience. The following points summarize the responses from stakeholders:

- *Create a steering group:* It was felt that having one project champion left the project vulnerable, but inclusion of a wider number of representatives from across the organization would improve visibility and buy-in to the process.
- *Present a united front:* Avoid the intervention being seen as a stand-alone project and involve as many stakeholder groups as possible (HR, H&S, OH, managers, senior managers and direct reports).
- *Integrate the project into existing organizational policies and practices:* The aim would be to link to an existing strategy on well-being or management, or to an existing management development programme. It could also be embedded into existing development programmes or induction processes.
- *Rename the research:* Not all organizations are receptive to the words 'stress' or 'well-being' and stakeholders felt it was important to consider the audience, renaming the intervention according to organizational culture, or perhaps to existing programmes or initiatives.
- *Choose participants strategically:* This may include focusing on organizational hot-spots or upon low-performing teams – addressing a need rather than using the intervention as a 'nice to have'.

In addition to the above, both managers and stakeholders felt the most important element to affect success of the intervention was support throughout, and after, the intervention process. Table 11.3 summarizes the suggestions for support both from stakeholders and managers.

It is noteworthy that, although both stakeholders and managers agreed that support following the intervention was key to embedding behaviour change, the research demonstrated that managers did not always receive this support. Three months after the intervention, responses from 38 managers about the support they had accessed showed that: almost a quarter of managers had not been able to

Table 11.3 Suggestions for support from stakeholders and managers

Support	Requested by
Before the intervention:	
Support to develop relationships with managers and communicate the aims of the project	Stakeholders
Gain buy-in from managers	Stakeholders
Gain buy-in from senior managers	Stakeholders / Managers
Encourage senior manager role modelling of behaviour	Managers / Stakeholders
Define support needs for the whole process	Stakeholders
During implementation:	
Provide support throughout the process	Stakeholders
Following the intervention:	
Further training and development, particularly around leadership and conflict management	Managers
Continued feedback, appraisal and action planning	Managers / Stakeholders
Supervision and mentoring from other managers	Managers / Stakeholders

access any kind of support following the intervention; less than 10 per cent had received support from HR or Health and Safety; and less than 5 per cent had received access to mentoring or further training and development. In contrast, the support these managers had received from their own managers and team members was more encouraging: over a third had got support from their managers, and a quarter from their team and colleagues, to help them make necessary behaviour changes.

Barriers to positive line manager behaviour and how to overcome them

During the workshop, all participating managers (n = 112) took part in an exercise to identify their own barriers to displaying the positive manager behaviours in the workplace. Managers were then asked, in a plenary format, to suggest how these barriers could be overcome, and therefore develop workplace solutions. Results from all the workshops were collated and analysed. Responses fell into four categories: individual/work level barriers; organizational/wider-level work barriers; team/relationship barriers; and personal barriers. Table 11.4, adapted from Donaldson-Feilder, Lewis, and Yarker (2009), summarizes the barriers and solutions under each theme.

Gaining manager buy-in to managing psychosocial risks

In addition to the process issues covered so far, the research showed that manager buy-in is a further key factor that will determine the success of this type of intervention and the degree to which line managers manage psychosocial risk.

Table 11.4 Barriers to displaying positive manager behaviour and suggested solutions for overcoming each group of barriers

Barriers	Suggested solutions
Individual/work level barriers • Workload • Short-term deadlines and demands • Conflicting pressures and multiple priorities • Lack of resource • Senior/line managers (pressurising, inconsistency, undermining, directionless)	• Planning and prioritizing • Delegating and finding extra resource, • Communicating with, and thanking the team • Encourage teamwork • Challenging upwards • Seeking advice from others • Recognize own emotions
Organizational/wider-level barriers • Organizational barriers (such as processes and bureaucracy) • IT issues, particularly excessive use of email • Impact of legislation, policy and government targets • Not being able to share information	• Prioritize work • Create a steering group to focus on issues • Make senior managers aware • Challenge the issues • Consult specialists • Make use of training and development • Communicate honestly with the team
Team/relationship barriers • Lack of capability within the team • Problematic behaviours and attitudes within the team	• Deal with poor performance • Make use of organizational policies • Communicate honestly with team • Use role modelling • Clarify self and team's objectives • Develop an evidence base
Personal barriers • Personal/home-life issues • Lack of confidence in own ability • Feeling under stress/undue pressure	• Recognise own behaviour/ emotions/ strengths and limitations • Talk to others and seek help • Take time out/off • Use training and development • Manage expectations of self and others • Communicate honestly with team

Barriers to line manager buy-in and how to overcome them

Gaining participation from line managers to health and well-being interventions within organizations is not always easy, particularly if longitudinal data gathering is required. In order to explore further the reasons why managers may not participate, a plenary discussion was held with stakeholders from the participating organization. Ideas were also sought about how to overcome these barriers to participation. The data was transcribed and content analysed and the outcomes of this exercise are included in Table 11.5, which has been adapted from Donaldson-Feilder, Lewis, and Yarker (2009).

Table 11.5 Barriers to manager participation and ideas on how to overcome these

Barriers to manager participation	Ideas on how to overcome this barrier
• Difficulties communicating with managers	• Gain senior management endorsement and participation • Target the programme at specific managers • Increase profile by linking with other initiatives • Communicate widely
• Managers too busy/pressured to focus on the intervention	• Focus on benefits of participation • Embed within existing initiatives/policies and practices
• Concerns about confidentiality of the process and implications of participation	• Clarify organizational objectives • Use an external provider for data capture • Focus on building trust and relationships
• Lack of knowledge about and belief in the 'concept of stress'	• Build the business case • Link to legislative requirements • Build an understanding of stress
• Non-attendance on the day, resenting being told to attend	• Integrate into management development/induction programme • Provide continuing professional development (CPD) points
• Lack of senior manager support	• See next section

How to gain buy-in from senior managers

During Phase Three, qualitative feedback on indicators of success in interventions from both managers and stakeholders showed that gaining senior management buy-in was key to success. To explore the question of how to gain buy-in, two focus groups were conducted with stakeholders. The emphasis was on previous experiences of success in gaining senior management buy-in to health and well-being interventions, rather than gathering thoughts on 'possible buy-in solutions'. The data from the focus groups were transcribed and content analysed. Five themes emerged as important, listed in order of frequency of mentions:

- *Having a clear business case*: Stakeholders stressed how important objective data was in order to demonstrate the business case and therefore the utility of interventions. Examples provided included demonstrating reductions in insurance premiums obtained by implementing stress management and demonstrating the cost of sickness related absence and therefore benefits of its reduction.
- *Linking the intervention to national goals*: These national goals included the UK governmental initiatives such as Dame Carol Black's report on the health and well-being of working age people (www.dh.gov.uk/en/Publicationsand-statistics/Publications/PublicationsPolicyAndGuidance/DH_083560); sector

specific goals such as the Boorman report on well-being of staff within the UK National Health Service (www.nhshealthandwellbeing.org/FinalReport.html); and national assessments such as IiP or employer of choice accreditations.

- *Clarification of legal responsibilities and risks*: Particularly effective links were made with HSE inspection and enforcement orders.
- *Having a champion within senior management*: Suggestions include having a senior manager with direct responsibility for health and safety, or including a senior manager on a relevant steering or working group.
- *Taking a multi-disciplinary approach*: It was agreed that senior managers were more likely to buy-in to initiatives where it was a joint or joined-up activity, either by linking with existing initiatives or by including HR, H&S and OH in the activity.

Overall learning from the intervention process

The third phase of this research provides a rich source of learning about interventions targeted at helping managers change their behaviour, in order to reduce psychosocial risk. Building on other research in this field (e.g. Biron, Gatrell, and Cooper, 2010; Nielsen and Randall, 2009) it also considers the range of process/contextual factors that affect the success of such interventions, including the following:

- *Manager behaviour change may be more effective for some than others*: While managers in the study were helped to adopt behaviours associated with reducing psychosocial risk, it appears that the intervention used was most effective for those who were not already showing these behaviours.
- *Organizational factors make a difference*: Stakeholders cited a range of process and contextual factors that they felt could improve the effectiveness of the intervention, including the integration into existing initiatives, its name and multi-disciplinary working.
- *Support is important*: Both managers and stakeholders identified support throughout and after the intervention as being important to its impact.
- *Manager buy-in is important and needs fostering*: In order to achieve success in this type of intervention, it is important to gain buy-in of the managers involved and there are barriers to this that need to be overcome. Gaining senior management buy-in is particularly important.

Chapter summary

This chapter has examined the vital role of line managers in managing psychosocial risks and the line manager behaviours that are important for preventing stress and improving well-being. It has looked specifically at an intervention to help managers behave in ways that reduce psychosocial risks and some of the process/

contextual issues that influence the effectiveness of this kind of intervention. Drawing from a review of the academic literature, a range of consultancy experiences and a five-year research programme working with managers to prevent and reduce stress, the authors identify not only what manager behaviours are important in this context, but also how managers can be helped to change their behaviour. A range of process issues are explored, looking at factors that can help and hinder managers in showing positive manager behaviour and making changes in order to manage psychosocial risks better.

References

Alimo-Metcalfe, B., & Alban-Metcalfe, R. J. (2001). The development of a new transformational leadership questionnaire. *Journal of Occupational and Organizational Psychology*, 74, 1–27.

Amick, B. C., & Celentano, D. D. (1991). Structural determinants of the psychosocial work environment: Introducing technology in the work stress framework. *Ergonomics*, 34, 625–646.

Arnold, J., Cooper, C. L., & Robertson, I. T. (1995). *Work psychology: Understanding human behaviour in the workplace*. London: Pitman.

Ashforth, B. E. (1994). Petty tyranny in organizations. *Human Relations*, 47, 755–778.

Baker, E., Israel, B., & Schurman, S. (1996). Role of control and support in occupational stress: An integrated model. *Social Science and Medicine*, 43, 1145–1159.

Baruch-Feldman, C., Brondolo, E., Ben-Dayan, D., & Schwartz, J. (2002). Sources of social support and burnout, job satisfaction and productivity. *Journal of Occupational Health Psychology*, 7, 84–93.

Bass, B. (1985). *Leadership and performance beyond expectations*. New York: Free Press.

Bass, B. M. (1998). *Transformational leadership: Industrial, military, and educational impact*. Hillsdale, NJ: Erlbaum.

Bass, B. M. (1999). Two decades of research and development in transformational leadership. *European Journal of Work and Organizational Psychology*, 8, 9–32.

Bass, B. M., & Avolio, B. J. (1994). *Improving organizational effectiveness through transformational leadership*. Thousand Oaks, CA: Sage Publications.

Bass, B. M., & Avolio, B. J. (1997). *Full range leadership development: Manual for the Multifactor Leadership Questionnaire*. Mind Garden, Palo Alto, CA.

Beehr, T. A., King, L. A., & King, D. W. (1990). Social support, occupational stress: talking to supervisors. *Journal of Vocational Behavior*, 36, 61–81.

Biron, C., Gatrell, C., & Cooper, C. L. (2010). Autopsy of a Failure: Evaluating process and contextual issues in an organizational-level work stress intervention. *International Journal of Stress Management*, 17, 135–158.

Bono, J. E., Foldes, H. J., Vinson, G., & Muros, J. P. (2007). Workplace emotions: The role of supervision and leadership. *Journal of Applied Psychology*, 92, 1357–1367.

Brossoit, K. B. (2001). *Understanding employee empowerment in the workplace: Exploring the relationships between transformational leadership, employee perceptions of employment and key work outcomes*. Unpublished doctoral dissertation. The Claremont Graduate University.

Burns, J. M. (1978). *Leadership*. New York: Harper & Row.

Cherniss, C. (1995). *Beyond Burnout: Helping teachers, nurses, therapists & lawyers recover from stress and disillusionment*. New York: Routledge.

CIPD (2008). *Annual survey report 2008: Absence management*. Available online at: www.cipd.co.uk/subjects/hrpract/absence/absmagmt. htm?IsSrchRes=1

Dekker, S. W. A., & Schaufeli, W. B. (1995). The effects of job insecurity on psychological health and withdrawal: A longitudinal study. *Australian Psychologist, 30,* 57–63.

Donaldson-Feilder, E. J., Lewis, R., & Yarker, J. (2009). Preventing stress: Promoting positive manager behaviour. CIPD Insight Report.

Duffy, M. K., Ganster, D. C., & Pagon, M. (2002) Social undermining in the workplace. *Academy of Management Journal, 45,* 331–351.

Duxbury, M. L., Armstrong, G. D., Drew, D. J., & Henly, S. J. (1984). Head nurse leadership style with staff nurse burnout and job satisfaction in neonatal intensive care units. *Nursing Research, 33,* 97–101.

Einarsen, S., Aasland, M. S., & Skogstad, A. (2007). Destructive leadership behaviour: A definition and conceptual model. *Leadership Quarterly, 18,* 207–216.

Fenalson, K. J., & Beehr, T. A. (1994). Social support and occupational stress: Effects of talking to others. *Journal of Organizational Behavior, 15,* 157–175.

French, W. L., & Bell, C. H. (1995). *Organization development: Behavioral science interventions for organization development*. Englewood Cliffs: Prentice Hall.

Gilbreath, B. (2004). Creating healthy workplaces: The supervisor's role. *International Review of Industrial and Organizational Psychology, 19,* 93–118.

Gilbreath, B., & Benson, P. G. (2004). The contribution of supervisor behaviour to employee psychological well-being. *Work & Stress, 18,* 255–266.

Harris, K. J., & Kacmar, K. M. (2005). Easing the strain: The buffering role of supervisors in the perceptions of politics-strain relationship. *Journal of Occupational and Organizational Psychology, 78,* 337–354.

Harvey, P., Stoner, J., Hochwarter, W., & Kacmar, C. (2007). Coping with abusive supervision: The neutralising effects of ingratiation and positive affect on negative employee outcomes. *Leadership Quarterly, 18,* 264–280.

Hauge, L. J., Skogstad, A., & Einarsen, S. (2007). Relationships between stressful work environments and bullying: Results of a large representative study. *Work and Stress, 21,* 220–242.

Hetland, H., Sandal, G. M., & Johnsen, T. B. (2007). Burnout in the information technology sector: Does leadership matter? *European Journal of Work and Organisational Psychology, 16,* 58–75.

Hogan, R., Curphy, G. J., & Hogan, J. (1994). What we know about leadership: Effectiveness and personality. *American Psychologist, 49,* 493–504.

Judge, T. A., & Piccolo, R. F. (2004). Transformational and transactional leadership: A meta-analytic test of their relative validity. *Journal of Applied Psychology, 89,* 755–768.

Keller, R. T. (2006). Transformational leadership, initiating structure, and substitutes for leadership: A longitudinal study of research and development project team performance. *Journal of Applied Psychology, 91,* 202–210.

Kelloway, E. K., & Barling, J. (2010). Leadership development as an intervention in occupational health psychology. *Work & Stress, 24,* 260–279.

Kile, S. M. (2000). *Health Endangering Leadership*. Unpublished doctoral dissertation. Bergen: Norway.

Kuoppala, J., Lamminpaa, A., Liira, J., & Vainio, H. (2008). Leadership, job well-being, and health effects – A systematic review and a meta-analysis. *Journal of Occupational and Environmental Medicine, 50,* 904–915.

Landeweerd, J. A., & Boumans, N. P. G. (1994). The effect of work dimensions and need for autonomy on nurses' work satisfaction and health. *Journal of Occupational and Organizational Psychology, 67*, 207–217.

Lee, R. T., & Ashforth, B. E. (1996). A meta-analytic examination of the correlates of the three dimensions of burnout. *Journal of Applied Psychology, 81*, 123–133.

Levy, P. E. (2003). *Industrial/organizational psychology: Understanding the workplace.* Boston, MA: Houghton Mifflin Co.

McDaniel, C., & Wolf, G. A. (1992). Transformational leadership in nursing service. *Journal of Nursing Administration, 22*, 60–65.

Moyle, P., & Parkes, K. (1999). The effects of transition stress: A relocation study. *Journal of Organizational Behavior, 20*, 625–646.

Nielsen, K., & Randall, R. (2009). Managers' active support when implementing teams: The impact on employee well-being. *Applied Psychology: Health and Well-being, 1*, 374–390.

Nielsen, K., Randall, R., Yarker, J., & Brenner, S. O. (2008). The effects of transformational leadership in followers perceived work characteristics and well-being: A longitudinal study. *Work and Stress, 22*, 16–32.

Nielsen, M. L., Rugulies, R., Christiansen, K. B., Smith-Hansen, L., & Kristiansen, T. S. (2006). Psychosocial work environment predictors of short and long spells of sickness absence during a two year follow up. *Journal of Occupational and Environmental Medicine, 48*, 591–595.

Nyberg, A., Bernin, P., & Theorell, T. (2005). *The impact of leadership on the health of subordinates.* The National Institute for Working Life.

O'Driscoll, M. P., & Beehr, T. A. (1994). Supervisor behaviors, role stressors and uncertainty as predictors of personal outcomes for subordinates. *Journal of Organizational Behavior, 15*, 141–155.

Offerman, L. R., & Hellmann, P. S. (1996). Leadership behaviour and subordinate stress: A 360 view. *Journal of Occupational Health Psychology, 1*, 382–390.

Parker, S. K., Chmiel, N., & Wall, T. D. (1997). Work characteristics and employee well-being within a context of strategic downsizing. *Journal of Occupational Health Psychology, 2*, 289–303.

Parker, S., & Williams, H. (2001). *Effective teamworking: Reducing the psychosocial risks.* Norwich, UK: HSE Books.

Rayner, C., & McIvor, K. (2006). *Report to the Dignity at Work Project Steering Committee.* Dignity at Work Report: Portsmouth University.

Rooney, J. A., & Gottlieb, B. H. (2007). Development and initial validation of a measure of supportive and unsupportive managerial behaviours. *Journal of Vocational Behavior. 71*, 186–203.

Rubin, R. S., Munz, D. C., & Bommer, W. H. (2005). Leading from within: The effects of emotion recognition and personality on transformational leadership behavior. *Academy of Management Journal, 48*, 845–858.

Saksvik, P. Ø., Nytrø, K., Dahl-Jorgensen, C., & Mikkelsen, A. (2002). A process evaluation of individual and organisational occupational stress and health interventions. *Work & Stress, 16*, 37–57.

Schat, A. C. H, Frone, M. R., & Kelloway, E. K (2006). Prevalence of workplace aggression in the U. S. workforce: Findings from a national study. In E. K. Kelloway, J. Barling & J. J. Hurrell (Eds.). *Handbook of workplace violence.* Thousand Oaks, CA: Sage.

Schaubroeck, J., Walumbwa, F. O., Ganster, D. C., & Kepes, S. (2007). Destructive leadership traits and the neutralising influence of an 'enriched' job. *Leadership Quarterly, 18*, 236–251.

Schaufeli, W. B., & Enzmann, D. (1998). *The burnout companion to study and research*. London: Taylor & Francis.

Seltzer, J., & Numerof, R. E. (1988). Supervisory leadership and subordinate burnout, *Academy of Management Journal, 31*, 439–446.

Sheridan, J. E., & Vredenburgh, D. J. (1978). Usefulness of leadership behavior and social power variables in predicting job tension, performance, and turnover of nursing employees. *Journal of Applied Psychology, 63*, 89–95.

Skogstad, A., Einarsen, S., Torsheim, T., Aasland, M. S., & Hetland, H. (2007). The destructiveness of laissez-faire leadership behaviour. *Journal of Occupational Health Psychology, 12*, 80–92.

Sosik, J. J., & Godshalk, V. M. (2000). Leadership styles, mentoring functions received, and job-related stress: A conceptual model and preliminary study. *Journal of Organizational Behavior, 21*, 365–390.

Stephens, C., & Long, N. (2000). Communication with police supervisors and peers as a buffer of work-related traumatic stress. *Journal of Organizational Behavior, 21*, 407–424.

Stogdill, R. M. (1950). Leadership, membership and organization. *Psychological Bulletin, 47*, 1–14.

Stogdill, R. M. (1963). *Manual for the Leader Behaviour Description Questionnaire*. The Ohio State University.

Tepper, B. J. (2000). Consequences of abusive supervision. *Academy of Management Journal, 43*, 178–190.

Tepper, B. J. (2007). Abusive supervision in work organizations: Review, synthesis and research agenda. *Journal of Management, 33*, 261–289.

Tepper, B. J., Carr, J. C., Breaux, D. M., Geider, S., Hu, C., & Hua, W. (2009). Abusive supervision, intentions to quit, and employees' workplace deviance: A power/dependence analysis. *Organizational Behavior and Human Decision Processes, 109*(2), 156–167.

Tepper, B. J., Duffy, M., Hoobler, J., & Ensley, M. (2004). Moderators of the relationships between co-workers organizational citizenship behaviour and fellow employee's attitudes. *Journal of Applied Psychology, 89*, 455–465.

Thomas, L. T., & Ganster, D. C. (1995). Impact of family-supportive work variables on work-family conflict and strain: A control perspective. *Journal of Applied Psychology, 80*, 6–15.

Thomson, L., Rick, J., & Neathey, F. (2004). *Best practice in rehabilitating employees following absence due to work-related stress*. London: HSE Books.

Van Dierendonck, D., Haynes, C., Borrill, C., & Stride, C. (2004). Leadership behavior and subordinate well-being. *Journal of Occupational Health Psychology, 9*, 165–175.

Yagil, D. (2006). The relationship of abusive and supportive workplace supervision to employee burnout and upward influence tactics. *Journal of Emotional Abuse, 6*, 49–65.

Yarker, J., Donaldson-Feilder, E., Lewis, R., & Flaxman, P. E. (2007). *Management competencies for preventing and reducing stress at work: Identifying and developing the management behaviours necessary to implement the HSE Management Standards*. London: HSE Books.

Yarker, J., Donaldson-Feilder, E. & Lewis, R. (2008). *Management competencies for preventing and reducing stress at work: Identifying and developing the management behaviours necessary to implement the HSE Management Standards: Phase 2*. HSE Books: London.

12 The impact of process issues on stress interventions in the emergency services

Viv Brunsden, Rowena Hill, and Kevin Maguire

Introduction

Emergency service personnel (i.e. police, fire and ambulance) can have high levels of exposure to both occupational and post-traumatic stress, with the interplay between these forms of stress generating a range of complex responses. This chapter considers the particular occupational context of the emergency services and the ways in which this context affects the nature and effectiveness of stress interventions, both at a strategic and interpersonal level. The nature and cultural context of emergency services working will be described, highlighting those specific organizational and occupational features that have the potential to impact on the take up, and effectiveness, of stress interventions. The emergency services provide a clear example of the need to understand both the occupational environment and specific cultural context of an organization in order to design and apply stress interventions successfully. This chapter will discuss a variety of process issues that ought to be considered when developing and delivering interventions in the emergency services.

Stress in the emergency services

The emergency services provide a complex occupational environment for both personnel and management. At one level they are similar to any other public sector organization; however, at another level, their frontline responding means that personnel are routinely exposed to unusual and traumatic stressors not evidenced in any other occupations. This exposure can interact with occupational stress, generated both by general and occupationally specific stressors, and create a unique and highly complex pattern of stress responses. The routine nature of stress exposure can be seen to generate particular stress responses and coping behaviours and these then affect employee relationships and the broader social organizational context. These effects can also reach beyond the organization itself, creating occupation-specific work-family relationships. These can, in turn, reduce the potential for employing preventative stress strategies, because of the inevitable stress exposure involved in emergency responding. Successful stress management strategies in the emergency services therefore tend to be responsive post-exposure interventions that take account of this unique occupational context.

Stress prevalence rates reported in emergency service audits have varied, but figures have been found as high as 19.3 per cent showing clinical signs of traumatic stress and 8 per cent meeting all post-traumatic stress disorder (PTSD) diagnostic criteria (Brunsden, Woodward, and Regel, 2003; Regel et al., 2001); other researchers have found even higher rates of 20 per cent having PTSD (see Durkin, 2006). Stepping away from trauma to occupational stress and related issues such as anxiety and psychological distress, the figures become even higher. Courtney, Francis, and Paxton (2010) found that almost a quarter of their sample of Australian paramedics suffered from above normal levels of anxiety and 40 per cent suffered from stress. Joseph, Brown, and Mulhern (2003) found one fifth of their sample of Irish firefighters showed high levels of psychological distress (as measured by the General Health Questionnaire) and suggested that this figure could be an underestimate of what might be found in the wider emergency services population, given that the most distressed individuals also tend to be the most avoidant. This avoidance is likely to be further exacerbated by emergency service culture and the reluctance to self refer on grounds of stress (Lawrence, 2003). This is supported by Brunsden, Woodward, and Regel (2003) who found that 76 per cent of their UK fire service sample reported physical ill-health symptoms associated with stress related illnesses but that only 3.8 per cent attributed those symptoms to stress.

That emergency services personnel are subject to a wide variety of stressors, both psychological and physiological, is well recognized in the literature (e.g. Hill and Brunsden, 2003; Brown, Mulhern, and Joseph, 2002; Baker and Williams, 2001; Regehr, Hill and Glancy, 2000). Despite this there can be a reticence to acknowledge and discuss stress within the services themselves (Beale, 2003). Within the emergency service stress literature the main focus has been on traumatic stress and, given the nature of the incidents that emergency personnel attend, this is perhaps unsurprising. However the relative neglect of other origins of stress may be problematic, as personnel are also exposed to a wide variety of non-traumatic stressors; for example chemical and biological hazards (Markowitz, 1989; Malek, Mearns, and Flin, 2003), bad weather conditions (Beale, 2003) or extreme heat (Brenner et al., 1998), and protracted or night-time operations (Beale, 2003). The need for 24-hour cover necessitates shift working (Ernst et al., 2004) which can itself be a stressor additionally creating tiredness and sleep disturbances (Courtney, Francis, and Paxton, 2010; Holmes, 2003; Murphy, 1999).

Personnel can also suffer stress arising from aggressive encounters with the general public (see Brunsden, Hill et al., in press; Brunsden, 2007); an analysis of British Crime Survey data suggested that protective service occupations, such as the Police and Fire and Rescue Service, are the most at risk of experiencing violence at work (Webster et al., 2007). Even when aggression is manifested as verbal abuse rather than physical violence, this can still create stress and anxiety (Communities & Local Government, 2006; Brunsden et al., in press). Non-traumatic stressors generate stress in and of themselves but are also likely to further complicate incidence of traumatic stress where this does occur.

Stress may also arise from a variety of standard occupational stressors as encountered in any other organization (e.g. see Brough, 2002; Brown, Fielding, and Grover, 1999) and there is some evidence that the presence of occupational stress can outweigh traumatic stress despite the emphasis on trauma in the emergency service's stress literature (Brunsden, Woodward, and Regel, 2003). However, traumatic stress must also be acknowledged because emergency services personnel are necessarily subject to this through their attendance at distressing emergency events (e.g. Hill and Brunsden, 2003; Brown, Mulhern, and Joseph, 2002; Beaton et al., 1999). Their traumatic stress is unusual, however, in that the traumatic exposure is expected and routine whereas most traumatic stress exposure is unexpected and unpredictable. This has led to recent suggestions of a need for a new diagnostic category of posttraumatic stress, specifically that of duty-related traumatic stress (Paton, 2006). The occupational implications of such a categorisation could be profound, given that a survey of UK Fire and Rescue Services found a psychologist to staff ratio of only 1:2600 (Durkin, 2006), perhaps suggesting that current occupational health services in the UK could be considered legally inadequate in terms of trauma support.

Studies related to stress in the emergency services have tended to focus on operational staff, i.e. those who actually attend the emergency events. However, these are not the only emergency service personnel who are exposed to stress. Control room staff can also suffer stress through standard occupational and organizational factors (see Brunsden, Robinson et al., in press; Brunsden, Woodward, and Regel, 2003; Wastell and Newman, 1996); but they may also experience traumatic stress vicariously, through staying on the phone with distressed callers during incidents and hearing their description of events. Because of their close working relations with operational personnel, control staff may also be vulnerable to stress crossover; a form of stress contagion whereby an individual assumes the stress of another because of role obligations and commitments (Wethington, 2000). Control staff are also vulnerable to distress arising from hoax calls and may also receive sexually harassing calls (Brunsden and Goatcher, 2009; Brunsden, Hill et al., in press). As yet however this occupational group has received little attention from academic psychologists (Brunsden, Hill et al., 2007; Brunsden, Robinson et al., in press).

Developing and implementing interventions in the emergency services

Cox's (1993) primary and secondary interventions (reproduced in Table 12.1) remind us how difficult it is to manage stress in the field of emergency work. It has even been argued that this is impossible where conventional control over both the workplace and workplace equipment cannot be achieved (Ash and Smallman, 2008). Thinking of person-environment fit theory (Caplan, 1987) for example, it is difficult to remove the road collision, fire, or hostile criminal from the emergency service worker (despite efforts in areas such as road safety, fire prevention and crime reduction).

Table 12.1 Primary and secondary stress management strategies from Cox (1993)

	Individual	Organizational
Primary	To reduce the risk factor or change the nature of the stressor	To remove the hazard or reduce employees' exposure or its impact on them
Secondary	To alter the ways in which individuals respond to the risks and stressors	To improve the organization's ability to recognize and deal with stress-related problems as they arise

The emergency services therefore tend to rely on individual-based primary and secondary strategies and organizationally-based secondary strategies. Commonly they are focussed on a combination of developing coping strategies, including greater control latitude (thus making them more 'stress resilient'); actively reducing strain levels ('destressing' initiatives); and monitoring for, and responding to, early identification of strain.

Process and context issues specific to the emergency services

It is tempting to think of the emergency services as a homogeneous population. Where research has been carried out on specific emergency services, the police dominate the literature, with far less research into the fire or ambulance services (Malek et al., 2003). However, often studies combine the services without considering any differences between them. These differences that make the emergency services distinct also define the context and process issues that any efforts to manage or reduce stress in the emergency services ought to take into account – these are discussed next.

Personality and individual differences

It has been suggested that emergency services personnel share similar personality characteristics, having "very different personalities from the average person ..." (Mitchell and Bray, 1990, p. 19). Mitchell (1983) suggested a distinct personality type seen in emergency service personnel, characterized by dedication, risk taking, a high need for stimulation, and a need to help others, terming this "the rescue personality". This notion of a rescue personality has been greatly debated (Wagner, 2005; Gist and Woodall, 1998) but there is certainly some evidence that emergency service personnel do differ from the general population in some ways. For example, Salters-Pedneault, Ruef and Orr (2010) found evidence that fire and police recruits both scored higher in terms of excitement seeking; however they also found differences between these two groups with police scoring higher on gregariousness, conscientiousness, deliberation and dutifulness. This suggests that there can be some personality differences across those attracted to the different forms of emergency service working. There can also be personality differences within emergency services, related to fulfilling different occupational roles within a service; for example, Kirkcaldy, Brown, and Cooper (1998) found, when exploring personality in a sample of police superintendents, that those who had primarily served in criminal detection prior to their promotion to superintendent had significantly higher scores for Type A behaviour.

Although operational personnel from the different services attend similar, or even the same events (Beale, 2003) and can experience very similar occupational realities, there can also be important differences between them. Pendleton et al. (1989) investigated stress and strain in firefighters, police and non-emergency public sector workers and found that firefighters reported the lowest strain levels whilst police officers reported the highest levels of stressor exposure. It is not clear whether this suggests perceptual differences amongst the groups or is a result of actual difference in levels of exposure, however such issues would need consideration when designing stress management interventions. Brough (2004) considered traumatic stress and organizational strain across the three services and found that whilst personnel within the ambulance and fire services experienced similar stress responses, there appeared to be a unique experience of occupational stress in participants from the police. The fire and ambulance services have also been found to show no differences from one another in their use of coping strategies (Young and Cooper, 1997).

Different occupational roles within services can also experience different stressors. In their study of the police, Brown and Campbell (1990) found that senior management's primary stressor was criticism from the media whereas constables were more likely to be stressed by factors such as time pressure, long hours, working with civilians and organizational politics, and sergeants by having to manage or supervise, working in isolation, and lack of consultation. Such differences may require different forms of response in terms of prevention and intervention.

Demographics and organizational cultures

There are also important differences in the demographics of the services and in their organizational cultures, both of which are likely to impact on the ways that stress is experienced. Despite recruitment drives which have targeted women and ethnic minorities, the fire and rescue services remain dominated by white male personnel (Hashem and Lilly, 2007). The police and ambulance services have however made some strides in this area and have better female representation, although even here females and minority ethnic groups are still under-represented (Sigurdsson and Dhani, 2010). This male dominance, combined with the "heroic" nature of emergency service work can lead to a macho organizational culture; sometimes labelled the three Ts of "testosterone, tattoos and taut-biceps" (Beale, 2003, p. 29). The pseudo-military nature of the emergency services can also feed in here (despite recent modernisation processes aimed at removing militaristic tones). When discussing the military, Greenberg, Langston and Scott (2006) noted that the military culture can be an important barrier to help-seeking and a similar problem can exist within the emergency services, particularly in relation to the self-reporting of stress (Lawrence, 2003). Organizational culture can therefore be seen as a key process issue that requires explicit consideration.

Industrial relations

A further difference between the emergency services relates to industrial relations. UK ambulance service personnel have a number of trade unions they can join which could be said to diffuse their collective identity. This is far from the case in the fire and rescue service where the Fire Brigades Union (FBU) could be described as the strongest single occupation public sector trade union and there is a strong collective identity (Brunsden and Hill, 2009). Following the 2002–2003 UK fire strike a modernisation process took place which changed many of the working conditions within the service (Bain, Lyons and Young, 2002), acting as a significant stressor, the repercussions of which are continuing. Similarly, the strike itself could be seen as a traumatising event which continues to generate stress reactions (see Brunsden and Hill, 2009). In contrast, the UK police are not allowed to join a trade union, or to strike, but instead have a federation to which all officers up to a certain rank are members, regardless of whether they opt to be. On the surface this suggests that police might enjoy more positive employer-employee relations than do firefighters. However, firefighters typically experience high job satisfaction (Guidotti, 2000; North et al., 2002; Smith, 2007) whereas the literature on police is more complex, contradictory and inconclusive (Getahun, Sims, and Hummer, 2008). The politicised context of emergency service working can affect the stress intervention strategies offered to personnel; for example it has been argued that fatigue management programmes, widely used in other high risk industries such as aviation and the petrochemical industry, are rarely implemented in the UK emergency services because the stress inducing factors of sleep loss and tiredness are necessarily entwined in issues of work hours, pay and secondary employment (Holmes, 2003).

Politicised activities, relations and cooperation

The very different industrial relations across the three services mean that stress audits can become highly politicised activities. Conducting stress audits necessarily requires the co-operation of both management and workers. During our own audits (e.g. Brunsden, Woodward and Regel, 2003; Eyre and Brunsden, 2003; Maguire and MacPherson, 1997; Regel et al., 2001; Woodward, Brunsden and Regel, 2000), variously commissioned by both emergency service management and by trade unions, it became clear that without both sides' cooperation the auditing would fail. Management necessarily had to approve the work to enable access, whereas the union necessarily had to endorse the initiatives in order to get any reasonable participation rate. Both sides have different agendas regarding the outcomes of auditing. It is inevitably in management's interests that audits show low levels of stress and high levels of job satisfaction. Conversely, if audits show high stress and low satisfaction trade unions can use these data to their own advantage in confronting management. This may lead both sides to exert pressure on personnel, consciously or unconsciously, to provide answers that skew the data in particular ways. This is further complicated if participating personnel do not trust that their individual response will not somehow be fed back to management. If this fear persists, regardless of whether it is founded in any actuality, then personnel may wish to represent their psychological health more positively than is actually the case. The vested interests of these various parties potentially render findings of audits as artefactual, saying more about the processes by which they were carried out, rather than indicating genuine stress. Despite this issue, audits are still worthwhile exercises as, even if taken very conservatively, the negative implications are still concerning.

Occupational identity

Within the emergency service literature there is evidence for a strong occupational identity (Fannin and Dabbs, 2003; Lee and Olshfski, 2002; Mitchell, 1983). This is a clear commitment to the occupational role, but also to the way in which that role is carried out and how it is viewed by members of the public. For example, firefighters have been found to have such strong role identity that they are never actually off duty (Lee and Olshfski, 2002). It has been suggested that emergency service personnel in operational roles have a need for control, a need to be needed, and a need to rescue (Brown, Mulhern, and Joseph, 2002; Regel et al., 2001). In addition, emergency personnel can have a denial of the need for assistance (Lawrence, 2003) and high empathy levels (Mitchell, 1983). Where this identity is threatened or prevented from operating in some way (for example, during change processes or industrial disputes) there can be negative and detrimental consequences for the emergency service personnel (Brunsden and Hill, 2009). Given this, consideration should be given to how this strong role identity affects the ways in which emergency service personnel experience stress. For example, if this identity leads personnel to experience a sense of enhanced responsibility, or an increase in guilt or a sense of failure, as has been suggested (Hill and Brunsden, 2009) this will have a confounding effect on stress symptoms. This strong group identity may also generate resistance towards stress interventions delivered by those seen as outsiders.

Gender and race representation

This strong role identity is complicated further by the well-reported observation that all emergency services are under-represented with regard to gender and race (Sigurdsson and Dhani, 2010), and thus the emergency service identity could be said to reflect a white male identity with all the concomitant characteristics of that identity. Certainly this white male dominance has led to accusations of a self-replicating, self-protectionist culture (Archer, 1999) within the emergency services. Although dated, Bennett and Greenstein's (1975) work is still relevant to the emergency services in terms of their conceptualisation of self-replication through the socialisation model (becoming similar through performing the job) and the predisposition model (being attracted to a job to find similar people to yourself). In order to increase person-environment and person-role fit, recently appointed and trained emergency service personnel tend to replicate the behaviours, attitudes and attributes of colleagues in their immediate contact. With such collective identities and tight co-worker networks within the emergency services (Beaton et al., 1999; Brunsden and Hill, 2009; Nixon et al., 1999), this has led to claims that women are subverting themselves and imitating male behaviours in order to become accepted into groups and maintain their role within those groups (Archer, 1999). Such imitation is problematic given the well-recorded evidence for male reluctance to seek or accept help (Galdas, Cheater, and Marshall 2005) and may detrimentally impact upon any interventions promoted by occupational health teams. Group members try to maintain their status and role within the group and sharing honest accounts of stress levels and psychological health difficulties is not conducive with this because of the attached stigma. Occupational health practitioners should consider the impact of stigma on both self-referral rates and the uptake of offered interventions (Hill and Brunsden, 2009; Lawrence, 2003).

Actual and fictive families

Close emergency service co-worker relations can lead to the development of fictive families; for example, the term 'the brotherhood' is widely used internationally to describe firefighter fictive families and the strong co-worker loyalties that exist (Regehr et al., 2005). These close relationships act as a stress buffer through the provision of strong social support (see Regehr, 2009), however they can themselves be a stressor in certain circumstances, for example if a colleague is injured or killed at work (Hill and Brunsden, 2003, 2009). Moos and Moos (1976) suggested that different types of social environments and living conditions can develop family processes and this allows fictive families to be considered in the same ways as 'traditional' families. The actual and fictive families become intertwined; Jackson and Maslach (1982) suggested the need for families of emergency service personnel to share their experiences in order to support and cope with the occupational demands that impact on family life.

Members of the emergency services become very close by the nature of their job, working in tight co-worker networks (Neale, 1991) that operate as both operational and emotional teams and which allow collective identities to form (Brunsden and Hill, 2009). This can lead to a reluctance to access interventions, with a commonly cited excuse that the team do not need this as they use each other as counsel. However, as Parkinson (1993) points out, this peer support is usually closer to defusing than debriefing, acting merely as social support rather than a coherent stress intervention *per se*. It is further suggested that this reluctance is heightened within emergency service workers as males are reluctant to share (Wester and Lyubelsky, 2005), isolating female colleagues and maintaining barriers to help-seeking. This exclusion is not just limited to female members of emergency services. Varvel et al. (2007) suggest that coping can become the exclusive function of a tight co-worker network to such an extent that it can also exclude family members. However they conclude that this is a forced situation due to shift work; an emergency service worker may not see their spouse or children for three consecutive days if the family is at work or school during the day and the emergency service worker is on shift. It may be therefore that this inevitable distancing from the family becomes another reason for the reliance on co-workers, rather than the co-worker relationship being the reason for excluding family members. Certainly a reliance on the immediate occupational team members is reflected in other research (e.g. see Bacharach, Bamberger, and Doveh, 2008).

However, the family and wider social support (outside of the immediate team) are also an important resource meriting consideration. Social support should be considered in any intervention for post-traumatic reactions (Landsman et al., 1990). Emergency service worker research has supported the importance of social support (Regehr, 2009), as emergency service personnel, their close occupational team and their family are inter-related systems (Schumm, Bell, and Resnick, 2001). This is a premise that health professionals need to consider, as it will likely impact on uptake of interventions, and on the groups that require and participate in these.

We have summarised a range of issues that can frame the success or failure of interventions for stress in the emergency services. Not only can the emergency services differ a lot from 'standard' occupations, but also there is a lot of variation within the emergency services themselves to consider them as three rather than one occupational groups. These differences provide the process and context issues that are important when developing and implementing interventions. We have highlighted a few that have emerged from our experience of working with this unique group. With these in mind, the next section discusses specific types of interventions that we have found are most relevant for these groups.

Interventions following stress exposure

As mentioned at the start of the chapter, although primary interventions are generally recommended (e.g. LaMontagne et al., 2007), in some organizations and for some occupations it is simply not possible to develop proactive actions. In occupations where extreme stressors such as those experienced by the emergency services are an inherent part of the job, primary interventions may imply changing the actual nature of the job. Therefore, often the only option is to focus on reactive secondary and tertiary interventions.

Health strategies employed within the emergency services can be differentiated into therapeutic treatments and reactive interventions. Therapeutic treatments are the province of the healing professions whereas reactive interventions can be seen as precursory actions to further deterioration in an attempt to prevent serious stress (which would require therapeutic treatment). Just to confuse the issue a little, however, reactive interventions can be carried out by members of the healing

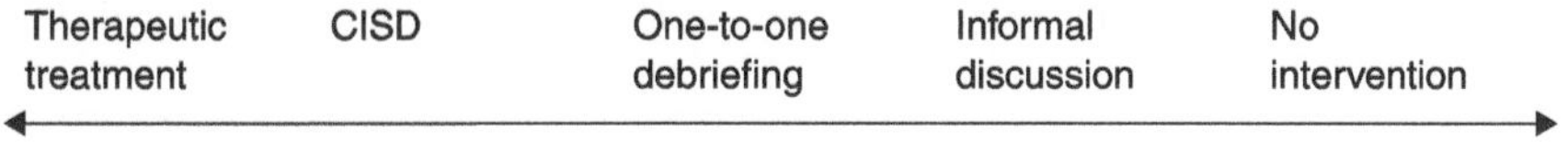

| Therapeutic treatment | CISD | One-to-one debriefing | Informal discussion | No intervention |

◄───►

Figure 12.1 The spectrum of interventions.

professions. In terms of reactive interventions related to traumatic stress Jeannette and Scoboria (2008) identified three levels: critical incident stress debriefing (CISD); one-to-one debriefing; and informal discussion. It is important to note however that CISD and, the critical incident stress management (CISM) process in which it is nested, are neither a therapy nor a substitute for one (Blaney, 2005, 2009; Mitchell, 2004). For completeness, we can add therapeutic treatment and no intervention as fourth and fifth levels, on a continuum of interventions based on their intensity (see Figure 12.1), which fully embraces Cox's (1993) definition of tertiary interventions.

Of Jeannette and Scoboria's (2008) three interventions, CISD, sometimes called psychological debriefing, is the one most associated with the emergency services. It is often considered as just one part of a CISM approach which is generally considered to be more effective than CISD alone (see Regel, 2007; Mitchell, 2004). The origins of CISD and CISM lie in crisis intervention theories dating back to 1944 (Regel, 2007) or even earlier (see Mitchell, 2004). Regel (2007) describes CISM as a:

> comprehensive, systematic and integrated multi-component crisis intervention package that enables individuals and groups to receive assessment of need, practical support and follow-up following exposure to traumatic events ... it facilitates the early detection and treatment of post-trauma reactions and other psychological sequelae.
>
> (p. 411)

Regel gives three elements in CISM that precede the actual CISD element: pre-crisis education, assessment, and defusing. There is also one element in CISM that comes after CISD; that of therapeutic intervention should PTSD occur. However, Mitchell (2004) went further in his detailing of CISM, listing 12 components rather than Regel's mere five (see Table 12.2). This confusion as to the exact nature of CISM, and indeed CISD, is common within the literature.

Mitchell (2003) notes that "everyone talks about 'debriefing' and means something different" (p. 56). Certainly different authors and practitioners use this term to describe what can be very different practices (see Brunsden, Woodward and Regel, 2003). This confusion is complicated further by the same practices also being referred to by different names. For example, Devilly and Cotton (2003, p. 144) refer to psychological debriefing as "emotional first aid" and Dyregov (1989) talks of psychological debriefing and CISD as if they were interchangeable terms. Regel (2007), however, shows preference for the term psychological debriefing over CISD, claiming the support of the British Psychological Society

248 *Addressing process and context in practice*

Table 12.2 Mitchell's 12 components of most CISM programmes (adapted from Mitchell, 2004)

Pre-event preparation including policy development as well as training etc.
Assessment procedures
Strategic planning procedures
Individual crisis intervention support actions
On-site support
Facility and plan for demobilisation from a disaster area
Crisis Management Briefing
Defusing on the same day
CISD
Other support services e.g. for family
Follow-up services
Referral services

---

in this. Other preventative programmes which appear to be CISD include psychological first aid (Vernberg et al., 2008) which has the slight distinction of being applied "in the field" and being intended for children as well as adults; in all other ways it sounds like CISD. The UK military's programme called Trauma Risk Management (TRiM) has similar echoes but is designed to be delivered individually as well as in a group (Greenberg et al., 2010). Mitchell (2004) tells us that CISD is not the most frequently used intervention; however, it is the most prominent and visible and is firmly associated with the emergency services.

Bearing in mind the differences already mentioned between the differing forms of intervention described as CISD, there are a number of principles which *generally* seem to characterize these. First, there is an agreement that the group debriefed should be homogeneous so that there is a greater shared understanding of the experience(s) with an implied likelihood of preparedness to listen, empathise, inform, and therefore to make progress. Therefore CISD tends to be carried out on the whole group who have attended, or dealt with, a specific potentially traumatic event; for example, a single fire and rescue service watch who had attended a fire involving fatalities.

Second, early intervention is also agreed as a general principle but the actual timing varies. Vernberg et al.'s (2008) psychological first aid aims to intervene as soon as possible but Dyregov (1989) states that CISD should not occur on the same day as the traumatic event and Regel's (2007) review found that it can be held anytime between 3 and 14 days after the event.

Third, Greenberg et al. (2006) identify the number of sessions as a key difference between CISD and TRiM (trauma risk management) stating that TRiM entails multiple sessions whereas CISD consists of a single session only. However, their position is out of sync with the views of Mitchell who can be considered the originator of CISD; he stated in 2004 that single session debriefings are *not* appropriate for CISD and goes on to list a host of organizations that do not endorse or approve single session CISD, including the International Critical Incident Stress Foundation. This is an important point of contention as Jones, Roberts and

Greenberg (2003) highlight research that suggests one-off sessions can cause more harm than good.

Fourth, one characteristic that remains unresolved is *who should deliver the intervention*. Traditionally these have been delivered by occupational health personnel or external consultants. However, in recent years there has been a shift within the emergency services to delivery from trained peers (see Barber, 2003; Durkin, 2006;). This has benefits in that trained peers can identify psychological risk factors that non-peers might not notice or appreciate (Jones et al., 2003).

Fifth, Jones et al. (2003) also note a key process issue, specifically that external practitioners lack in-depth organizational understanding. This results in employee hostility to outsiders and poor receipt of interventions. The suspicion and distrust of external practitioners is further complicated by the strong role identity seen in emergency services personnel, aligned with the culture of suspicion created by emergency services' unusual industrial relations. This is then exacerbated by the unusual nature of the occupational role in terms of the sights, sounds and smells that personnel have to face and deal with; limiting the discussions they feel they can have with someone who has not experienced these. This can be because of a disbelief that someone with no such experience could ever truly empathise or understand; but also because of an unwillingness to burden others with what they themselves have faced.

Interventions which use trained peers are therefore becoming increasingly popular within the emergency services. This has led to an increased uptake of the TRiM process (Greenberg et al., 2010) which was specifically designed to be delivered by trained peers. However, emergency services are also developing their own versions of CISD and CISM, adapting and modifying these to fit their own local organizations. One example of this is what has been termed "the Tyne and Wear approach" (see Barber, 2003; Lawrence and Barber, 2004;) which originated in Tyne and Wear Fire and Rescue Service but which has subsequently been adopted more widely. Trained peers provide both the initial diffusing and the debriefing following a traumatic event. However, as well as having access to these peer-led interventions there is also support available from chaplains, occupational health workers, a psychologist and a psychiatrist. This additional support can be accessed by the whole group by agreement, individuals can self-refer, or the peer supporters can refer individuals for therapeutic treatment if required. The peer trauma team also offers support to one another considering case studies to explore best practice and also debriefing one another to prevent the development of secondary trauma or burnout. The peer supporters are voluntary applicants who go through a selection process and then undergo extensive training (this can be in house but may also lead to external formal qualifications).

In both its original and modified formulations CISD has not been without its critics. Raphael, Meldrum and MacFarlane (1995) noted that CISD had rarely been systematically evaluated with no randomised controlled trials (RCT) being reported. However, whether RCT are either appropriate or ethical in the case of

CISD is highly debatable. The different formulations of CISD mean that different intervention processes are being confused and would suffer unfair comparison if RCTs appeared to provide an authoritative voice on the matter. Further the real world conditions mean that the allocation of groups are hardly random and are certainly not controlled. This is self-evident given that CISD is employed because of unpredictable and chaotic events (Deahl, 2000). Deahl (2000) also notes the ethical problems in having the non-intervention group required by RCT because denying one group the opportunity for debriefing may be detrimental, particularly given that many individuals find it substantially helpful at the time.

Much of the criticism of CISD stems from the Cochrane Review (Rose, Bisson, and Wessely, 2002). However, as Devilly and Cotton (2003) note, this review evaluated only single-session interventions meaning it did not consider the majority of CISD programmes, or reflect CISD as originally intended (Mitchell, 2003). It is fundamental to any rigorous evaluation to compare like with like, but this has rarely been the case with CISD evaluations.

Aside from trauma interventions, the emergency services also need to manage other stressors and strain. This includes not only the types of stress seen in any organization but also those peculiar to emergency service work. Within the emergency services even those stressors seen in other organizations take on additional significance because they can ultimately lead to harm in the field, risking the safety of both emergency service personnel and the communities they serve. Thus stressors have greater importance and urgency than among other working populations and it therefore behoves emergency services to look carefully at issues such as support, workload and communication. Removing and reducing generic stressors also reduces the effect that the work-peculiar stressors will have (see Fletcher, 1991).

Stress prevention

Notions of stress resilience focus interventions on preparing the emergency worker, as exemplified by emergency services' efforts to create the "safe worker" (Ash and Smallman, 2008) by way of appropriate personnel selection, instruction and training. Training reduces cognitive load (Paas, 1992) and so general training for regular activities can help, as well as training specific to the work stressors faced by emergency service personnel. The latter form of training carries twin benefits of developing specific strategies for coping with those particular stressors and automatizing responses (although greater problems can then arise if these automatized responses are disrupted by the nature of the emergency; see Hill and Brunsden, 2003). Research suggests that this strategy development gives greater choice to the worker, increasing control (Karasek, 1979), and leaving more cognitive capacity to face other less predictable challenges. For example, Kagan, Kagan, and Watson's (1995) study of emergency medical workers found that training in interpersonal coping and developing interpersonal awareness (both important in emergency situations) were associated with lower levels of anxiety and depression. Similarly, Michie and Williams (2002) found that problem-solving training helped reduce strain levels. Training and personal development can also help personnel to become more stress-resilient in other ways. Wang et al. (2010) found that greater feelings of self-esteem, which partly comes from feelings of competence (Warr, 1987), were predictive of lower levels of depression in police officers. Training can therefore be seen as an important stress management tool, however it should also be remembered that training for emergency service personnel poses special challenges since, if it is to be realistic, training can in itself be a source of danger (Cooper and Cotton, 2000).

Stress monitoring, the minimum required by law, has been an important plank in emergency services' intervention strategies. However, because there are no rigorously precise ways for an employer to assessing levels of strain, legal standards are rarely specified. Advice from the Health and Safety Executive (HSE) in Great Britain is that employers, as well as encouraging symptom reporting, should check sickness records (HSE, 1999). This identifies patterns across shifts and watches, as well as individual problems. There is also less formal but continual monitoring by way of the continual interaction of colleagues and family who can notice small changes; allied with referral processes to get affected individuals to the correct professional, such informal strategies can be highly effective. Any consequent reactive help from healing professionals might also improve later stress resilience. It should however be kept in mind that, regardless of pre-exposure stress management strategies, exposure is still inevitable and that post-exposure interventions are likely to always dominate stress policies within the emergency services.

Conclusions

It is clear that environmental exposure, the cultural context and the resultant co-worker, inter- and intra-familial relations, all interact to affect the success (or otherwise) of interventions within the emergency services. The emergency services thus serve as a useful example to demonstrate the need for knowing and understanding an organization before attempting to develop or deliver interventions. Although this chapter's focus has been on the emergency services there is a wider message here, in that all organizations will have their own occupational, environmental, and social contexts that have the potential to affect and distort the utility of specific interventions. It has been suggested that the only successful way to understand a culture is to live in it, thus allowing understanding of linguistic nuances and practices (Gomm and Hammersley, 2001). Where external practitioners enter organizations as an outside consultant or "external expert" they need to ensure that they are able to appreciate the organization's specific context and gain an understanding of that before blithely attempting to conduct audits or administer interventions. Without this deeper knowledge there is the potential not only to alienate the workforce and obtain inaccurate audit data but also to deliver interventions in ways that could risk doing more harm than good.

References

Archer, D. (1999). Exploring "bullying" culture in the para-military organization. *International Journal of Manpower, 20*(1/2), 94–105.

Ash, J., & Smallman, C. (2008). Rescue missions and risk management: highly reliable or over committed? *Journal of Contingencies and Crisis Management, 16*(10), 37–52

Bacharach, S.B., Bamberger, P.A., & Doveh, E. (2008). Firefighters, critical incidents, and drinking to cope: The adequacy of unit-level performance resources as a source of vulnerability and protection. *Journal of Applied Psychology, 93*(1), 155–169.

Bain, G., Lyons, M., & Young, A. (2002). *The Future of the Fire Service: Reducing Risk, Saving Lives: The Independent Review of the Fire Service.* London: Office of the Deputy Prime Minister.

Baker, S.R., & Williams, K. (2001). Relation between social problem solving appraisals, work stress, and psychological distress in male firefighters. Stress and Health. *Journal of the International Society for the Investigation of Stress, 17*(4), 219–229.

Barber, G. (2003). Critical Incident debriefing: a personal perspective. *Fire Safety, Technology and Management, 8*(3), 31–34.

Beale, A. (2003). Researching stress in emergency first responders. *Journal of Fire Safety, Technology & Management, 8*(2), 25–30.

Beaton, R., Murphy, S., Johnson, C., Pike, K., & Corneil, W. (1999). Coping responses and posttraumatic stress symptomatology in urban fire service personnel. *Journal of Traumatic Stress, 12*(2), 293–308.

Beehr, T.A., Johnson, L.B., & Nieva, R. (1995). Occupational stress: coping of police and their spouses. *Journal of Organizational Behavior, 16*(1), 3–25.

Bennett, R.R., & Greenstein, T. (1975). The police personality: a test of the predispositional model. *Journal of Police Science and Administration, 3*(4), 439–445.

Blaney, L. (2005). Leaping into the inferno: CISM works! *Journal of Fire Safety, Technology & Management, 9*, 19–26.

Blaney, L. (2009). Beyond 'knee jerk' reaction: CISM as a health promotion construct. *Irish Journal of Psychology, 30*(1–2), 37–58.

Brenner, I., Shek, P.N., Zamecnik, J., & Shephard, R.J. (1998). Stress hormones and the immunological responses to heat and exercise. *International Journal of Sports Medicine, 19*(2), 130–143.

Brough, P. (2002). Female police officers' work experiences, job satisfaction and psychological well-being. *Psychology of Women Section Review, 4*, 3–15.

Brough, P. (2004). Comparing the influence of traumatic and organizational stressors on the psychological health of police, fire, and ambulance officers. *International journal of Stress Management, 11*(3), 277–244.

Brown, J.M., Fielding, J., & Grover, J. (1999). Distinguishing traumatic, vicarious and routine operational stressor exposure and attendant adverse consequences in a sample of police officers. *Work and Stress, 13*, 312–325.

Brown, J., Mulhern, G., & Joseph, S. (2002). Incident-related stressors, locus of control, coping, and psychological distress among firefighters in Northern Ireland. *Journal of Traumatic Stress, 15*(2), 161–168.

Brown, J.M., & Campbell, E.A. (1990). Sources of occupational stress in the police. *Work and Stress, 4*(4), 305–318.

Brunsden, V. (2007). Award-winning research reveals truth of attacks on firefighters. *FIRE, 100*(1229), 41–43.

Brunsden, V., & Goatcher J. (2009). Firefighters' experiences of attack. *Violence in Public Places and Institutions.* University of Central Lancashire. 25–27 June 2009.

Brunsden, V., & Hill, R. (2009). Firefighters' experience of strike: an interpretative phenomenological analysis case study. *Irish Journal of Psychology, 30*(1–2), 99–115.

Brunsden, V., Hill, R., McTernan, J., & Shuttlewood, A. (in press). Under fire: firefighters' coping with aggression and violence. *International Fire Service Journal of Leadership and Management.*

Brunsden, V., Robinson, J., Goatcher, J., & Hill, R. (in press). "Brick, Ball, Hoax call": The dissemination of psychological research on attacks on firefighters through multi-media art-works. In P. Vannini (ed.). *Popularizing Research: engaging new genres, media, and audiences.* New York: Peter Lang Publishing.

Brunsden, V., Woodward, L., & Regel, S. (2003). Occupational stress and posttraumatic reactions in fire-fighters and control room staff. *Fire Safety, Technology and Management, 8*(3), 11–14.

Brunsden, V., Woodward, L., & Wilson, S. (2007). Stress in union officials: an issue for managerial concern? *International Fire Service Journal of Leadership and Management, 1*(2), 17–22.

Caplan, R. (1987). Person-environment fit theory and organizations: commensurate dimensions, time perspectives, and mechanisms, *Journal of Vocational Behavior, 31,* 248–267.

Communities and Local Government. (2006). *Tackling Violence at Work: good practice guidance document for Fire and Rescue Services.* London: HMSO.

Cooper, M., & Cotton, D. (2000). Safety training – a special case? *Journal of European Industrial Training, 24*(9), 481–490.

Courtney, J.A., Francis, A.J.P., & Paxton, S.J. (2010). Caring for the carers: fatigue, sleep and mental health in Australian paramedic shiftworkers. *Australian and New Zealand Journal of Organizational Psychology, 3*(1), 32.

Cox, T. (1993). *Stress Research and Stress Management: putting theory to work* (HSE Contract research report No. 61/1993). London: HSE HMSO.

Deahl, M. (2000). Psychological debriefing: controversy and challenge, *Australian and New Zealand Journal of Psychiatry, 34,* 929–939.

Demerouti, E., Geurts, S.A.E., Bakker, A.B., & Euwema, M. (2004). The impact of shiftwork on work-home conflict, job attitudes and health. *Ergonomics, 47*(9), 987–1002.

Devilly, G., & Cotton, P. (2003). Psychological debriefing and workplace: defining a concept, controversies and guidelines for intervention, *Australian Psychiatrist, 38*(2), 144–150.

Durkin, J. (2006), Isn't it NICE to be ignored when you're stressed? *Counselling at Work, Summer,* 13–16.

Dyregov, A. (1989). Caring for helpers in disaster situations: psychological debriefing. *Disaster Management, 2*(1), 25–30.

Ernst, A. T., Jiang, H., Krishnamoorthy, M., & Sier, D. (2004). Staff scheduling and rostering: a review of applications, methods and models. *European Journal of Operational Research, 153*(1), 3–27.

Eyre, A., & Brunsden, V. (2003). *Essex County Fire & Rescue Service: Stress Management Programme, Progress Report.* Essex: Essex County Fire & Rescue Service.

Family Safety and Health (2006). Study shows how shiftwork may strain family well-being. *Family Safety and Health, 65*(3), 4.

Fannin, N., & Dabbs, J.M. (2003). Testosterone and the work of firefighters: fighting fires and delivering medical care. *Journal of Research in Personality, 37*(2), 107–115.

Fletcher, B. (1991). *Work, stress, disease, and life expectancy*. Chichester: John Wiley and Sons.

Galdas, P.M., Cheater, F., & Marshall, P. (2005). Men and health help-seeking behaviour: literature review. *Journal of Advanced Nursing, 49*(6), 616–623.

Getahun, S., Sims, B. & Hummer, D. (2008). Job Satisfaction and Organizational Commitment Among Probation and Parole Officers: A Case Study. *Professional Issues in Criminal Justice, 3*(1), 39–56.

Gist, R., & Woodall, S. J. (1998). Social science versus social movements: the origins and natural history of debriefing. *The Australasian Journal of Disaster and Trauma Studies, 1*. Retrieved January 15 2011, from www.massey.ac.nz/~trauma/issues/1998-1/gist1.htm

Gomm, R., & Hammersley M. (2001). Thick ethnographic description and thin models of complexity. *Presented at the Annual Conference of the British Educational Research Association*, University of Leeds, England. 13–15 September 2001.

Greenberg, N., Langston, V., Everitt, B., Iversen, A., Fear, N., Jones, N., & Wessely, S. (2010). A cluster randomised control trial to determine the efficacy of Trauma Risk Management (TRiM) in a military population, *Journal of Traumatic Stress, 23*(4), 430–436.

Greenberg, N., Langston, V., & Scott, R., (2006). How to TRiM away at post traumatic stress reactions: Traumatic Risk Management – Now and in the Future. In *Human Dimensions in Military Operations – Military Leaders' Strategies for Addressing Stress and Psychological Support* (pp. 35-1 – 35-6). Meeting Proceedings RTO-MP-HFM-134, Paper 35. Neuilly-sur-Seine, France: RTO.

Guidotti, T.L. (2000). Stress in firefighters. In G. Fink. *Encyclopaedia of Stress: Volume II*. San Diego, CA: Academic Press.

Hashem, F., & Lilly, J. (2007). Career aspirations and desirability: minority ethnic young people and their perceptions of the fire service as a career. *International Fire Service Journal of Leadership and Management, 1*(2), 23–27.

Health & Safety Executive. (1999). *Understanding Health Surveillance at Work: An introduction for employers*. London: HMSO.

Hill, R., & Brunsden, V. (2003). Surviving disaster: Firefighters as victims – preliminary findings. *Fire Safety, Technology and Management, 8*(3), 21–24.

Hill, R., & Brunsden, V. (2006). I love a firefighter; an IPA analysis of the experiences of operational FRS staff's relatives. *Presented at the Fire Service College Research Event*. Moreton in the Marsh, Gloucestershire. 14–16 November 2006.

Hill, R., & Brunsden, V. (2008). *Identifying the Needs of Relatives of Highlands and Islands Fire and Rescue Service Firefighters*. Inverness: Highlands and Islands Fire and Rescue Service.

Hill, R., & Brunsden, V. (2009). 'Heroes' as victims: role reversal in the fire and rescue service. *Irish Journal of Psychology, 30*(1–2), 75–86.

Holmes, A. (2003). Sleepless knights: sleep loss and its contribution to occupational stress in the UK fire service. *Fire Safety, Technology and Management, 8*(2), 7–10.

Jackson, S.E., & Maslach, C. (1982). After-effects of job-related stress: Families as victims. *Journal of Occupational Behaviour, 3*, 63–77.

Jeannette, J.M., & Scoboria, A. (2008). Firefighter preferences regarding post-incident intervention. *Work and Stress, 22*(4), 314–326.

Jones, N., Roberts, P., & Greenberg, N. (2003). Peer-group risk assessment: a post-traumatic management strategy for hierarchical organizations, *Occupational Medicine, 53*, 469–475

Joseph, S., Brown, J., & Mulhern, G. (2003). Stress and coping fire-fighters: implications for occupational health psychology. *Fire Safety, Technology and Management*, 8(2), 17–20.

Kagan, N.I., Kagan, H., & Watson, M.G. (1995). Stress reduction in the workplace: the effectiveness of psychoeducational programs. *Journal of Counselling Psychology*, 42(1), 71–78.

Karasek, R. (1979). Job demands, job decision latitude, and mental strain: implications for job redesign. *Administrative Science Quarterly*, 24(2), 285–308.

Kirkcaldy, B., Brown, J., & Cooper, C.L. (1998). The demographics of occupational stress among police superintendents. *Journal of Managerial Psychology*, 13(1–2), 90–101.

Lamontagne, A. D., Keegel, T., Louie, A. M., Ostry, A., & Landbergis, P. A. (2007). A Systematic Review of the Job-stress Intervention Evaluation Literature, 1990–2005. *International Journal of Occupational and Environmental Health*, 13, 268–280.

Landsman, I.S., Baum, C.G., Arnkoff, D.B., Craig, M.J., Lynch, I., Copes, W.S., & Champion, H.R. (1990). The psychosocial consequences of traumatic injury. *Journal of Behavioural Medicine*, 13(6), 561–581.

Lasky, R. (2004). Pride and ownership: the love for the job. Our two families. *Fire Engineering*, 157(9), 91–92.

Lawrence, L. (2003). Stress in the fire service: a therapist's perspective. *Fire Safety, Technology and Management*, 8(3), 35–37.

Lawrence, L., & Barber, G. (2004). Debriefing in the fire service. *Counselling at Work*, Winter, 11–13.

Lee, S-H., & Olshfski, D. (2002). Employee commitment and firefighters: it's my job. *Public Administration Review*, 62(s1), 108.

Maguire, K., & MacPherson, D. (1997). *Norfolk Fire Service Stress Audit: Final Report*. Nottingham: Nottingham Trent University.

Malek, M.D.A., Mearns, K., & Flin, R. (2003). Stress and well-being in firefighters: a review of the literature. *Fire Safety, Technology and Management*, 8(3), 1–6.

Markowitz, J.S. (1989). Long-term psychological distress among chemically exposed firefighters. *Behavioral Medicine*, 15, 75–83.

Menendez, A.M., Molloy, J., & Magaldi, M.C. (2006). Health responses of New York city firefighter spouses and their families post-September 11, 2001 terrorist attacks. *Issues in Mental Health Nursing*, 27, 905–917.

Michie, S., & Williams, S. (2003), Reducing work related psychological ill health and sickness absence: a systematic literature review, *Occupational and Environmental Medicine*, 60, 9.

Mitchell, J.T. (1983). When disaster strikes...the critical incident stress debriefing process. *Journal of Emergency Medical Service*, 8, 36–39.

Mitchell, J.T. (2003). Crisis intervention and CISM: a research summary. Retrieved November 2010: http://tinyurl.com/378ccvq

Mitchell, J.T. (2004). Characteristics of successful early intervention programs. *International Journal of Emergency Mental Health*, 6(4), 175–184.

Mitchell, J.T., & Bray, G.P. (1990). *Emergency Services Stress: Guidelines for preserving the health and careers of emergency services personnel*. Englewood Cliffs, NJ: Prentice-Hall.

Moos, R.H., & Moos, B.S. (1976). A typology of family social environments. *Family Process*, 15(4), 357–371.

Murphy, S.A., Beaton, R.D., Cain, K., & Pike, K. (1999). Gender differences in fire-fighter job stressors and symptoms of stress. *Women's Health*, 22(2), 55–69.

Neale, A.V. (1991). Work stress in emergency medical technicians. *Journal of Occupational Medicine*, *33*(9), 991–997.

Nixon, S.J., Schorr, J., Boudreaux, A., & Vincent, R.D. (1999). Perceived sources of support and their effectiveness for Oklahoma city firefighters. *Psychiatric Annuals*, *29*(2), 101–105.

North, C.S., Tivis, L., McMillen, J.C., Pfefferbaum, B., Cox, J., Spitznagel, E.L, Bunch, K., Schorr, J., & Smith, E.M. (2002). Coping, functioning and adjustment of rescue workers after the Oklahoma City bombing. *Journal of Traumatic Stress*, *15*(3), 171–175.

Paas, F. (1992). Training strategies for attaining transfer of problem-solving skill in statistics: a cognitive-load approach. *Journal of Educational Psychology*, *84*(4), 9–434.

Parkinson, F. (1993). *Post-Trauma Stress*. London: Sheldon Press.

Paton, D. (2006). Posttraumatic growth in disaster and emergency work. In L. G. Calhoun & R. G. Tedeschi (Eds). *Handbook of Posttraumatic Growth: Research and Practice*. (2006). New Jersey: Lawrence Erlbaum Associates, Inc.

Pendleton, M., Stotland, E., Spiers, P., & Kirsch, E. (1989). Stress and strain among police, firefighters, and government workers: a comparative analysis. *Criminal Justice and Behavior*, *16*, 196–210.

Raphael, B., Meldrum, L., & MacFarlane, A.C. (1995). Does debriefing after psychological trauma work? *British Medical Journal*, *310*, 1479–1480.

Regehr, C. (2009). Social support as a mediator of psychological distress in firefighters. *Irish Journal of Psychology*, *30*(1–2), 87–98.

Regehr, C., Hill, J., & Glancy, G.D. (2000). Individual predictors of traumatic reactions in firefighters. *Journal of Nervous & Mental Disease*, *188*(6), 333–339.

Regehr, C., Dimitropoulos, G., Bright, E., George, S., & Henderson, J. (2005). Behind the brotherhood: rewards and challenges for wives of firefighters. *Family Relations*, *54*(3), 423–435.

Regel, S. (2007). Post-trauma support in the workplace: the current status and practice of critical incident stress management (CISM) and psychological debriefing (PD) within organizations in the UK. *Occupational Medicine*, *57*, 411–416.

Regel, S., Woodward, L., Horsley, R., & Brunsden, V. (2001). *Audit of Operational Stressors & Post Traumatic Reactions in Firefighters & Control Room Staff*. Nottingham: Centre for Trauma studies, Nottinghamshire Healthcare NHS Trust.

Rose, S., Bisson, J., & Wessely, S. (2002). Psychological debriefing for preventing post traumatic stress disorder (PTSD) (Cochrane Review). *The Cochrane Library*, *2*, Oxford: Update Software.

Salters-Pedneault, K., Ruef, A.M., & Orr, S.P. (2010). Personality and psychophysiological profiles of police officer and firefighter recruits. *Personality and Individual Differences*, *49*(3), 210–215.

Schumm, W.R., Bell, D.B. & Resnick, G. (2001). Recent research on family factors and readiness: implications for military leaders. *Psychological Report*, *89*, 153–165.

Sigurdsson, J. and Dhani, A. (2010). *Police Service Strength. England and Wales, 31 March 2010*, 14/10. Home Office, Retrieved November 2010: http://tinyurl.com/39usjuc

Smith, (2007). *Job Satisfaction in the United States*. Chicago: NORC/University of Chicago. Retrieved November 2010: http://tinyurl.com/32h96ak

Varvel, S.J., He, Y., Shannon, J.K., Tager, D, Bledman, R.A., Chaichanasakul, A, Mendoza, M.M., & Mallinckrodt, B. (2007). Multidimensional, threshold effects of social support in firefighters: is more support invariably better? *Journal of Counselling Psychology*, *54*(4), 458–465.

Vernberg, E., Steinberg, A., Jacobs, A., Brymer, M., Watson, P., Osofsky, J.D., Layne C.M., Pynoos, R.S., & Ruszek, J.I. (2008). Innovation in disaster mental health: psychological first aid. *Professional Psychology: Research and Practice, 39*(4), 381–388.

Wagner, S. L. (2005). The "Rescue Personality": Fact or Fiction? The Australasian Journal of Disaster and Trauma Studies, 2005–2: www.massey.ac.nz/~trauma/issues/2005-2/wagner.htm (last accessed January, 2012).

Wang, Z., Inslicht, S., Metzler, T., Henn-Haase, C., McCaslin, S., Tong, H., Neylan, T., & Marmar, C. (2010). A prospective study of predictors of depression symptoms in police. *Psychiatry Research, 175*, 211–216.

Warr, P. (1987). *Work, Unemployment and Mental Health*, Oxford, Clarendon.

Wastell, D., & Newman, M. (1996). Information system design, stress and organizational change in the ambulance services: A tale of two cities. *Accounting, Management and Information Technologies, 6*(4), 283–300.

Webster, S., Patterson, A., Hoare, J., & O'Loughlin, A. (2007). *Violence at Work: Findings from the 2005/06 and 2006/07 British Crime Survey*. Bootle: Health & Safety Executive.

Wester, S.R., & Lyubelsky, J. (2005). Supporting the Thin Blue Line: Gender-Sensitive Therapy with Male Police Officers. *Professional Psychology: Research and Practice, 36*(1), 51–58.

Wethington, E. (2000). Contagion of stress. *Advances in Group Processes, 17*, 229–53.

Wagner, S. L. (2005). The "rescue personality": fact or fiction? *The Australasian Journal of Disaster & Trauma Studies, 2*. Retrieved January 15 2011, from www.massey.ac.nz/~trauma/issues/2005-2/wagner.htm

Woodward, L., Brunsden, V., & Regel, S. (2000). *Audit of Operational Stressors & Post Traumatic Reactions in Firefighters & Control Room Staff; report of the pilot study*. Nottingham: Centre for Traumatic Stress – Research and Practice, Nottingham Trent University.

Young, K.M., & Cooper, C.L. (1997). Occupational stress in the ambulance service: a diagnostic study. *Health Manpower Management, 23*(4), 140–147.

13 The development of smart and practical small group interventions for work stress

John Klein Hesselink, Noortje Wiezer, Heleen den Besten, and Erna de Kleijn

Practitioners need smart and practical organizational interventions that can be easily applied to implement quick changes in small groups of workers with high levels of work stress. To establish the effectiveness of interventions, most practitioners rely on scientific evidence and standards which are often difficult to apply or can be impractical in smaller settings. In this chapter, suggestions are given for a more straightforward intervention approach to change the organizational environment of small groups of workers with alternatives for measurement and the use of control groups.

Introduction

Rapidly changing organizational contexts rarely offer opportunities for adequate reflection on a detailed evaluation of work stress interventions. Often, external practitioners and consultants are hired to address identified problems. The topic of stress at work is complex, and finding the appropriate intervention approach with sound methods to evaluate its effectiveness is often impracticable. Before starting an intervention practitioners would need to be certain that the intervention is likely to be effective and demonstrate that it can generate the expected outcomes. However, these criteria are not easy to meet, especially when intervention groups are small and control groups are difficult to find. Practitioners who work with small groups in organizations may turn to the scientific literature to find solutions for smart, practical, and evidence-based organizational interventions for reducing work-related stress in small groups of employees. They may discover, however, that this research area is not very extensive yet, certainly when the focus is changing organizational and work-related aspects in order to reduce exposure to occupational stressors. The scientific and practical reasons for this lack of information are explored in the following section.

Reasons why practical information on interventions is scarce

One reason that explains the scarcity of information in scientific journals on smart, practical and evidence-based organizational interventions is that the history of organizational stress intervention research itself is not very old or extensive. In

1992 for instance, an important conclusion of the American Psychological Association (APA) and the National Institute of Safety and Health (NIOSH) conference was that three decades of fundamental research had generated much knowledge on the aetiology and consequences of occupational stress, but very few studies focused on preventing or reducing exposure to occupational stressors with appropriate interventions. More intervention research was deemed necessary (Keita and Sauter, 1992). In the following years, the number of work stress intervention studies steadily increased and generated new research on the effect of individual-level interventions, aiming to help individual workers to cope with their work (see, for example, Richardson and Rothstein, 2008; van der Klink et al., 2001). These studies also indicate that organizational-level interventions are scarce, and that it is difficult to draw firm conclusions about their effectiveness. Besides, these systematic reviews are methodologically orientated and meant to evaluate the effect of individual-level and organizational-level interventions, but do not address "how" and "why" interventions produce their effects (Griffiths, 1999; Kompier and Kristensen, 2000; Kristensen, 2005). Only a minority of intervention studies concentrate on organizational-level interventions, tackling problems at their source. The development of methods for small group preventative interventions that change organizational factors lags behind. But the field is developing and organizational-level interventions are emerging. For instance, Brun, Biron, and Ivers (2008), Cox et al. (2000), Cox, Randall, and Griffiths (2002), Klein Hesselink et al. (2001), Kompier, Cooper, and Geurts (2000) offer needed practical guidelines to implementing organizational interventions to reduce exposure to psychosocial risks and improve well-being.

Another reason explaining the scarcity of practical guidelines for small-scale organizational interventions in the research literature is that scientific journals require research with sound methodology and unequivocal evidence. For practical organizational work stress interventions in small groups of employees, these methodological requirements and this type of evidence are incompatible and impossible to apply, as will be seen in the course of this chapter. In accordance with these methodological requirements, the ideal way to test intervention effectiveness is by using an experimental research design, which allows to establish the causal effects between the intervention and the changes observed in outcomes. Originating from the natural sciences, such as physics, this experimental research design is often impractical and difficult to apply for evaluating the effectiveness of interventions in organizational settings (Cox et al., 2007; Griffiths, 1999).

In addition, organizational work stress interventions face practical barriers. In her article *Organizational Interventions: Facing the limits of the natural science paradigm*, Griffiths (1999) characterizes the differences between a social experiment and a natural science experiment. This latter method includes scientific standards like testing cause-and-effect hypotheses with random allocation of subjects to treatment and control groups. In this laboratory-like situation, principles like temporal priority, control of potentially confounding variables and random assignment to conditions are used to discern whether desired changes occur as a result of the manipulation of some important variable or the introduction of a particular treatment. An experiment in "real" life is called a "social experiment". In real life settings, however, the conditions for an experiment rarely apply. Griffiths (1999) gives five reasons why these conditions are difficult to meet (see also Cox et al., 2007; Semmer, 2006):

1. researchers are often guests in an organization, not autocrats who have been given "carte blanche" to do whatever they see fit,
2. causal relationships are not simple; they are embedded in an extremely complex context and are therefore not simple to establish,
3. the effects are seldom caused by a single factor,
4. it is nearly impossible to allocate participants randomly in a control and intervention group in an organizational context, and
5. even if groups can be allocated randomly, there are countless other factors besides the intervention being investigated that may affect or explain the results.

Cox et al. (2007) and Griffiths (1999) provide several more reasons why the effects of a social experiment are difficult to establish:

6. reductions in exposure to risks can be observed only in the long term, during which period many confounding factors may affect the results,
7. organizational stress interventions usually imply a chain of cause and effect combinations and each relationship has different context variables and possible disruptions, and
8. performing a social experiment requires considerable time and also commitment by the participants.

This may not be convenient to participants, who may already have to cope with work demands. Consultants therefore often choose to implement an intervention without evaluating its effects on outcomes.

Another barrier to performing intervention studies in practical situations is the lack of focus on organizational change. Reviews by Richardson and Rothstein (2008) and van der Klink et al. (2001) show that organizations have a preference for interventions directed at the individual (individual-level) instead of interventions directed at changing the work environment (organizational-level). Organizations are often reluctant to change existing production or working conditions because this may involve considerable effort and uncertainty. Kompier et al., (1998) and Kompier and Cooper (1999) discuss four reasons why organizations prefer individual rather than organizational-level interventions:

1. values and norms of the management of organizations are often directed at "blaming" the individual worker,
2. occupational psychology and medicine are traditionally focused on individual differences,

3   random assignment of cases is easier to achieve with individuals than with organizations, and
4   stress research is traditionally directed at subjective outcomes rather than organizational outcomes like productivity, sickness absence, and accident rates.

Why is it important to concentrate on organizational interventions?

The stress literature distinguishes between primary, secondary, and tertiary intervention (e.g. Lamontagne et al., 2007; van der Klink et al., 2001). Primary intervention is proactive and preventative and means that risks are eliminated at source. For instance, if employees do not lift or carry heavy loads, back problems related to these risks cannot develop. In the case of occupational stress, this implies reducing exposure to work-related stressors to acceptable levels so that negative effects do not emerge. Secondary intervention aims to modify and improve individuals' response to stressors (e.g. by modifying their perceptions of stressful situations). Tertiary intervention is reactive and aims to minimize the consequences of stress. In the research literature and also in legislation and prevention there is a clear preference for primary organizational-level interventions since they eliminate the sources of stress instead of the consequences.

No intervention published does not mean no intervention implemented

The fact that there are not many published articles on organizational-level interventions to ameliorate work-related stress does not mean that these interventions do not take place. Evidence from a Dutch study shows that many more interventions are implemented in practice than can be read about in the published literature (Klein Hesselink et al., 2010; van Hooff, van den Bossche and Smulders, 2008). In addition, the need for interventions is acknowledged by both employers and employees. Based on data from the 2009 wave of the Netherlands Working Conditions Survey (Koppes et al., 2010) and a representative sample of 22,768 employees, Figure 13.1 shows that employees in organizations of various sizes and a range of sectors consider that stress interventions are necessary. Specifically, 80 per cent of all employees considered it necessary for their organization to take measures regarding their workload and occupational stress, whereas only 11 per cent reported that no measures were taken (or that they were not aware of any measures). The need for measures to reduce work-related stress was felt more strongly in large organizations and the education, health and public care, and public administration sectors.

Employers' views reflect employees' views. Of the Dutch employers questioned, 46 per cent indicated that work stress is an important risk (Oeij et al., 2009). As in the employee survey, there was a strong positive relationship between employers' reporting of work stress and company size. In addition, many companies in education (85 per cent), health and social care (72 per cent), and public

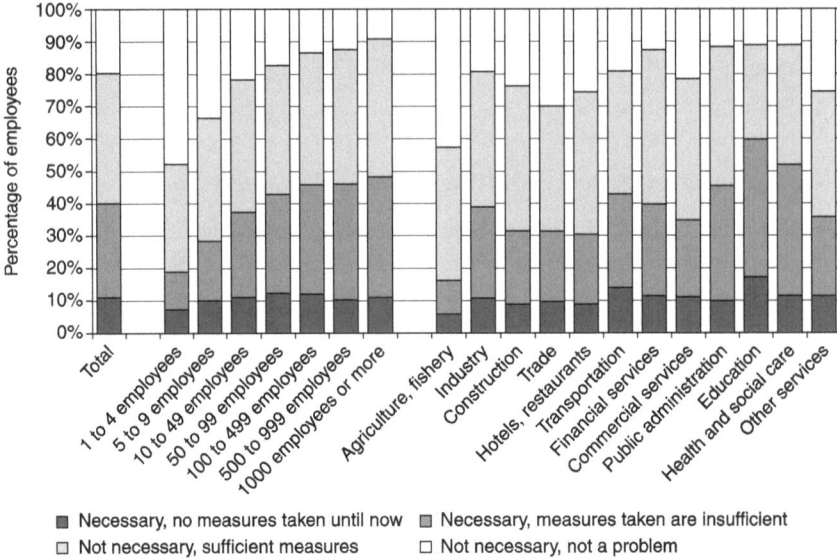

Figure 13.1 Answers to the question: "Do you consider it necessary that your organization takes interventions (measures) against work pressure and work stress?" by company size and sector.

Source: Koppes et al., 2010.

administration (72 per cent) reported work stress as an important risk in the company. Both employers and employees reported that tackling work stress is important for organizations and it is necessary to address it. There is no reason to believe that these findings are restricted to Dutch organizations.

How can we evaluate organizational interventions without robust experimentation?

A sound experimental design will always be necessary to establish cause-effect relationships between interventions and their effects. Employers can benefit from this type of result, and are also sometimes prepared to participate in this type of study. However, most employers' interests lie in interventions which are easy to apply *and* likely to be effective. They understand the need for sound methods to assess intervention outcomes, but are often unwilling or unable to participate in studies with strict methodological requirements. In other words, when you have a headache, you need a medicine, not an experiment to test if the medicine works. Therefore, there is a need to develop methods to evaluate organizational interventions in small groups of employees in a smart and practical way.

In the past decade, researchers such as Cox and his colleagues (2000, 2002, 2007), Griffiths (1999), Kompier and Kristensen (2000), and Kristensen (2005) started to challenge the view that the effectiveness of organizational work stress interventions can only be established by means of randomized control trials and quasi-experimental research designs. They state that experimental and control groups in organizations differ such that direct comparison is hardly possible. According to Kristensen (2005), three research questions are important in occupational intervention studies:

1. is the intervention carried out as intended?,
2. did the intervention lead to the intended change of exposure to stressors? (in a laboratory setting the researcher is, by definition, in control of the exposure, so this question is not relevant in this setting but is highly relevant in field settings), and
3. did the exposure to stressors bring the intended effect on the outcomes?

To answer the last question, aetiological studies are necessary, which require large samples, a clear endpoint (health/disease), randomization of participants in intervention and control groups, quantitative methods, and representative samples. In practice, however, for most organisational interventions, a large aetiological study is not necessary, because the effects of stressors on outcomes are already known from existing theory and research. Especially in small groups, this type of design is often not practical or viable. Yet, in order to evaluate the effectiveness of interventions in small groups, it is important to know if the endpoint (e.g. the exposure to stressors or stress levels) has improved. To obtain such results in small-scale interventions, mixed methods or case study approaches are more appropriate.

There are studies which are both rigorous and useful for organizations, and which allow them to implement interventions and evaluate their process and outcomes (Bond and Bunce, 2001; Bourbonnais et al., 2006; Brun, Biron, and Ivers, 2008; Nielsen et al., 2006; Nielsen et al., 2009). In each of these studies, comparisons with control groups were possible and reductions in exposure to stressors and improvements in health outcomes were found (either mental health or physical health, or well-being in a more general sense). Since this type of design is not always possible in small groups, mixed methods can be used fruitfully to evaluate the intervention outcomes. Nielsen et al. (2010), for instance, conclude that mixed method designs are needed to integrate process and outcome evaluation and increase the generalizability of interventions. Cox et al. (2007) use the term 'fit for purpose' to indicate that the correct approach to obtain good quality data has to be judged against the purpose of obtaining data. This is not an alternative to a scientific approach but a suggestion for a more broadly conceived and eclectic scientific framework for intervention evaluation.

In this chapter we advance on this line of thinking and present five organizational work stress interventions in small groups of employees. We suggest that although strict methodological requirements to claim causality can not always be met in small groups, effective interventions can still be developed, implemented, and evaluated. If a control group is available, it should be used to understand the effects of the intervention. If not, there are other ways to evaluate intervention effectiveness.

Our research and experience

In our work as consultants we assist organizations with implementing organizational change and redesign work environments to improve work quality and performance. Besides being involved in quasi-experimental evaluation studies (Klein Hesselink, 2002), we are also involved in practical interventions developed/based on the principles discussed in the previous section. We thus often experience the tension between rigid standards for careful experimentation and scientific evaluation on one hand, and the demands for smart and practical interventions, preferably conducted with high employee participation, on the other.

As consultants, we are involved in the process of implementing the interventions but aim to keep employees as much responsible as possible for the changes in their working situation. The lack of scientific rigour is compensated by the use of a mixed method design with additional interviewing and a participative approach in which information between the consultants and the employees is shared and communicated explicitly. The measurement of health and organizational outcomes is not valued strongly by the participating employees. Rather, their primary focus is on changes in their working situation. In terms of Kristensen's (2005) three research questions that are important in occupational intervention studies:

1 is the intervention carried out as intended?
2 did the intervention lead to the intended change of exposure?, and
3 did the changes in exposure to stressors lead to the intended changes in outcomes?

The focus on process issues in organizational interventions, which is central in this book, is also a central issue in our work.

In this section, we describe five intervention case studies, draw conclusions and provide recommendations on how interventions in small groups can be applied in practice. In five groups of employees within two large organizations, we assisted employees with the design, implementation, and evaluation of a work stress intervention. In four cases, all employees and/or managers were invited to be actively involved in the design of the intervention by participating in the steering committee and completing the surveys. The average number of employees in each department was 15. The interventions were carried out in 2005–2006. Three interventions were implemented in a mental health hospital and two in a local government department, each located in a large city in the Netherlands.

The interventions consisted of four integrated components:

1 participation of employees in the steering committee to determine the nature of the stress-related problems and agree ways to address these,

2   implementation of a guided intervention,
3   evaluation of changes in work stress by means of a survey completed before and after the intervention by employees and managers, and
4   evaluation of the efficacy of the intervention by means of an evaluation discussion in the steering committee with, again, as many employees as possible involved.

The design of the case studies differed in each case. In the two local government department cases, the Nova Weba survey (Kraan et al., 2000; Vaas et al., 1995), which consists of 120 questions, was completed by an intervention and a control group. For the control group, a department with similar characteristics (i.e. number of employees or job type) to the intervention group was selected. In the three mental health hospitals, no control group was used. Surveys in these three groups were tailored for the specific work stress situation in each department by means of collecting the complaints about the nature of the work stress problem in the first meeting of the steering committee. The results of the first survey were used as input to design the intervention.

As highlighted by Tvedt and Saksvik (in this volume), all interventions involved organizational change. This implies that all interventions involved the whole group or department and focused on issues that affected all employees. For that reason, all interventions in the five cases were guided by steering committees which were chaired by a manager and also involved employees. All interventions followed the same framework and consisted of the steering committee:

1   identifying bottleneck situations,
2   prioritizing these situations,
3   exploring the ideal situation,
4   designing the intervention, and
5   evaluating the results of the intervention on the basis of their personal experiences and results of the survey.

The manager chairing the steering committee was responsible for the implementation of the interventions.

Five case studies

Here we describe the interventions and examine their effects. At the end of each case description we offer some comments in terms of lessons learned. In the final paragraph of this chapter, we give our general conclusions and recommendations.

Case 1: Streamlining front and back office

The first intervention was conducted in a Human Resource Management department of a local government department of a large Dutch city. Our intervention involved the work content of a small group of 11 Human Resource (HR) consultants.

266 Addressing process and context in practice

Although formally all consultants belong to the same department, each has a different working area within the local government organization. The HR consultants experienced work-related stress mainly because of the overload of administrative tasks. The careful and correct execution of these tasks is fundamental to the job of an HR professional, but seemed to predominate over their primary core task to assist departments in attaining an optimal personnel staffing. The consultants were voluntarily split into an intervention (four individuals) and a control group (seven individuals). An administrative assistant who took over a number of administrative tasks from the consultants was assigned to the intervention group. This intervention was expected to reduce the consultants' workload and therefore their emotional exhaustion. The department managers agreed that if the intervention was successful, an administrative assistant would be appointed permanently to support all consultants. This meant that the control group did not feel resentful.

To evaluate the effects of this intervention, the Nova Weba survey was used to measure well-being at work (Kraan et al., 2000; Vaas et al., 1995). Workload is central in this model and is measured with "job demands" and "coordination problems".[1] "Job content" and "task complexity" are used to measure causes of high workload. Job control is considered a mediator and is measured with "autonomy", "contacts at work", "organizing tasks", and "information supply". "Exhaustion" was added to this survey to measure consequences of work stress.[2] The items were scored as 0 = "no" and 1 = "yes" and the mean score across all items was calculated for each variable. The scores in Figure 13.2 range from 0 = "always no" on all items to 1 = "always yes" on all items. The intervention was evaluated by

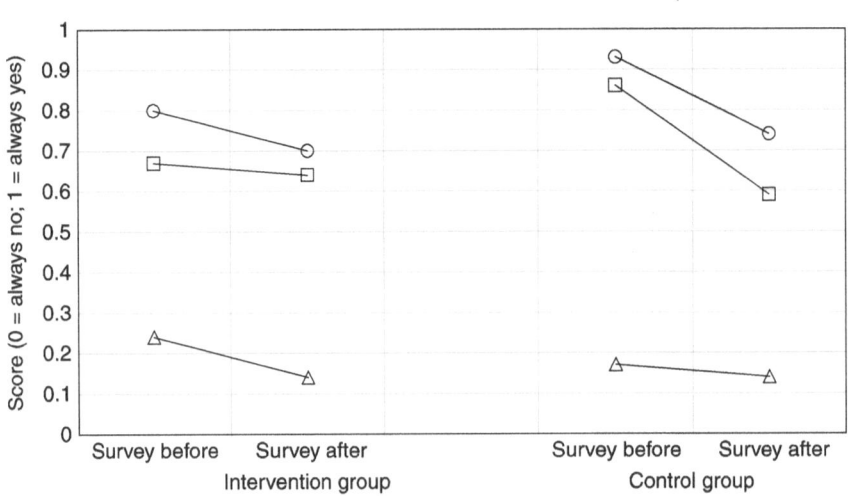

Figure 13.2 Scores on three work-related stress concepts of the intervention (n = 3 before and n = 4 after the intervention) and control group (n = 6 before and n = 7 after the intervention) (results not significant).

comparing the scores of the intervention and control groups. The intervention and control groups both received the same survey before and after the intervention.

After the intervention, coordination problems hardly diminished in the intervention group but, unexpectedly, decreased in the control group (Figure 13.2). The same pattern was found for job demands, where larger improvements were found in the control group compared to the intervention group. Job control is not depicted in Figure 13.2, but also showed a tendency contrary to what was expected. After the intervention, job control had increased for the control group and decreased for the intervention group. Given these results, the effect on exhaustion was expected to be stronger in the control group than in the intervention group, but that too was not the case. Exhaustion decreased more in the intervention than in the control group.

Explanations for these results were explored during steering committee meetings and an interview with the administrative assistant. Indeed, it was discovered that the administrative assistant adopted part of the HR consultants' administrative tasks. In the course of doing this, she reorganized the administrative process and improved the relationships with the administrative back office, allowing intervention group HR consultants to focus entirely on their core tasks. Job demands for the intervention group did not decrease because administrative tasks were replaced by consulting tasks. However, for the control group, this reorganization meant a decrease in their job demands. The appointment of the administrative assistant also resulted in a new type of coordination problem for the intervention group. The dependency on the administrative assistant for the execution of all administrative tasks hindered the consultants in solving easy problems by themselves. An unexpected intermediate effect was the increase in information supply. This affected the levels of exhaustion in both groups and could be explained by the appointment of a new manager who was responsible for the department and all HR consultants and who shared a lot of information with them. The final meeting between the steering committees and the consultants after the intervention showed that participants in the intervention group were satisfied with the solution of appointing the administrative assistant. Satisfaction with the job increased in the intervention group, as did the satisfaction with the quality of their work, and they became more involved in their core consultation task.

Lessons learned

This case illustrates the complexity of designing and evaluating organizational interventions in a small group. Unexpected side effects of the intervention (the assistant reorganizing the administrative process) and other developments (a new manager who facilitated information sharing) may have interfered with the outcomes. The number of respondents was too small to detect significant results. The use of mixed methods (surveys, participation in the steering committees, and interviews), however, helped us to understand the changes observed. Both the control group and the intervention group participated in the decision making on how to implement the intervention, creating high involvement among all employees.

268 *Addressing process and context in practice*

Case 2: Secretaries from front office to back office

Six secretaries in a mental hospital who were responsible for all reception and administrative tasks in their departments participated in a project to reduce workload. A steering committee identified and prioritized bottleneck situations and visualized the ideal situation. The committee was chaired by the department manager and all secretaries participated as members. Three bottleneck situations were identified: understaffing, inefficient work processes, and lack of training. Interventions were implemented by the manager to reduce the workload stemming from the first two bottlenecks whereas training was postponed to a later phase. Regarding the understaffing problem, two part-time receptionists were recruited to staff a front office. The intervention to improve work processes consisted of six actions:

1. creating a quieter work environment,
2. developing and implementing a checklist containing all administrative procedures,
3. implementing new patient intake procedures,
4. streamlining the information on therapies conducted in the hospital to the general practitioners of the patients,
5. standardizing the format of treatment registration, and
6. allocating more influence over the organization of the workflow to the manager.

By the time these interventions were implemented, the organization had also implemented a new electronic patient registration database system.

Based on the exploration of bottleneck situations, the researcher together with the secretaries designed a one-page survey to measure the time spent on specific tasks and workload, defined in terms of overtime and completed tasks. The survey covered a period of two weeks before and two weeks after the intervention. It was completed at the end of each working day. Seven questions covered the number of minutes spent each day on seven kinds of tasks (see Figure 13.3) and two questions covered daily overtime and the percentage of time spent on uncompleted tasks. Per task and workload indicator, the minutes spent by all secretaries were added up for all workdays in the two-week period and divided by the number of hours worked in these two weeks. The percentages are given in Figure 13.3 for the two week period before and after the implementation of the intervention.

Following the intervention, there was an increase in the number of hours the secretaries were able to dedicate to secretarial work. However, the amount of hours spent on secretarial work after the intervention was still less than 100 per cent, since they had to stand in for the part-time reception assistants in their absence. Time spent on registration and team support tasks decreased, in favour of accomplishing more secretarial tasks. According to the group evaluation, this decrease was a direct result of the intervention. Overtime decreased dramatically from 24 to 7 hours per week for each secretary. The daily work was completed more often after the intervention, although in 50 per cent of cases daily work was

Small and practical interventions 269

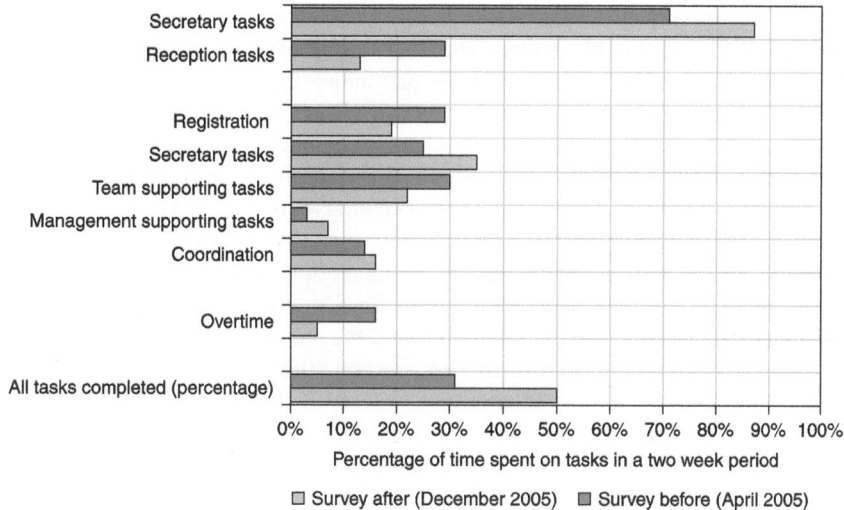

Figure 13.3 Percentage of hours worked of six secretaries in a two-week period before and a two-week period after the implementation of the interventions.

not completed as scheduled. At the evaluation session, the secretaries indicated that the uncompleted work was generally of lower priority and most high priority tasks were completed on time. They also indicated that they regarded the increase in staffing as the most important change in their work, followed by the standardization of the intake format and treatment registration, the quieter working environment, and the use of checklists for administrative procedures. All interventions were viewed positively by the secretaries, who perceived a substantial reduction in their workload.

Lessons learned

The sample was small and it was not possible to have a control group and therefore a case control experimental design. However, the members of the steering committee were directly involved, helping to design and administer the questionnaire and to evaluate the intervention. They were satisfied with the effects of the intervention, especially the reduction in overtime and the increase in the number of completed tasks.

Case 3: Centralization of work division

Twenty-one civil servants working as client mentors in a social security department of a local government department participated in a project to reduce work stress due to an overload of administrative tasks in the assignment of social

270 *Addressing process and context in practice*

benefits. Administrative tasks in this job are complex and can hinder activities related to the second goal of the social security law, work reintegration. Two types of intervention were implemented. The first was the streamlining of new requests for social security fees by assigning the distribution task to an experienced client mentor. This mentor became responsible for all intakes, digital administration, building up the physical document file, and assignment of clients to the appropriate client social servant. The second intervention streamlined the monthly group of benefit clients for reintegration. The intervention required that all clients were selected and invited by the team manager, with a request for delivering essential documents on solicitation. In collaboration with the financial department, the team manager checked the need for documents and planned appointments with clients in client mentors' agendas, and mentors subsequently met with the client and carried out follow-up meetings and reintegration activities.

The effectiveness of this intervention was evaluated using the Nova Weba survey (Kraan et al., 2000; Vaas et al., 1995). As described in Case 1, the scores in Figure 13.4 range from 0 = "always no" to 1 = "always yes" on all questions. All 21 mentors of the municipal social security department participated and 24 mentors of another department with comparable characteristics also participated as the control group.

Contrary to expectations, job demands hardly declined in the intervention group. This was explained at the steering committee by the fact that although the number of administrative tasks decreased, the overall workload remained the same. Tasks that were eliminated as part of the intervention were substituted by

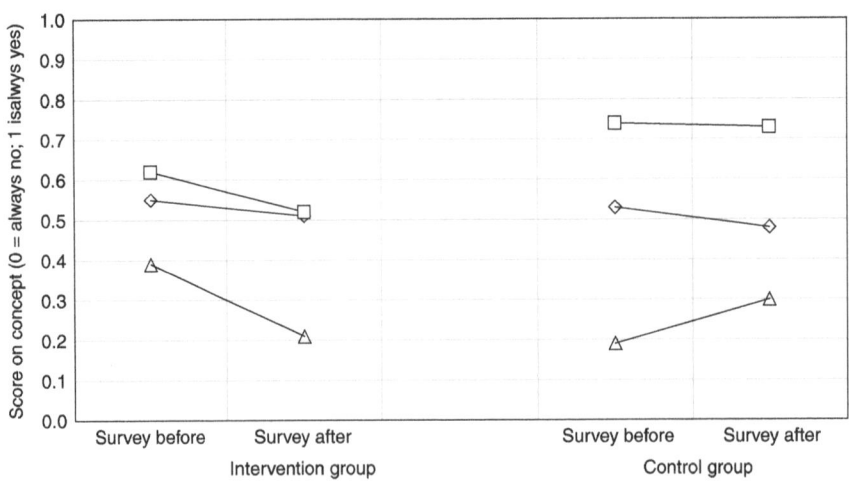

Figure 13.4 Scores on three work-related stress concepts of the intervention (n=12 before and n=9 after the intervention) and control group (n=14 before and n=10 after the intervention) (results not significant).

Small and practical interventions 271

other tasks, which were more related to the core mission of the group's mentoring task. Coordination problems declined in the intervention group but not in the control group. At the steering committee evaluation discussions, all participants showed enthusiasm about the results of the intervention. Cases were processed much faster, piled up work was eliminated, and perceived workload declined drastically. Less experienced mentors were linked with an experienced colleague to increase support. Exhaustion decreased more substantially in the intervention group than in the control group.

The scores for job content and task complexity were extremely high for both groups, indicating the complexity of performing the social security mentoring task. Work tasks continued to be complex, but this was not perceived as a problem by the consultants. The differences in the scores before and after measurement in the control group were small, with the exception of improvements in the organizing of tasks, which, however, did not explain the experience of work stress in both groups.

Lessons learned

It was not possible to establish the significance of the changes due to the small group size, but the intervention was viewed positively by the steering committee at the intervention evaluation meeting. As in the previous cases, the use of mixed methods (surveys and interviews) helped to explain the effects of the intervention.

Case 4: Improvement of the ICT helpdesk service

Eight managers of treatment departments in a mental hospital complained about the poor quality of the software for entering and retrieving personnel and patient information. The helpdesks did not function optimally and enquiries were too often postponed or not answered at all. The managers were primarily responsible for entering and updating the information on personnel and patients in these systems. The hospital board decided to take action to achieve better results in entering and retrieving information. A steering committee chaired by the head of the information and communication technology (ICT) department and consisting of the eight department managers was established. Complaints concerned:

1. the helpdesk staff inaccessibility,
2. the lack of communication of the helpdesk staff on promised software and hardware solutions and improvements,
3. the complexity and poor usability of the personnel database programme, and
4. the complexity and poor usability of the patient treatment database programme.

The interventions agreed by the steering committee and implemented by the ICT staff included:

272 *Addressing process and context in practice*

1   merging three helpdesks into one,
2   improving the helpdesk service quality,
3   improving the personnel database programme manual and the usability and accessibility of the programme, and
4   improving the patient treatment database programme manual and the usability and accessibility of the programme.

Nearly all interventions were implemented over a period of ten months, with the exception of the update of the patient database manual. Before and after the interventions were implemented, a short and tailored two-page survey was completed by five of the eight patient department managers, listing all contact with the helpdesks. All managers completed the survey two weeks before and two weeks after the intervention was implemented. Figure 13.5 presents the results of the questionnaire. Contacts were calculated as the mean number of contacts and calls per day.

The total number of helpdesk calls decreased after the implementation of the interventions. This was interpreted by the steering committee as an indication that the service level of the new helpdesk had improved. In the new arrangements, managers did not have to contact the ICT department several times for the same problem. In particular, improvements were made in accessing the general and the patient program helpdesk. The questions on the patients' program were redirected to the general helpdesk as a part of the intervention. Therefore, the one-helpdesk policy was successful also because information could be retrieved more easily. The number of first contacts with the wrong helpdesk (after the intervention the

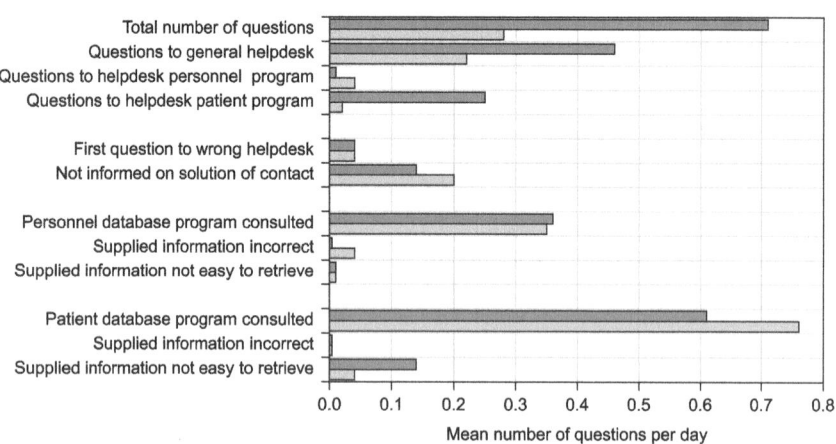

Figure 13.5 Mean number of contacts per day of five managers, as recorded in two surveys, one completed in a two-week period before and the second survey completed in a two-week period one year later, after the implementation of the interventions.

redirection to an ICT professional) and the number of times callers were not informed about the solution did not improve. The number of calls/queries on the personnel database program did not change before and after the intervention. At the last steering committee meeting where the results of the two surveys were discussed, participants confirmed that the functioning and accessibility of the general helpdesk had improved. However, managers did not agree on the improvement of the personnel and patients' treatment database programs. The most positive outcome of the intervention was that ICT staff had adopted a more service-orientated attitude. This, however, was not the case for the group of staff responsible for the database programs, although the steering committee valued their cooperation highly. It was decided to start a second intervention cycle to improve the situation further. For this second cycle, the supervisors of the personnel and patient treatment registration programs were also invited to the steering committee.

Lessons learned

The steering committee evaluated the intervention cycle as relevant and were satisfied with the results, but also agreed that an additional intervention cycle was necessary to further improve the situation. It is possible that these types of more complex and fundamental interventions involving many clients cannot be carried out in one cycle. Rather, an iterative approach may be necessary to assess the direction of the improvements. The steering committee agreed, however, that the approach was successful and should be continued.

Case 5: Improving the top-down information and communication

Eight managers of treatment departments of a mental hospital participated in a project to improve the information and communication streams from the top of the organization to the departmental employees. Bottleneck situations and suggestions for interventions were identified on the basis of the results of a personnel satisfaction survey that was administered in the organization a few months before the intervention was developed. The survey showed that the department managers could not respond to the regularly received requests for information from staff and patients, because they did not receive information from top management. This resulted in staff spending much time collecting the information themselves. One of the most important problems was that the cluster managers at the intermediate level between the top of the organization and the department managers were overloaded and were not able to spread the information from the top to the department managers adequately. Clusters are groups of departments in the organization that combine the treatment of similar groups (for example, children/adolescents, adults, and the elderly).

In this case, the steering committee was not charged with the implementation of the interventions; rather, this responsibility was allocated to the top managers. Four types of interventions were developed and implemented:

274 *Addressing process and context in practice*

1. the intermediate management level was reorganized, re-staffed, and extended from three to seven managers,
2. information dissemination was reorganized and improved with new management memos, a new regular company newsletter, and an intranet site with all relevant information, issued directly from the top of the organization to the personnel,
3. meeting formats and minute distribution was harmonized in the entire organization and bound to strict rules of communication, and
4. any dysfunctional dissemination of information in the form of rumours or gossip was discouraged by managers' immediate and direct interference. All interventions were implemented speedily, with the exception of the intranet site.

Two surveys were completed by the eight patient department managers, two weeks before and 13 months after the implementation of the interventions. A short and tailored survey was developed at a steering committee meeting with the input of its members. In the surveys, the managers recorded all requests for information to and by them and also recommendations regarding the preferred actions for responding to these requests. Figure 13.6 shows the results of the survey where the requests and actions were recalculated to the mean number of events per day.

The total number of formal requests for information was not reduced (i.e. 0.75 per day before and after the intervention). However, there was a reduction in the

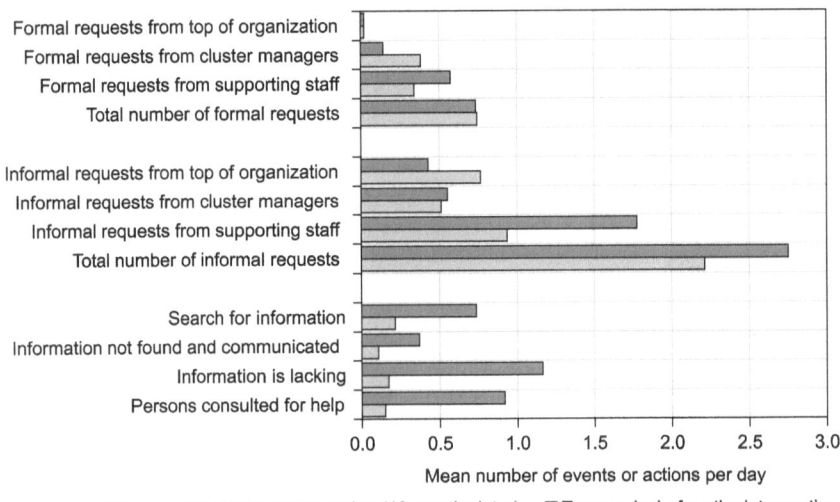

Figure 13.6 Mean number of requests for information per day from or to eight department managers and subsequent actions taken by these managers, as recorded in two surveys, one completed in a two-week period before and the second survey completed in a two-week period one year later, after the implementation of the interventions.

number of formal requests from supporting staff, indicating they received more information. Intermediate level managers asked for formal information more often, possibly because they were newly appointed and their number of requests increased from three to seven. The total number of informal requests for information also decreased from 2.7 to 2.2 events per day. Again the supporting staff issued fewer requests, but requests for information from the top of the organization increased, perhaps because of the newly acquired task of disseminating the information and the need for collection of this. In the first reporting period, information had to be searched for all formal requests. In the second period, the average of 0.75 requests per day decreased to 0.21, indicating that the managers could now find the needed information more easily. Overall, these results suggest that information retrieval improved. There were fewer cases where information was lacking, and fewer staff were consulted. Another important improvement, not shown in Figure 13.6, was that the mean number of minutes spent searching for information decreased from 24 to 15 minutes per event.

Overall, most departmental managers agreed that the reorganization of the intermediate management level contributed to better information supply and improved communication streams. Information dissemination by means of the new company newsletter strongly contributed to this. However, the issuing of the management memos was delayed. The harmonization of meeting formats and distribution of minutes improved communication, and the discouragement of informal gossip was also successful.

Lessons learned

Broad organization-wide participation of all employees in the intervention is not always required. Other sources of information in the organization can also be used for identifying problematic situations and suggestions for improvement. In this case this was the employee satisfaction survey. But there is a need for empirical evaluation and this case shows that it can be achieved in a small group. In this case, the top management selected a small group of managers to guide the interventions to improve the communication and information streams.

Discussion

Our main conclusion from the five cases described in this chapter is that small and practical interventions directed at the redesign of the sources of work stress, by means of improvement of organizational conditions, can be achieved successfully in several ways. In this last section of the chapter we discuss some issues central to our approach.

Causes and consequences of occupational stress

Our interventions were directed at improving organizational conditions (i.e. the causes of work-related stress). Stress was also the keyword used by the organizations for inviting us as consultants to assist in solving the problems. The working conditions to be dealt with were selected by the steering committees and the employees who were directly exposed to the stressful conditions. In two cases, we used the Nova Weba survey, which was designed to measure stressful conditions at work. The consequences of work-related stress were not measured in three of the five cases, because employees did not view them as important factors to be targeted. These factors were sometimes informally assessed in the final steering committee meetings after the intervention. In two cases, we looked more explicitly at work-related stress outcomes, since they were part of the Nova Weba survey. Therefore, we cannot assess whether the interventions in our five cases influenced work stress-related outcomes, such as health and well-being or organizational performance, productivity and economic improvement. Our assumption is that organizational level interventions can also help to improve these stress-related outcomes.

The numbers of respondents in our groups are too small to measure stress consequences in a reliable way. Hence, our primary interest was to demonstrate that we can influence risk factors in organizations and reduce the risks of stress. This focus on organizational change was also shared by the organizations and the employees participating in the steering committees. Participants focused on the sources rather than the consequences of stress, which they viewed as key for a successful intervention. The primary focus of the committees was on the direct and short-term factors that contributed to strain and loss of motivation in their work situation, rather than on long-term health issues. This is in line with the concept of "fit for purpose" suggested by Cox et al. (2007). Although in strict scientific terms the research designs were less than optimal and sample sizes were small, it was still possible to target problematic factors, develop interventions to tackle them, and evaluate the effects on outcomes perceived as relevant by the participants.

On the methods used to evaluate the interventions

The five cases differ in many aspects, but share the participative approach of designing the implementation in collaboration with the employees. Employees in the steering committees were also the target group for the interventions, so they themselves designed the changes in their work environment. The participating consultants facilitated the discussions, moving the focus away from collecting complaints to developing constructive suggestions for improvement. In this way, the interventions were tailored to the participants' specific needs, and the use of evaluation surveys that matched the interventions proved to be advantageous. Explicit and measurable targets were set and evaluated. This approach is practical because interventions are developed and implemented jointly with the employees, and it is small and focused, because we kept only the most necessary instruments for assessment (small surveys, and often no use of a control group). This helped to improve ownership of the changes and sustainability of the results.

In all cases, a mixed-methods approach was used to evaluate the outcomes and effectiveness of the interventions. Information on the effects of the interventions was gathered by means of surveys that were completed before and after the implementation of the interventions, and by means of evaluation discussion sessions at the steering committees. This approach provided an in-depth understanding of the daily work-related stress problems experienced by the employees and proved to be very helpful for understanding the effects of the interventions. Without the evaluation discussions, we would not have been able to ask the relevant questions or understand the survey results, which were, in some cases, different from what we expected.

Using pre-existing vs. tailored surveys

Surveys were used to evaluate the effects of the interventions. In three of the five cases, tailored surveys were developed, whereas in the other two cases, pre-existing ones were used. The use of off-the-shelf surveys guarantees precise assessment of the concepts to be measured, because the concepts are developed in earlier studies, they consist of statistically closely related items, and because the concepts have been linked in earlier studies with theoretically associated concepts.

In the three cases where tailored surveys were used, their development was integrated into the intervention process. The surveys helped to define indicators of the identified problems and also, when completed before the intervention, they helped to assess whether these problems were important and relevant to the employees.

The use of tailored surveys is preferred in intervention studies, but this does not mean that detailed and precise feedback cannot be achieved using pre-existing off-the shelf measures. Available surveys and validated concepts in these surveys may not be relevant to the specific identified issues that the intervention targets. If the employees in the steering committees do not perceive the link between the content of the survey and the target problem of the intervention as relevant, this may impede survey responses and participation in the project. For these reasons, the face validity of the tailored surveys was more important than other forms of (statistical) validity. Because of the potential lack of some types of validity of tailor-made questionnaires, mixed method designs are suggested to integrate process and outcome evaluation and increase the generalizability of interventions (Nielsen et al., 2010). Here also, the term "fit for purpose" applies (Cox et al., 2007) since the need in these cases was to obtain data appropriate enough for intervention evaluation. The final discussions in the steering committees yielded many reasons for the obtained effects and suggestions for future improvements.

The use of a case control group design

Researchers and practitioners recommend that interventions should be evaluated using a pre- and post-test, and a control group design (Campbell and Stanley, 1966; Cook and Campbell, 1979; van der Klink et al., 2001). Of course, the application of a case control group design is fundamental for robust and reliable comparisons. In intervention research in small groups or organizations, however, an appropriate control group is very often difficult to find (Cox et al., 2007; Griffiths, 1999; Semmer, 2006) and the question remains as to whether the control group is adequate for comparisons (Griffiths, 1999). Departments within an organization can differ widely in functions, work tasks, policies, individuals, culture, managers, history, etc. Accurate comparisons related to the estimation of outcome effects of organizational interventions as used in quasi-experimental designs are often difficult (Griffiths, 1999; Kristensen, 2005; van der Klink et al., 2001). This can be solved by randomly selecting a sufficient number of participants in the intervention and control groups. But what does "sufficient" mean? In case control, the following are crucial: group comparisons, sample size and statistical power (Cohen, 1977). In our examples, the numbers of individuals in the intervention and control groups were small. Besides, in an organizational setting, there is always the risk of a "spill-over effect"; it is hardly possible to keep the control and intervention groups completely separated, as is shown in Case 1 (streamlining front and back office). In addition, the use of tailored surveys can make comparisons between the intervention and control groups problematic because the control group may not understand what the questions of the survey imply since they were developed for the intervention group.

On the effectiveness of interventions (comparison of the cases)

Interventions should be designed to achieve success. In nearly all cases success was achieved, but it is important to identify the common factors of this success. External validity of case studies can be assessed by applying a case study comparison design on predefined characteristics (Kompier and Kristensen, 2000). However, we did not start our study by selecting our cases on predefined characteristics. In fact the opposite was true: the organizations allowed us a closer look at the possible alternative actions. The ultimate goal in the mental hospital study even went one step further, because we wanted to design a self-help instrument for organizations to tackle work-related stress. Although we did not succeed, we learned much about the role of steering committees in designing an intervention. However, a comparison of the cases was still possible and provided important lessons regarding the effectiveness of interventions.

In the first three cases (streamlining front and back office, secretaries from front office to back office, and centralization of work division), the interventions consisted of a redesign of the employee tasks. In all cases, the redesign of tasks was evaluated as successful by the participants although it is unclear whether the intervention caused a decrease in their actual workload. Discussions with the members of the steering committees were fruitful because they helped to qualify the changes in workload, which might not have been detected with pre-existing surveys.

In cases 4 (improvement of the ICT helpdesk service) and 5 (improving the top down information and communication), work stress originated from the fact that retrieving information was difficult and time-consuming. The intervention in case 4 was partly successful because the service level of the helpdesk personnel improved, but not the use of the patient treatment database program. The intervention in case 5 reduced the time spent on retrieving information and was evaluated positively by the respondents. In both cases, however, the results of the intervention could be improved, for instance by organizing a new intervention cycle to address omissions or insufficient solutions in the former cycle. It can be expected that a new intervention cycle can be even more efficient because all parties are aware of the bottleneck situations and are experienced in intervention implementation. It is not necessary to restrict interventions to one intervention cycle. The advantage of a second intervention cycle is that it starts with the knowledge and experience learned in the previous cycle. It can also be carried out with less preparation and information gathering, because knowledge and experience from the preceding intervention cycle can be transferred to the next one. In the total quality literature this is referred to as quality cycles (Kaplan and Norton, 1996; 2001).

In our cases the steering committees were the forum where the evaluation criteria were developed and used for the final appraisal of intervention effectiveness. The mixed-method approach used in these examples allowed participants in the intervention groups to provide clear and detailed feedback on the changes, and on their experiences and evaluations of the process. This type of information is often richer and can be more informative than the use of surveys.

Conclusions

When an organization contacts a consultant or a researcher with a request for help to ameliorate a work-related stress problem, a range of alternative solutions can be considered. Consultants often help organizations to develop interventions without putting much effort into assessing the situation. Researchers, on the other hand, often tend to use a heavy arsenal of off-the-shelf surveys and rigorous scientific designs. Problems may arise with either approach. Consultants may not be able to provide scientific credentials for their practice-informed approaches. They often evaluate the effectiveness by means of short tailored surveys containing questions on participants' perceptions of the intervention and the implementation process. Researchers often end up with long evaluation reports containing detailed information on the identified problems, which can be sometimes difficult to decipher and apply in real-life settings. Optimizing both approaches may lie somewhere in-between and in the notion of "fit for purpose" (Cox et al., 2007). We hope that we have illustrated how smart and practical small group organizational interventions for work-related stress can be usefully developed and implemented, even when the research design is less than optimal, thus fulfilling organizations' needs for practical solutions.

Notes

1 Job demands is measured with five items, for example "do you have to work very fast?" Coordination problems is measured with eight items, for example: "do you experience delay in your work because you have to wait for other people's output?"
2 Job content is measured with eight items, for example "do you need craftsmanship to perform your job?". Task complexity is measured with six items, for example "do you

have to memorize a lot of information during your work?" Information supply is measured with 11 items, for example "do you receive information in time to do your job?" Autonomy is measured with 10 items, for example "can you decide for yourself the pace of your work?" Contacts at work are measured with six items, for example "can colleagues help you to finish your work?" Organizing tasks is measured with five items, for example "do you discuss in your team how tasks will be divided among team members?" Exhaustion is measured with eight items, from the UBOS (Schaufeli and Van Dierendonck, 2000)), for example "In the morning I'm tired just by thinking about the day ahead of me."

References

Bond, F. W., & Bunce, D. (2001). Job control mediates change in a work reorganization intervention for stress reduction. *Journal of Occupational Health Psychology, 6*, 290–302.

Bourbonnais, R., Brisson, C., Vinet, A., Vézina, M., & Lower, A. (2006). Development and implementation of a participative intervention to improve the psychosocial work environment and mental health in an acute care hospital. *Occupational and Environmental Medicine, 63*, 326–334.

Brun, J.-P., Biron, C., and Ivers, H. (2008). *Strategic approach to preventing occupational stress.* Montréal (Québec): Institute de recherche Robert-Sauvé en santé et en sécurité du travail.

Campbell, D. T., & Stanley, J. C. (1966). *Experimental and Quasi-experimental Designs for Research.* Chicago: Rand McNally.

Cohen, J. (1977). *Statistical Power Analysis for the Behavioral Sciences* (Revised Edition). New York: Academic Press.

Cook, T. D., & Campbell, D. T. (1979). *Quasi-experimentation: Design and Analysis Issues for Field Settings.* Boston: Houghton Mifflin.

Cox, T., Griffiths, A., Barlowe, C., Randall, R., Thomson, L., & Rial-González, E. (2000). *Organizational interventions for work stress: a risk management approach (Contract research report 286/2000).* Sudbury: HSE Books.

Cox, T., Randall, R., & Griffiths, A. (2002). *Interventions to control stress at work in hospital staff (Research report 435/2002).* Sudbury: HSE Books.

Cox, T., Karanika, M., Griffiths, A., & Houdmont, J. (2007). Evaluating organizational-level work stress interventions: Beyond traditional methods. *Work & Stress, 21*(4), 348–362.

Griffiths, A. (1999). Organizational Interventions: Facing the limits of the natural science paradigm. *Scandinavian Journal of Work, Environment & Health, 25*, 589–596.

Kaplan, R. S., & Norton, D. P. (1996). *The Balanced Scorecard.* Harvard Business School Press.

Kaplan, R. S., & Norton, D. P. (2001). *The Strategy-focussed Organization. How Balanced Scorecard companies thrive in the new business environment.* Harvard Business School Press.

Keita, G. P., & Sauter, S. L. (1992). *Work and Well-Being: an Agenda for the 1990s.* Washington DC: American Psychological Association.

Klein Hesselink, D. J., Klink, J. J. L., van der, Vaas, S., Houtveen, J. H., & Frielink, S.J. (2001). *Maatregelen werkdruk en werkstress. Catalogus ontwikkeld in het kader van arboconvenanten: stand der wetenschap 2001 (catalogue on organizational and individual interventions for the Dutch working conditions convenant approach).* Doetinchem: Elsevier bedrijfsinformatie bv.

Klein Hesselink, D. J. (2002). *Werkstress en verandering: verslag van een quasi- experimenteel evaluatieonderzoek van de cursus anders werken: een methode om werkstress bij werknemers op uitvoerend en laag leidinggevend niveau te verminderen. [Quasi-experimental effect evaluation study of a work stress intervention course for employees to reduce work stress complaints and improve their working situation].* Thesis, Hoofddorp: TNO Arbeid.

Klein Hesselink, J., Houtman, I., Hooftman, W., & Bakhuys Roozeboom, M. (2010). *Arbobalans 2009: Kwaliteit van de arbeid, effecten van maatregelen in Nederland (Working conditions report 2009: Quality of work, and effects of work interventions in The Netherlands).* Hoofddorp (The Netherlands): TNO Kwaliteit van leven.

Kompier, M. A. J., Geurts, S. A. E., Gründemann, R. W. M., Vink, P., & Smulders, P. G. W. (1998). Cases in stress prevention: the success of a participative and stepwise approach. *Stress medicine 14*, 155–168.

Kompier, M. A. J., & Cooper, C. L. (Eds.). (1999). *Preventing stress, improving productivity. European case studies in the workplace.* London: Routledge.

Kompier, M. A. J., Cooper, C. L., & Geurts, S. A. E. (2000). A multiple case study approach to work stress prevention in Europe. *European Journal of Work and Organizational Psychology 9*(3), 371–400.

Kompier, M. A. J., & Kristensen, T. S. (2000). Organizational Work Stress Interventions in a Theoretical, Methodological and Practical Context. In: J. Dunham (Ed.), *Stress in the Work Place: Past, Present and Future* (pp. 164–190). London: Whurr Publishers.

Koppes, L. L. J., Vroome, E. M. M., de Mol, M. E. M., Janssen, B. J. M., & van den Bossche, S. N. J. (2010). *Nationale Enquête Arbeidsomstandigheden 2009: Methodologie en globale resultaten. (Netherlands Working Conditions Survey 2009: Methodology and overall results).* Hoofddorp: TNO.

Kraan, K., Dhondt, S., Houtman, I., Nelemans, R., de Vroome, E. (2000). *Handleiding NOVA-WEBA: een vragenlijst om arbeidsorganisatorische knelpunten op te sporen: hernieuwde versie [A new update of the manual of the Nova-Weba instrument to detect bottleneck situations in the design of work in organizations].* Hoofddorp: TNO Arbeid.

Kristensen, T. S. (2005). Intervention studies in occupational epidemiology. *Occupational and Environmental Medicine, 62*, 205–210.

LaMontagne, A. D., Keegel, T., Louie, A. M., Ostry, A., & Landsbergis, P. A. (2007). A Systematic review of the Job-stress Intervention Evaluation Literature, 1990–2005. *International Journal of Occupational and Environmental Health, 13*, 286–280.

Nielsen, K., Fredslund, H., Christensen, K. B., & Albertsen, K. (2006). Success or failure? Interpreting and understanding the impact of interventions in four similar worksites. *Work & Stress, 20*, 272–287.

Nielsen, K., Randall, R., Brenner, S.-O., & Albertsen, K. (2009). Developing a framework for the 'Why' in change outcomes: The importance of employees' appraisal of changes. In: P.O. Saksvik (Ed.), *Prerequisites for healthy organizational change* (1st ed.) (pp. 76–78). Bentham.

Nielsen, K., Randall, R., Holten, A.-L., & Rial-González, E. (2010). Conducting organizational-level occupational health interventions: What works? *Work & Stress, 24*(3), 234–259.

Oeij, P. R. A., Vroome, E. M. M., Sanders, J. M. A. F., & Bossche, S. N. J. (2009). *Werkgevers Enquête Arbeid 2008 Methodologie en beschrijvende resultaten (Employers survey on work and employment 2008: methodology and descriptive results).* Hoofddorp (The Netherlands): TNO Work & Employment.

Richardson, K. M., & Rothstein, H. R. (2008). Effects of Occupational Stress Management Intervention Programs: A Meta-Analysis. *Journal of Occupational Health Psychology, 13*(1), 69–93.

Schaufeli, W. B., & Van Dierendonck, D. (2000). *Utrechtse Burnout Schaal (UBOS): Testhandleiding.* [Utrecht Burnout Scale. Test Manual]. Amsterdam: Harcourt Test Services.

Semmer, N. K. (2006). Job stress interventions and the organization of work. *Scandinavian Journal of Work, Environmental and Health, 32*(6), 515–527.

Vaas, S., Dhondt, S., Peeters, M. H. H., & Middendorp, J. (1995). *De weba-analyse (work situation analysis by means of the weba instrument).* Alphen aan den Rijn: Samsom bedrijfsinformatie.

van der Klink, J. J. L., Blonk, R. W. B., Schene, A. H., & van Dijk, F. J. H. (2001). The Benefits of Interventions for Work-Related Stress. *American Journal of Public Health, 91,* 270–276.

van Hooff, M., van den Bossche, S., & Smulders, P. (2008). *The Netherlands Working Conditions Survey: Highlights 2003–2006.* Hoofddorp: TNO (www.tno.nl/downloads/TNO-KvL_NEA_Brochure_2007_Eng.pdf).

Part 3
Policy implications

14 Implementation of the Management Standards for work-related stress in Great Britain[1]

Colin Mackay, David Palferman, Hannah Saul, Simon Webster, and Claire Packham

Introduction

In 2000 The Health and Safety Commission (HSC), the body responsible for health and safety matters in Great Britain (GB), set targets for the overall reduction in the burden of occupational health including that of the contribution from work-related stress. As a result, a 10-year priority programme was devised to meet these targets. A key part of the intervention programme was the development of the concept of a series of "Management Standards" that would allow organizations to both gauge their performance and to facilitate continuous improvement. The Management Standards (MS) themselves each consist of both a desirable "state to be achieved" for individual stressors and a process by which organizations can assess and manage their exposure to such stressors. This work is described earlier in Mackay et al. (2004) and Cousins et al. (2004).

The design of the stress priority programme was predicated on action being taken across GB industry; that is, applied to a large proportion of the GB workforce (*national level*). It is therefore a *population* approach rather than being targeted at the conspicuously ill (existing "cases") or particularly high-risk occupational groups. The requirement to reach the national targets necessitated the development of an Implementation Logical Model (ILM). The ILM made a number of assumptions at a population level about:

1   an organization's awareness and willingness to use the MS,
2   the uptake of the MS approach,
3   their proper implementation, and
4   the appropriate choice and effectiveness of interventions at an organizational level.

Progress towards the targets was driven by the extent to which these assumptions were met.

At an *organizational level*, the Management Standards enable the assessment of current exposure to stressors and then the development and implementation of action plans to reduce risk, followed by subsequent re-assessment. The change

1 © HMSO

pathway requires the workforce and management to participate together in risk assessment and action planning.

To test the process, an implementation plan based on five priority sectors (those with the highest incidence of stress-related health problems and associated sickness absence) was developed (Sector Implementation Plan 1 or SIP1). This required the collaboration of (initially) 100 organizations and was intended to enable a full evaluation of the Management Standards process to be undertaken. Especially important was to understand any obstacles in implementation – and thus to refine the process – and to gauge the likely impact on overall targets via the ILM. It was envisaged that a second Sector Implementation Plan (SIP2) and a Wider Implementation Plan (WIP), to cover non-target sectors, would follow.

In this chapter we would like to reflect upon the impact of the programme, via the ILM, upon the national targets and some of the learning points in trying to use such an approach (*population level*). Second, we wish to describe, in terms of process issues, the learning points that accrued from a number of different types of evaluation that were done of SIP1, and, to a lesser extent of SIP2 (*organizational level*).

Background

The health of the workforce and particularly their "mental health" is an issue of critical national importance (Black, 2008). Work-related stress, depression and anxiety have costs for employee, employer and society as a whole. Stress is responsible for lost economic productivity (including human error and consequent accidents) and is a major cause of sickness absence in that it accounts for a higher proportion of long spells than other common health conditions. Elevated rates of psychological work demands, including high work pace and conflicting priorities at work have been consistently associated with increased risk of psychiatric disorders in a number of prospective studies and it seems reasonable to assume that there may be a causal link between them.

Two developments in the early 1990s gave rise to an increased need for practical approaches to the management of work-related stress in the UK and wider European and global context: first, the rise in the apparent scale of the problem (Stansfeld et al., 2008) and, second, the requirements of European legislation (Leka, Cox, and Zwetsloot, 2008, Leka et al., 2011). In their respective ways, and in line with European Commission recommendations (European Commission, 1996), both developments pointed to the efficacy of a risk management paradigm that emphasizes prevention as the preferred strategy for tackling occupational health issues.

Stress is an inherent aspect of modern work and domestic life for many, one so great that, although not the sole or necessarily primary cause, it is implicated in over half of human morbidity and mortality and has become one of the most serious health issues of modern times. The debate has included the consideration of work and well-being and the "good jobs" agenda (Constable et al., 2009; Crawford et al., 2010) and how these concepts can be incorporated in an integrated approach.

In this chapter we discuss implementation process issues at both the national and organizational level.

The "Revitalising Health and Safety" targets

In June 2000 the UK Government and HSC set a number of national targets for the health and safety system as part of their "Revitalising Health and Safety" (RHS) initiative. The following headline targets were to be achieved over the ten-year period up to 2010:

- A 30 per cent reduction in the number of working days lost per 100,000 workers from work-related injury and ill-health.
- A 20 per cent reduction in the incidence rate of cases of work-related ill-health.
- A 10 per cent reduction in the incidence rate of fatal and major injury incidents.

As a major cause of ill-health, work-related stress is a large contributor to the burden of overall sickness absence and disease incidence and therefore a major component of the eventual ten-year RHS strategy. Baseline comparator data were those drawn from 2000–2001 survey of work-related ill-health. Meeting the targets would mean that by 2010 there would be around 34,000 fewer people first reporting a spell of stress, depression and anxiety and around 1.2 million fewer days would have been lost.

The Management Standards approach

In 1993, the Health & Safety Executive (HSE) commissioned a review of the literature on work-related stress (Cox, 1993) with the intention of it informing the HSE's approach to dealing with this problem. That review recommended the adoption of a risk management approach. It introduced a framework for a problem-solving approach to work-related stress and, within that framework, also suggested a possible taxonomy of psychosocial hazards or stressors. These ideas were incorporated into the subsequent HSE guidance *Stress at Work: A Guide for Employers* (HSE, 1995). The risk management approach to work-related stress was further advocated in *Tackling Work-Related Stress* (HSE, 2001) and, finally, given operational form in 2004 when HSE launched its *Management Standards for Work-Related Stress* (MS). These standards took the form of a set of procedures to help employers meet their legal obligations in relation to the prevention and control of work-related stress that have at their core the notion of risk management (Cousins et al., 2004; Mackay et al., 2004).

Following a period of extensive research and consultation with stakeholders concerning the way forward in tackling work-related stress, the MS were launched in November 2004. The approach is preventative and is underpinned by the rationale that "collective protective measures should be given priority over individual protective measures" (Mackay et al., 2004, p. 101). Taking what is known as a "population strategy" (Rose, 1992) the approach is intended to be applied to populations rather than to those individuals identified as "high risk", or conspicuously "ill", with the aim of shifting entire populations towards the desired *states to be achieved* as set out in each of the standards (see below for further elaboration). This rationale reflects HSE's remit, which is to prevent ill health among those at work and is founded in the belief that a large number exposed to a small risk generates a greater burden to the population than a small number exposed to a conspicuous risk. Therefore it is hoped that making small changes to address the risk factors associated with work-related stress could bring substantial improvement to the health of the working population (Mackay et al., 2004, p. 107). Of course, this strategy carries with it the risk that it will bring little obvious benefit to the majority of individuals who would not have developed work-related stress symptoms anyway.

The MS approach (details of which can be accessed on the HSE website together with downloadable tools; www.hse.gov.uk/stress) consists of both a set of standards and a risk management process by which they can be implemented. The standards themselves are based on a typology of six domains of psychosocial risk identified through extensive research, including a review for HSE by Cox (1993) and then extensive consultation with stakeholder groups. The six domains are:

1 Demands (workload, skills, abilities)
2 Control (over pace of work, development opportunities, work patterns)
3 Support (feedback and support from managers and between colleagues, employees are aware of support and how to access it)
4 Relationships (promotion of positive behaviours, avoiding conflict and dealing with unacceptable behaviour)
5 Role (clear and compatible requirements and responsibilities)
6 Change (timely consultation and support during organizational change).

Each of the six stressor areas is accompanied by a description of the desirable *states to be achieved* which are seen to reflect high levels of health, well-being and organizational performance. These states provide a level of performance to be worked towards, or in other terms, good job design practice. The basis of the MS approach is to compare the states to be achieved with the actual conditions that currently exist within an organization, using a gap analysis set within a risk assessment framework (see below). This aims to help employers identify the underlying causes of workplace stress and think about how they might be prevented through practical improvements through organizational level interventions. It does this by using risk assessment, which is itself based on a standard business problem-solving model.

Assessing and managing risks

The aim of conducting a risk assessment is to identify what in a workplace could potentially cause harm to people so that suitable precautions and solutions might be developed in order to prevent the harm from occurring. Employers are required to undertake risk assessments for health as well as safety risks and to take measures to control and address the identified risks so identified. This approach is proactive as it emphasizes prevention and seeks to involve employees through active discussion and partnership working with their employers to develop locally relevant understandings and practical ways to address the issues. The MS as a risk management strategy for work-related stress is intended to be a flexible approach that can be tailored to suit the needs of different employee groups and organizations. Cox et al. (2000) suggest that it is important to be mindful that the risk assessment method for workplace stress is not an "off the shelf recipe to be mechanically followed" (p. 20), rather they argue that what is essentially required is adherence to the underpinning logic and principles of the method. Employers were encouraged to go beyond simply complying with their legal health and safety responsibilities; to prevent and manage workplace stress by developing workplaces that are conducive to staff well-being. The five steps approach to risk assessment is well understood.

In the MS, Step 1 is represented by each of the six individual factors described above. To assess exposure to these, at Step 2, we have previously described an indicator tool (Cousins et al., 2004) to help organizations characterize risks. This tool has robust psychometric properties (e.g. Edwards et al., 2008, Kerr, McHugh, and McCrory, 2009) that have been demonstrated in recent empirical studies (e.g. Bartram, Yadegarfar, and Baldwin, 2009). What is perhaps novel about our approach is that the data collected at this step (use of the indicator tool is not mandatory) are then used in Step 3 to enable a fuller understanding of the nature of risks to be established. It was assumed at this point the "states to be achieved" for each of the six standards would be used to guide discussion as to where there might be deficiencies in job content and job design that could be rectified by appropriate interventions. Crucially, at this step, active workforce participation and involvement is key, in that workers are best able to articulate the precise nature of the risks from poor job design to which they are exposed. Organizations have found that focus groups are one way of doing this. In such a forum it is envisaged that employers and employees would work together to prioritize reasonably practicable interventions that could be then implemented to address the problems as identified (Step 4). Step 5 is designed to monitor and review progress on the impact of solution intervention. The MS process is shown in Figure 14.1 (see the HSE website (www.hse.gov.uk/stress/standards/index.htm) for further details).

The overall generic change journey was devised around a high level process of:

- Securing engagement from organizations in target sectors who fully support the implementation of the MS approach (especially in the pilot phase – Sector Implementation Plan 1 – see p. 296).
- HSE and partner agencies to provide training, support and guidance for participating organizations in targeted sectors and increased awareness of solutions and business benefits.

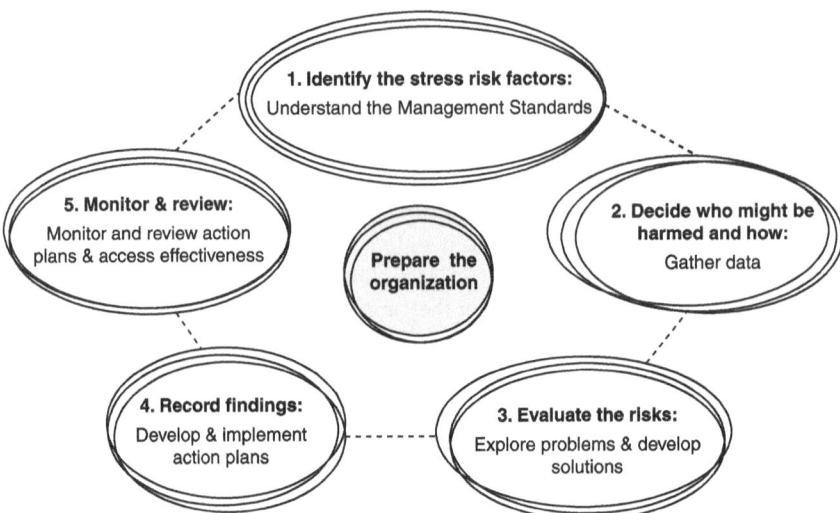

Figure 14.1 The Management Standards approach.

- Achieving "buy-in" from board level senior management (and equivalents) to support the adoption of the MS (especially resourcing adequately).
- Management and workforce working collaboratively to collect data, identify potential stressors, and to discuss implications.
- Management and workforce then working together to discuss practical ways to address problems and identify and implement interventions (control measures).
- Monitoring the extent to which interventions have been properly implemented and have led to improvements and benefits in key performance indicators.
- At a national level, realization and measurement of the benefits in terms of a reduced incidence of stress-related ill health and a concomitant reduction in associated sickness absence.

Overall intervention strategy

The approach to reducing the number of cases of work-related stress was to attempt to shift the entire distribution of exposure to the risk in a favourable direction. This approach is described in detail in Rose (1985, 1992) and is known as a population strategy (also see Anderson, Huppert, and Rose, 1993; and Rose, 1989).

The population strategy attempts to control the determinants of incidence of disease, to lower the mean level of risk factors and to shift the whole population in a favourable direction (from a less desirable to a more desirable state). Hence the "states to be achieved" as incorporated into the standards. The idea is predicated

The Management Standards for work-related stress in Great Britain 291

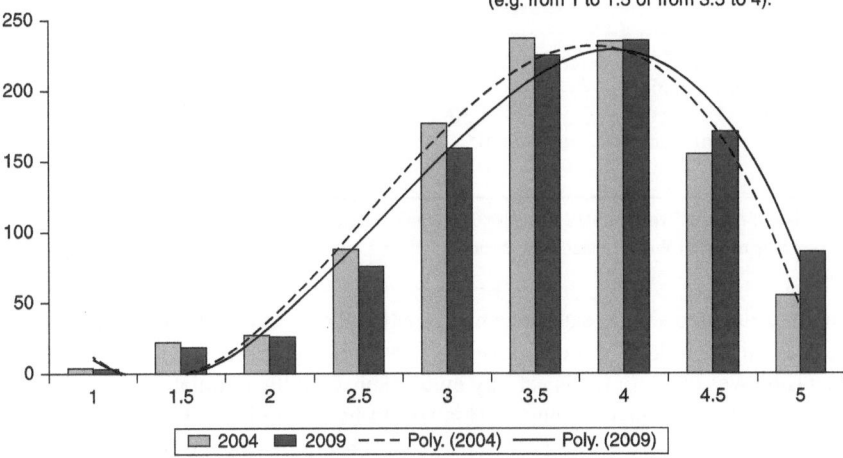

Figure 14.2 An illustration of what a 20% improvement in work characteristics (*demand*) would look like – on the x axis, 1 represents poor management of *demand* whilst 5 represents good performance.

on the fact that a large number of people exposed to a small risk may generate a greater population burden than a small number exposed to a conspicuous risk, and, conversely, if large populations are exposed, a small change in a risk factor may bring about substantial improvements in the health of the working population (Rose, 1992). Using, as an example, the Management Standard for Demand, Figure 14.2 shows schematically what a 20 per cent positive shift (using the HSE indicator tool questions) between the years 2004 (pre launch) and 2009 would look like.

The population approach to prevention does rely on one critical assumption when applied in this context. Adams and White (2005) showed that if the association between exposure and harm is non-linear, then the strategy may be harmful to some individuals. Evidence from the Whitehall II study, presented in Rydstedt, Ferrie, and Head (2006), shows a linear association between work stressors and health, supporting the argument that it was reasonable to apply the population approach to tackling workplace stress.

Selecting a target population

Although the population concept was pre-eminent in reality, a combined strategy was thought to have the best chance of success: the population approach plus the targeting of "high risk" sectors (Health, Education, Central Government, Local Government, and Finance) that had been shown to have the highest incidence

292 *Policy implications*

Table 14.1 Incidence rate and average days lost per worker due to work-related stress, depression and anxiety

Industry (SIC92 classification)	Incidence rate*	Average number of days lost per worker
Public administration and defence	1.40	1.30
Financial intermediation	1.50	0.82
Education	1.30	0.85
Health and social work	0.94	0.81
All industry	0.78	0.50

* Number of cases of work-related stress per 100 employed in last 12 months.
Source: Self-reported Work-related Illness survey 2001/02.

rates and elevated sickness absence in GB national data as described below. Given the scale of the task of reducing levels of work-related stress as set in the targets, a decision was taken to focus activity on this subset of the population where there was greatest potential for change. This would have two benefits – first, lessons could be learned before any potential roll-out to the rest of the population; and second, it would allow HSE to achieve more with limited resources available. It was assumed that the MS could have the biggest impact where there were high levels of work-related stress. In segmenting the potential target audience, industry sector was used as a categorization. This was partly driven by the availability of data and partly by a recognition that industry sectors share similar working conditions, similar characteristics and would form natural communities for sharing information and best practice on the management of risk.

The 2001/02 *Self-reported Work-related Ill-health* (SWI) survey was used as the primary data source for determining which industries should be selected. The SWI survey recorded industry via the SIC 1992 industry classification system. In determining the target sectors, HSE took into account the incidence of work-related stress (i.e. the number of new cases per year) and the number of working days lost to work-related stress. Four industry sectors stood out with incidence rates and days lost figures which were higher than average. Table 14.1 shows the relevant figures for these four industries and how they compare with the national GB averages.

For practical reasons including the simplification of communication routes, HSE referred to five separate industry categories: splitting public administration into its two major components, namely central government and local government. However for statistical calculations and target decisions the original category was retained.

Development of an intervention logic model

Having determined the overall intervention strategy the next step was to consider the scale and develop an idea of the resources needed to achieve the desired outcomes. The process was mapped out using an intervention logic model (ILM), a common tool used to plan interventions that is centred around the idea of a

The Management Standards for work-related stress in Great Britain 293

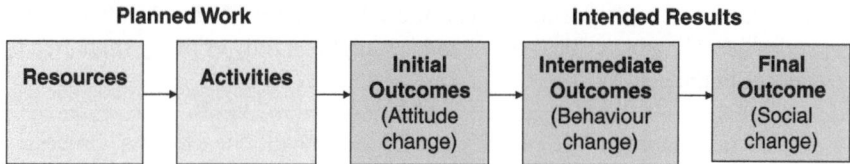

Figure 14.3 Outline of an intervention logic model.

"pathway to change" (McLaughlin and Jordan, 1999; Rogers, 2000). In an ILM, each stage between the allocation of project resources and the final outcome (in this case a reduction in cases of work-related stress) is mapped out along a change pathway which demonstrates clear logical links between each stage and any assumptions that need to be tested along the way. Figure 14.3 below summarizes the basic structure of the ILM as used.

If the desired final outcome can be quantified then the ILM can help determine the scale of the resources and inputs needed to meet the outcome, including decisions on appropriate target audiences, staff resources and costs. Intervention logic models are used by working backwards along the pathway shown in Figure 14.3. The user would start with the final outcome and work backwards until they had a clear idea of the activities and resources needed to achieve this. Full details on how HSE resourced the work described is available from the first author.

In the case of the MS intervention, the final outcome would be a reduction in cases of work-related stress and associated sickness absence. Figure 14.4 shows a simple version of the intervention logic model applied to this intervention.

The ILM was used to generate estimates and assumptions about what HSE would need to do in order to meet the final outcomes. The "change pathway" involved raising awareness of the MS approach, increasing organizational uptake and assisting organizations in implementing the approach properly. An ILM is, in essence, a hypothesis about how the change process should work, based upon the best currently available knowledge and experience and, in reality, the eventual

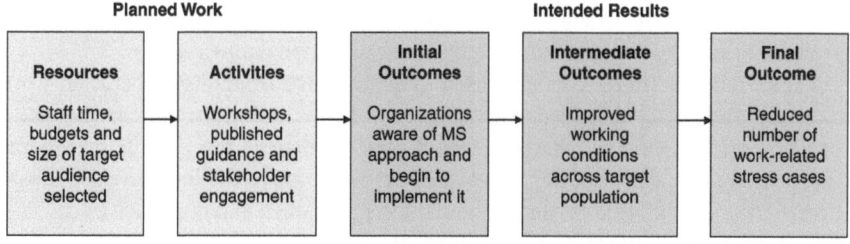

Figure 14.4 Intervention logic model summary for Management Standards intervention.

model was much more complex than the one described here. A fuller picture includes the following critical steps (and assumptions fully or partly supported by evidence) in the pathway:

- Working conditions are causally related to ill-health via a "stress" process.
- The MS incorporate these conditions as fully as possible.
- The mechanism of *delivery* to bring about changes in working conditions is optimal given cost (resource)/benefit constraints.

In Mackay et al. (2004) and Cousins et al. (2004) we explore the evidence supporting these statements. Some further assumptions were made:

- Optimal *awareness* of the standards in industry is achieved, principally in terms of timeliness and coverage (especially to target sectors).
- Sufficient *uptake* of the standards to ensure that a significant proportion of industry was actively using them (or an "equivalent" approach).
- That uptake would be followed by *proper use* of the MS approach.

As part of the overall RHS initiative considerable communication resources were devoted to the design and dissemination of material on the MS through a variety of media and methods to achieve the above aims. A series of surveys of both employers and employees were undertaken to measure the extent to which awareness, uptake and proper use were happening, and, in particular, with the rates of usage of the MS process (FiT3 programme). Finally it was assumed that the following would occur:

- Organizations would properly identify the risk to which their employees were exposed and would be able to choose the appropriate control measures.
- That these control measures would be fully and properly implemented (with associated monitoring and review of impact).
- Impact would be sufficiently quick to influence targets during the timescale of the programme.

The aim was that actions from the above across GB industry would lead eventually to an improvement in working conditions across the target population and ultimately a reduction in work-related stress cases across this population. It should be noted that although this process appears linear for simplicity, in practice it is necessarily iterative. As will be described later, HSE put in place a series of systems of monitoring each of the steps in this change pathway.

Data from this study were also used to quantify the relationship between work characteristics and subsequent incidence of ill health in order for HSE to make a decision on the scale of change it would seek to incur in the population. Scores representing a shift in working characteristics were converted into standard deviation (SD) units and used in models predicting prevalence and incidence of disease. For illustration, a 1 SD shift in working characteristics is approximately equivalent to shifting the distribution such that the new mean score is at the old 80th percentile cut-off. In other words before such a shift, 20 per cent would be above this cut-off and after such a shift, 50 per cent would be above that same cut-off.

Models were produced to relate some of the MS factors to psychiatric morbidity, for example, as measured by the General Health Questionnaire (GHQ) (Goldberg, 1972). The results of this modelling indicated that an average shift of around 0.5 SD in each of *control, job demands and social support* predicted a reduction in the prevalence of psychiatric morbidity (GHQ caseness) of 20 per cent.

Working conditions would be influenced within those organizations where employers had embraced the concept of the MS approach (or equivalent) and were implementing it successfully. The policy team responsible for deciding on the scale of the initial intervention chose aspirational targets of 80 per cent awareness amongst those employers in the target sectors, with 80 per cent uptake of the approach from those organizations who were made aware of it and 75 per cent of these using the approach properly (i.e. not just paying lip service to the approach through fear of reprisals if seen to ignore it).

These aspirational targets were chosen by again working backwards through the ILM. Starting with a population of around nine million and assuming the awareness and uptake levels stated, the model predicts that just over four million workers would have their working conditions improved, thus in theory meeting the targets. The calculations are summarized in Table 14.2 which shows a simplified version of the change pathway within the Intervention Logic Model. To test the assumptions behind awareness, uptake (adoption) and proper use, a number of surveys were subsequently carried out.

Implementation and delivery mechanisms

To implement the change pathway the overall strategy was rolledout in three parts – two targeted at the key five sectors known as Sector Implementation Plans 1 and 2 (SIP1 and SIP2) and a Wider Implementation Plan (WIP) to cover all the

Table 14.2 Change pathway for the Management Standards intervention

Target audience	9 million workers across 5 priority industry sectors
Awareness	Organizations employing 80% of the workforce (7.1m workers) within these sectors made aware of the Management Standards
Uptake	Assumption that 80% of organizations becoming aware of the Management Standards will adopt the approach (covering 5.7m workers)
Proper use	Assumption that 75% of organizations using the Management Standards will use the approach properly and sustain their activity (covering 4.3m workers)
At-risk cases	Approximately 1% of workers in these organizations at risk of developing work-related stress (43,000 workers)
Estimated incidence reduction	Assumption that 40% of potential work-related stress cases avoided by using the Management Standards approach (17,000 cases saved)

remaining sectors and industries in the UK. It was envisaged that they would run consecutively so that learning from each stage could be incorporated into the subsequent steps.

Structure of Sector Implementation Plan 1 (SIP1)

The objectives of SIP1 were as follows:

- Identify those sectors in which there is a high incidence/high risk of work-related stress that lend themselves to a risk management approach to risk assessment and reduction.
- Gather evidence to demonstrate the extent to which the incidence of stress-related ill health can be reduced via a risk management approach.
- Engage a volunteer sample of 100 "willing" businesses from identified priority sectors to take part in an HSE led intervention programme designed to facilitate the effective management of work-related stress.
- Test the reliability and usability of the MS methodology especially barriers and facilitators to implementation within organizations.
- Gather empirical evidence of variables impacting upon the success of the introduction of a risk management based approach to inform the trajectory of the HSE's broader stress management programme i.e. "what works, where, under what circumstances and in what way?"
- Use participating organizations as potential case studies.

The basic design of SIP1 consisted of 20 organizations from each of the five sectors (originally the so-called "Willing 100"). They agreed to enter into a compact with HSE such that for implementing the MS within their organization at an agreed level ("proper use") and to an agreed time (approximately 18 months) they would be given free consultancy to help them through the process and at the same time immunity from any potential legislation enforcement activity during the timescale of the project. It was not intended that these organizations would represent a statistically valid subset of the overall sectors but that they would provide useful data on the "implementability" of the overall process that could be fed into subsequent parts of the programme.

Communication and engagement strategy

The overall communication strategy for the stress programme was based on the UK Office of Government and Commerce (OGC) guidance for Managing Successful programmes adapted in the light of HSE's policies and discussions with the main stress priority programme stakeholders. The key messages were:

- Work-related stress is a serious problem for organizations and GB industry as a whole. Tackling it effectively can result in significant benefits for organizations.

- There are things organizations can do to prevent and control work-related stress.
- Although the stress programme is not enforcement led and is aimed at helping fulfil legal duties, the law requires organizations to take action to reduce risks to health.

Over the course of the programme these key messages were developed, regularly reviewed and tailored for specific audiences – segmented into audience type, key messages, activity/channels/product, timescale, evaluation and budget.

SIP1 Initial engagement activity

Stakeholder workshops were held for each of the five sectors in which organizations present were asked if they would be willing to participate (to become part of the "Willing 100"). The aim was to identify approximately 100 organizations – around 20 for each sector. This was carried out in the spring of 2005 after the launch of the MS. The list of potential volunteers was oversubscribed especially from the health sector. At this stage the aim was to recruit a reasonable cross-section of the primary sectors rather than obtain a truly representative (statistical) sample.

SIP1: HSE support structures

Various internal HSE administrative and support structures were set up to liaise between the various members of the HSE programme team and to also interface with participating organizations – effectively in a "troubleshooting" role. Further details of exactly how this component of the programme operated and performed is available from the authors. The multidisciplinary teams comprised policy specialists, occupational and health psychologists, field inspection staff, specialists in communication and engagement, and statisticians. Each participating organization within a particular sector had the following assigned to it:

- An HSE inspector (or in the case of the finance sector a Local Authority inspector) who became known as a "Stress Partner".
- A Sector Account Manager (SAM) from the stress policy group whose role was to manage the various interactions between the team and organization, "troubleshoot" and track progress.
- A technical support officer – usually an organizational psychologist.
- An ACAS Consultant. The Arbitration Conciliation Advisory Service (ACAS) is a UK Government agency that provides expert advice on dispute resolution, employee engagement and, generally, best practice in industrial relations matters.

SIP2: Structure of Sector Implementation Plan 2

Sector Implementation Plan 2 was designed to be the main phase of delivery of the overall programme of implementation in the five key target sectors building on lessons learned from SIP1. In fact because of the delays in starting, SIP1 was still underway when SIP2 was begun and only partial learning from organizations who had been quick to implement the MS approach could be used. In response to feedback from stakeholders it was decided to try to improve knowledge and awareness of the MS approach in those target sectors – the aim was to help them understand the rationale behind the methodology and provide them with the skills and commitment to take the process forward in their organizations unaided by outside structured/pre-planned help (as in SIP1). Organizations in the target sectors that collectively employed 80 per cent of employees in that particular sector were subject to high level stakeholder engagement and then invited to attend a series of workshops. It was intended that two participants from each organization would attend – a senior human resource manager and a staff or union representative (to underline the importance of worker involvement). In the event representatives from a range of functions across organizations attended including senior management, occupational health and from health and safety type functions.

The original programme plan for the idea was to take the learning from SIP1 and do a wider dissemination to a large proportion of those other organizations (the bulk of them) also from the five key sectors. The intervention chosen was a free, one-day workshop with a very practical content based on the use or exercises based on each stage of the MS process. This core content was supplemented by case studies largely drawn from organizations who participated in SIP1 and from sector partners and stakeholders to provide encouragement that the process was "doable" and resulted in the desired organizational outcomes (much of this material is now on the HSE website). The design of the content went through various iterations and testing in pilot workshops.

Part of the rationale for the SIP2 workshops was that most of the key sectors had high levels of sickness absence (more properly *absence attributed to sickness*) from common health problems notably stress, depression and anxiety and musculoskeletal disorders. It was envisaged that the subsequent interventions undertaken by workshop delegates in their organizations would contribute to overall reductions in sickness absence in their sectors. The SIP2 workshops were delivered on a sector-by-sector basis, and as far as possible to cover as much as the country (geographically) as was possible. There was an expectation on the part of HSE that organizations that attended would then begin to implement the MS approach (or an "equivalent" approach).

During 2006–2007 HSE ran 67 of these workshops (covering 800 organizations with a combined workforce of over seven million). Further so-called "*Masterclasses*" were run in 2007–2008 to cover specific aspects of the process especially where organizations had, or were, experiencing difficulties and obstacles to implementation (such as data handling). HSE also provided follow-up support in the form of an infoline service and advice via a website. Insofar as this represented a significant resource commitment, and in order to maintain momentum and capitalize on this investment, a programme of follow-up inspection work was developed. A large number of nominated inspectors were trained to undertake this work (approximately 141 HSE inspectors and 100 Local Authority inspectors who, between them, are responsible for monitoring the key sectors). To further support these inspectors an *inspection tool* was developed to help them assess and record duty holders' progress with risk assessment and control both in terms of rate of progress and also the quality of their implementation ("proper use"). The focus of these inspections was the extent to which duty holders could demonstrate progress on each of the five steps in terms of management commitment and workforce participation. A copy of this tool is available from the authors. These follow-up inspections began in April 2007. The implications of the data derived from this monitoring work is discussed in a subsequent section.

Structure of Wider Implementation Plan (WIP)

It was envisaged in the original design of the programme that a third element comprising those remaining sectors outside the five key sectors – essentially the remaining approximately 19 million workers – would be covered by a lighter touch Wider Implementation Plan (WIP). In the event because of resource constraints and changes in priorities a properly formulated implementation plan for the active engagement of these sectors did not take place.

Process evaluation and impact assessment

In the remainder of this chapter we reflect on the current state of knowledge about the impact of the MS approach (in both specific and generic senses) and the benefits and the experiences that the approach has had for end user organizations so far. We also consider the extent to which the original national targets on incidence and sickness absence have been met. We focus upon the following specific questions:

1. What is the extent and quality of existing evidence about how the MS approach has been implemented thus far (as of 2011)?
2. What is the extent and quality of existing evidence about whether using the MS (or equivalent approach) produces a sustainable reduction in incidence of stress-related problems in an organization?
3. What activities (control measures) have proved effective in tackling work-related stress?
4. What gaps are there in the existing body of evidence and how might these be filled effectively in the future?
5. What has been the impact of implementing the MS at a national (population) level and, in particular, on the Revitalising Health and Safety (RHS) targets set in 2000?

To do this we will reflect on evidence from current literature where the MS approach has been adopted (or equivalent), especially relevant studies post 2004, and from experiences from the Stress Priority programme described above. We will touch on a number of process issues that are germane to the successful implementation of a programme such as we have described but will leave other authors of chapters elsewhere in this book to dwell on the detailed implications in terms of process design. But it is worthy of note that each of the building blocks behind the MS process as finally implemented has been extensively tested (for example the individual components of the risk assessment cycle).

Evaluation: outcome or process?

There are two main types of evaluation: outcome evaluation, also known as summative evaluation, aims to tell us what impact (if any) a particular intervention has had in terms of meeting its objectives. Process or formative evaluation, on the other hand, can provide evidence about how and why a particular intervention works or fails to work. This distinction represents the difference between asking whether an intervention has been effective and why – very often evaluation will seek to answer both questions by combining elements of both types of evaluation. The fundamental aim of outcome evaluation is to assess whether any observed outcomes were a result of the intervention in question.

Griffiths (1999) argues that the nature of the population approach implies that organizational interventions for tackling work-related stress do not need to be demonstrated to be particularly powerful for improvements to be obtained (also see Mackay et al., 2004, p. 107). However, it is important to be able to demonstrate that any improvements made are a direct result of the intervention rather than due to other factors. For workplace stress interventions targeted at the organizational level in particular, it is extremely difficult to obtain the meaningful counterfactual that is required of robust outcome evaluation research; therefore the current evidence base does not facilitate a reliable assessment of whether using the MS approach produces a sustainable reduction in work-related stress within an organization. It is also not possible to identify from existing data the specific activities that have proved effective in tackling work-related stress. To make sound assessments of this type would require an empirical examination of the type of interventions that have been developed using the MS approach; how they have been implemented in the "real world" and, use of longitudinal data to assess what the long term outcomes have been. Long term outcomes might include levels of reported stress; overall sickness absence data; organizational productivity and levels of staff engagement, although it is likely that a significant period of time would need to elapse before population level shifts in health trends could be identified.

There is now a wealth of "process" evidence ranging from anecdotal insights and examples of good practice that have been provided by end users of the approach, to systematically researched case studies which often combine the analysis of interview data and organizational documents. It is important to bear in mind that the evidence drawn from process evaluations is merely suggestive of associations between actions and outcomes and any conclusions are likely to be context dependent. Nevertheless, these accounts do provide useful insights into the processes occurring at the micro level, which may help to explain the success or failure of stress programmes that have been implemented as part of a MS approach.

Micro level evidence

Following the launch of the MS in 2004, there were considerable efforts by the HSE to encourage uptake of the approach, particularly within those industries identified as having high levels of work-related stress, related ill health and sickness absence. The process evaluation of SIP1 is described in two separate publications (Cox et al., 2007; Tyers et al., 2009); both involved the development of organizational case studies which included interviews with key stakeholders involved in implementing the MS. From these preliminary studies early indications emerged of how employers and employees were likely to engage with the MS and the types of barriers and enablers they might encounter. In particular the process was found to be lengthy to implement and it was often difficult to maintain momentum throughout the implementation process. The importance of line manager competency and of securing full and sustained commitment from senior management were found to be important factors in ensuring the best chance of successful implementation.

Cox et al. (2007) conducted a process evaluation with a subsample of 11 of the organizations that took part in the SIP1 initiative. These organizations volunteered to take part in the additional study and their participation involved providing documentary evidence and taking part in interviews. The study identified some key process issues relating to the implementation of the MS approach, for example the importance of employee involvement and communications. The study highlighted the significance of securing senior support for the approach, which could be achieved through their participation in a stress steering group or taking on the role of senior level project champion. This finding corroborates the findings from the initial contacts with SIP1 organizations that also emphasized the importance of securing commitment from senior staff. Cox et al. (2007) also identified the value of integrating the approach with existing policies and procedures; ensuring that sufficient resources are set aside for implementation and the importance of getting the organization ready for change (Nielsen et al., 2010).

The initial idea was that lessons from SIP1 would feed into a second wave of implementation (SIP2); however due to timetable changes, this proved unfeasible and in the end SIP2 had to begin before SIP 1 was fully complete. This second phase focused upon engaging with the target audience on a bigger scale; through workshops, master classes, a telephone helpline and non-enforcement inspections. It was designed to reach 1,500 organizations to help them implement the MS. The process evaluation of SIP2 (Broughton et al., 2009) involved a telephone survey of workshop participants and case studies which combined interviews, focus groups and documentary analysis. The findings suggest that many of the case study organizations were finding it difficult to engage with the notion of stressors on an organization-wide basis and often focused considerably upon individual stress factors and reacting to individual absence records. In keeping with the philosophy of the MS approach, some of the organizations were choosing to tailor the approach to suit their own needs rather than interpreting it rigidly. This demonstrates a more flexible interpretation of the approach than had originally been promoted during SIP1. SIP2 reiterated the process issues involved in implementation that are identified consistently throughout the existing evidence, such as senior manager and staff involvement and support, integration with existing policies and procedures, and allocation of resources – all of which could act as barriers or enablers for the organizations.

Although the evaluations of SIP1, the Cox et al. (2007) review, and SIP2 provide some useful information about the processes involved in implementation, a degree of caution must be taken when considering the findings. The organizations involved are unlikely to be "typical" of the target audience because participation was voluntary and their experiences were shaped by a special set of circumstances which are not representative of the "real world". For example during SIP1, HSE provided dedicated support through its Stress Partners and ACAS, to help the organizations fully implement the approach. Participating organizations were exempt from formal enforcement action on work-related stress issues throughout the process, providing that they implemented the approach properly. During SIP2, all the organizations involved were already "tuned in" to the issue of work-related stress and some had already begun implementing the MS approach. Evidence from inspector visits indicated that some of the organizations that had attended the workshops were able to follow the process to the end but overall there was a large range in the speed at which organizations were able to progress – mainly for the reasons highlighted in the SIP1 evaluation reports. Despite these issues, the findings provided some useful early insights into the process issues relating to engagement with the approach and successful implementation and these early findings have frequently been echoed in subsequent process focused assessments of the MS approach.

Research for HSE by Cox et al. (2009) used a Delphi methodology to consult with a range of experts in occupational health about their views on the MS approach. The Delphi approach tends to be most useful for consultation with a view to reaching consensus on an issue and it is not a well established social research method. However, the research is a good starting point for considering the utility of the MS approach so far. The study reiterated some of the findings from the SIP evaluations, for example identifying the importance of staff involvement and of the competencies of managers to implement the approach successfully. The experts were mindful of the difficulty of assessing cause and effect since organizational changes are likely to have a delayed impact. Of particular concern to the experts was their recognition of a discrepancy between assessment and action because many felt that although the MS approach enables organizations to identify areas of weakness in performance, it doesn't necessarily facilitate the development of strategies for improvement in these areas. The Delphi panel were concerned that there is a distinct lack of a sound evidence base about what makes a good intervention for preventing work-related stress. They suggested that a more prescriptive approach regarding how to address the problems identified in a risk assessment would be useful and could involve the provision of sector- or organization-specific examples or case studies. The lack of organizational case studies has since been addressed by HSE who now provide a range of examples on their website which provide potential end users with models of good practice. However the MS approach is intended to be a flexible rather than rigid approach so it is likely to be difficult to provide the kind of prescriptive advice to organizations suggested by the Delphi experts.

Micro level evidence: Case studies

Egan et al.'s (2009) review of organization level stress interventions found that the explanations for poor outcomes tended to be revealed more effectively through the commentaries rather than through any quantitative data collected. Griffiths (1999) suggests that it may simply be the case that "the most useful information cannot be collected through quantitative approaches". Mellor et al. (under review) have undertaken first hand research using qualitative case study methods to examine the processes involved in implementing the MS approach. They carried out five case studies using a combination of semi-structured interviews and documentary analysis, with particular consideration of the activities that occurred at each of the five stages of risk assessment advocated by HSE. Interestingly this study comprised private as well as public sector organizations, which have tended to be the focus of discussion so far. These findings again echo some of the consistent findings relating to enablers, for example: the importance of securing senior management commitment, sufficient resource allocation, active and committed line managers, the creation of a multidisciplinary steering group, worker involvement, and manager training on stress and risk assessment. Typical barriers included: lack of manager availability to carry out actions following the risk assessment, having to persuade managers that stress is an issue, and having to keep them motivated and interested in it. The findings also echo concerns raised by the experts in Cox et al.'s (2009) Delphi study, about whether guidance for the design of interventions following the risk assessment is suitable and sufficient. The organizations in Mellor's research did not tend to view stress as an independent issue and most were implementing the MS as part of a wider well-being strategy rather than as a stand-alone initiative. More specifically Mellor's research underlines the problem of evaluating organizational interventions as the findings suggest that the case study organizations felt that it was difficult to demonstrate the impact of any interventions they had put in place in response to the risk assessment. They commonly attempted to measure success using key performance indicators such as HR, occupational health and sickness absence data and, while most did see a decline in reported stress and sickness absence, it was widely felt that more guidance was needed on how to conduct

304 *Policy implications*

robust evaluations in order to establish whether positive outcomes could be attributed to their interventions.

Macro level evidence: Impact on targets and outcome measures

Overall in the ten years since the HSC targets were set and the stress programme begun, is it possible to detect any (significant) improvements at the national (GB) level? Macro level quantitative data from a number of large scale surveys can tell us about the incidence prevalence of work-related stress symptoms in the GB and whether these have changed significantly over time. According to the latest data from the Labour Force Survey (LFS), the incidence rate of self-reported stress, depression and anxiety for 2008/2009 was estimated at 760 cases per 100,000. This represents a significant decrease since 2001/2002 when prevalence was estimated at 890 cases per 100,000. The data also suggests that there has been a decline in the number of working days lost to cases of work-related stress, depression and anxiety which have fallen from an estimated 12,919 in 2001/2002 to 11,420 in 2008/2009 (HSE 2009, p. 18) – see also discussion below. Although these figures imply that a positive shift may have occurred during the last decade, it is not, of course, possible to link these macro level trends directly to the launch and implementation of the MS in 2004.

Data from the *Psychosocial Working Conditions Survey* and the *Self-Reported Working Conditions Survey* also provide some evidence about outcomes for work-related stress and will be discussed in turn below. First, the *Psychosocial Working Conditions Survey*, an HSE-funded module in the ONS (GB: Office of National Statistics) Omnibus Survey, was set up specifically to monitor changes in the psychosocial working conditions related to Demand, Control, Managerial Support, Peer Support, Role, Relationships and Change in GB workplaces. These are the stressors that make up the MS typology and represent the working conditions which HSE is aiming to improve through the approach. The most recent report of the Psychosocial Working Conditions survey compares the results over a six year period from 2004 to 2009 (Packham and Webster, 2009). The first year (2004) was completed just before the launch of the MS and so provides a useful baseline against which to compare data from subsequent years. The data suggests that in general there has been no significant change to psychosocial working conditions overall during this period. However there does appear to have been an improvement in the specific conditions of "Peer Support" and "Managerial Support" but none in "Control". These data are shown in Figure 14.5 below. The data also suggests that there was little change during the six year period in terms of the proportion who reported that they have discussed stress with their line manager or who were aware of stress initiatives in their workplace. It was hoped that these questions would help elicit information about the level of activity on work-related stress, with the assumption that activity would increase in organizations that were using the MS, however no such increase has occurred yet.

HSE estimates that once the MS process is initiated by an organization it could be at least 18 months before any benefits are realized from interventions that are

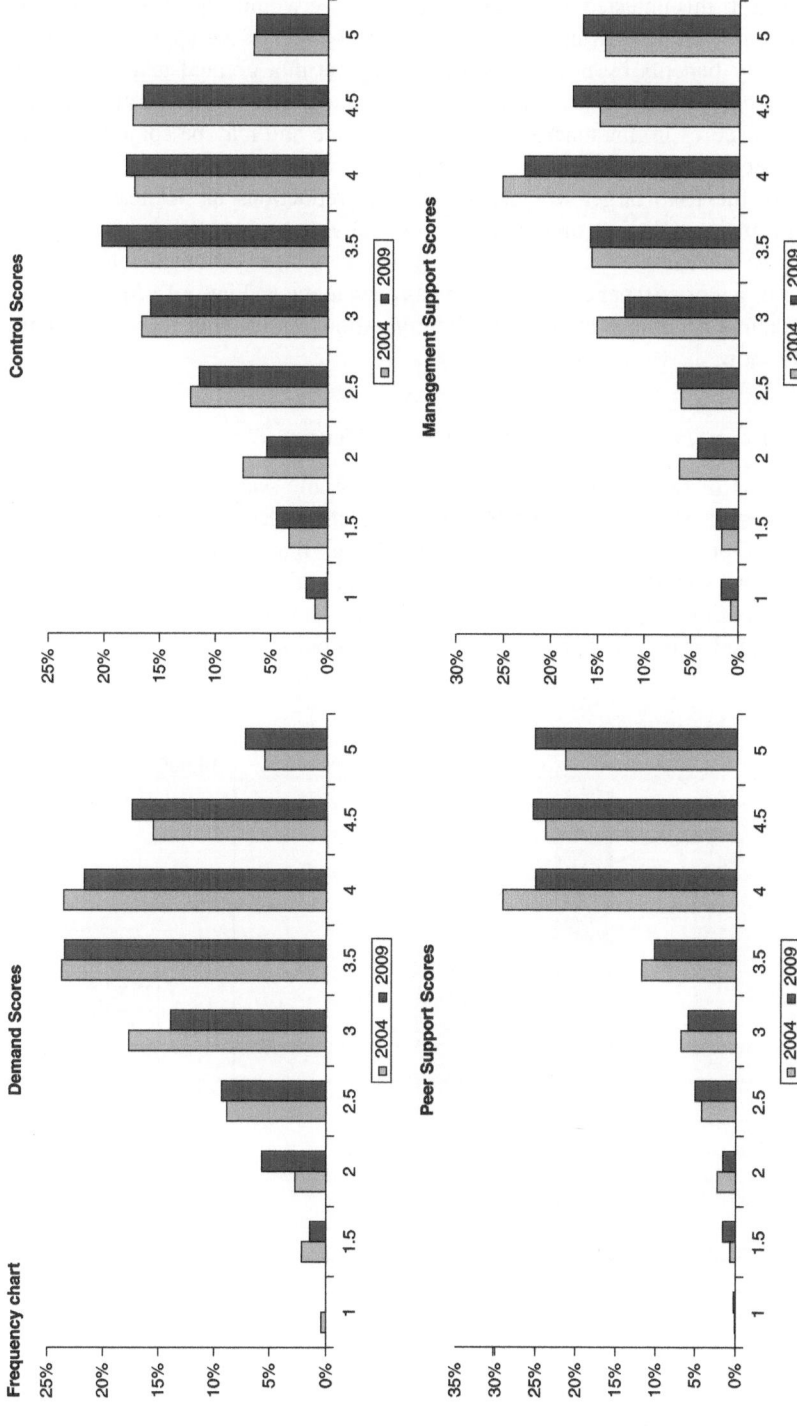

Figure 14.5 Population shifts in selected psychosocial work characteristics 2004–2009 using data from the HSE indicator tool questions.

306 Policy implications

introduced. If this untested assumption is correct, we would expect organizations that implemented the MS immediately after the launch of the approach to have realized any benefits by now. However it may be that gradual take up of the approach has precluded a significant impact upon the psychosocial working conditions scores in this macro level survey to date and may be something that will not become apparent unless measured over a much longer period of time. The second of the RHS target was concerned with reductions in sickness absence and was an integral part of the workshops delivered under SIP2 ("*Managing sickness absence and work-related stress*"). The very latest data available suggest that this target of a 30 per cent reduction has been met (see Figure 14.6). But there are of course a number of differing interpretations as to what these data may actually show.

Reflections on lessons learned: Organizational level

The success of the overall stress priority programme was predicated at action being taken at organizational level (see next section also). It is clear from the various components of the evaluation methodology that assumptions about how organizations would respond to efforts to implement the MS process were, if not incorrect, optimistic. Here we make some general comments about factors that either enabled or impeded progress with implementation.

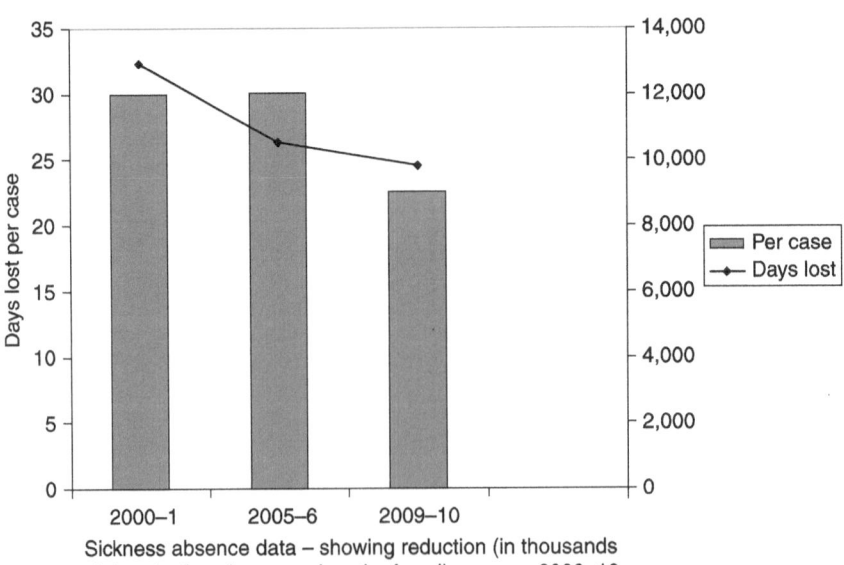

Figure 14.6 Ten-year changes in days lost from work-related stress. Data derived from the SWI (Self-reported, Work-related Illness Survey) Survey.

1   Organizations were often committed to taking part in SIP1 but found it difficult to integrate into work plans that had been agreed for example in the previous financial year – this was very difficult to reconcile with HSE's timing. Senior management (or equivalent depending on sector) commitment was absolutely vital – preparatory work to help with questions (typically, "what's in it for us?") raised by the board was found helpful here.
2   Some organizations (or levels within them) were uncomfortable with language drawn from the health and safety world (e.g. "risk assessment").
3   Many large organizations were found to have initiatives in place that replicated some features of the MS process – it was found that these could be recruited without starting the whole process *de novo*. On the other hand some of these were part of annual staff surveys and there was often the issue of "survey fatigue".
4   Some organizations found that (even with HSE and ACAS support) they had difficulty in adapting the generic guidance material to their specific needs ("flat-pack") and wanted a much more prescriptive approach (see also point 6).
5   In an earlier piece of research done for HSE (Jordan et al., 2003) it was demonstrated that where key corporate units (Human Resources, Occupational Health, Health and Safety) cooperated in tackling occupational health issues then outcomes tended to be better. Conversely, where activity was uncoordinated, the reverse was the case. Of particular note was the difficulty some organizations have in collecting, analysing and interpreting organizational data, that is, of course, critical to the MS process. In fact, as a post-workshop activity special data handling seminars ("*Masterclasses*") were held for those participating organizations that were struggling with this area. The critical learning point is that a key individual or group needs to take "ownership" of the project and inject sufficient resource.
6   The successful completion of the process is critically dependant on workforce engagement – the guidance recommends the use of (facilitated) focus groups to complete this step. Some participants who had historically found active worker participation in decision making difficult found this critical step hard to complete with the result that action planning and decisions on what interventions were appropriate was not done properly. Linked to this was a perception on the part of some organizations that they needed a "recipe book" that would assist them in choosing appropriate interventions. This is perhaps a reflection of not trusting the workforce to be able to specify what would make real improvements to their jobs or management concern about what the outcome might be in terms of resource commitment or perceived negative impact on productivity. In reality most of the interventions that were implemented proved to be low cost.
7   A generic issue that became apparent in many organizations was the issue of line manager competency to address work-related stress issues at both an individual (cf. each of the six standards specifies that "*systems should be in place to deal with individual concerns*") and team level. This seems to be a

reflection of the training that managers are given, that, in many organizations, reflects the technical requirements of their post rather than the ability to address how they interact appropriately with their staff. To an extent this has been addressed by the work HSE has co-funded on line manager competencies (see Lewis et al., this volume).

8   Finally, and perhaps most disappointingly at an organizational level, the MS themselves (that is *the states to be achieved* – to guide progress in developing specific interventions) appear to be used only occasionally, perhaps reflecting an overemphasis on the initial stages of the risk assessment process rather than on the need to focus on and implement context-specific interventions at the job level.

Reflections on lessons learned: National level

The overall thrust of this programme we have described was to both reduce the prevalence of work-related stress (and associated sickness absence) and, in so doing, achieve country (GB) level targets. The approach has been to take a public health model of prevention and apply it to a workplace health issue – in doing so various untestable assumptions were made about the extent to which this could be successful given the constraints of the overall programme. From the outset it was recognized that the targets were both ambitious, and, to some extent, arbitrarily specified. In retrospect, it perhaps should have been better to do some preliminary testing of the power of the MS approach to bring about sustained improvements over a specified time period; but under the constraints of the overall policy requirements and associated timescale such testing was not possible. So the eventually designed ILM was, in a sense, based on largely untested assumptions about the power and "implementability" of the MS even though the underlying science was known to be sound and that individual (key) components of the model had previously been tested and validated.

At the outset a considerable amount of front loading of engagement of key stakeholders was done so that the chosen sectors were primed and that awareness of the launch of the programme was assured. Although considerable effort was expended in this launch and post-launch activity, it is clear from extensive surveys done on HSE's behalf that *awareness* and eventual *uptake* (key parts of the ILM) were patchy. In retrospect it seems that assumptions about how quickly this would happen across the target sectors were overestimated. One can also speculate that the wrong sectors were chosen for the initial implementation work and that, perhaps, those where incidence of cases was low (indicating where some activity had already taken place) or where existing management systems were more conducive to MS implementation might have provided a better test bed.

The second part of the implementation plan was based around workshops ("*Managing sickness absence and work-related stress*"). It may be that the linking of these two issues was not the right approach; that delegates became, on their return to their organizations, focused on sickness absence *management* rather than more fundamental commitment to prevention through job design and organizational change. These areas of speculation are perhaps reflected in the national data we have so far. Also, in retrospect, the design of the content of the workshops could have been improved. The focus was very much (because of time constraints) upon the early stages of the MS process – very little of the material covered action planning and the choice of interventions; monitoring and review and organizational learning aspects were only touched upon. Perhaps the overall lesson here is that organizations became more comfortable and proficient at sickness absence *management* but were less convinced about the importance of *primary prevention* approaches. Structural changes in job content (towards "good jobs") would seem to need much more time to be achieved as evidenced from data in changes in work characteristics. More detailed analysis, which is currently ongoing, should help us to disentangle the drivers towards improvement. The national level approach was also predicated very much upon a partnership approach between key stakeholders and other Government agencies and whilst there were some very good examples of how this worked effectively, greater planning and time would have allowed these collaborations to work more synergistically.

Conclusions

In this chapter we have briefly described a national programme aimed at reducing the burden of work-related stress from both a national and organizational perspective. We have attempted to draw some general lessons for those who may wish to repeat such an exercise. The overall conclusion must be that such programmes are "doable" but consume large resources in terms of time and effort. This commitment is easy to underestimate both nationally and organizationally. A better approach focusing more widely on improving the quality of working life for employees by seeking to enhance their well-being is likely to have a positive impact on other common health conditions beyond work-related stress and might be easier to implement. The experts in Cox et al.'s (2009) Delphi study were in favour of making the MS approach broader to address well-being issues and the findings from some case studies certainly highlighted the similarity between enablers and barriers for well-being programmes and for organizational stress interventions. On the other hand, the uncertainty about the relationship between stress and well-being is also a case for keeping them as separate issues.

There is now plenty of evidence about why and how the MS have been implemented from the abundance of case studies and anecdotal examples of good practice, however there is much less evidence about whether the approach has been successful at the micro and macro levels because it is difficult to attribute positive outcomes to the implementation of the MS approach. A need for further long-term data collection, both qualitative and quantitative, has been identified by numerous commentators including Cox et al. (2007, p. 60), Jordan et al. (2003), and LaMontagne et al. (2007). This would improve the generalizability of the findings and enable us to better understand the impact of the approach upon end user organizations. A carefully controlled matched comparison study could provide an estimate of the counterfactual where, for example, within one organization the MS approach is adopted in certain parts of the organization (test group) but not others (control group) so that before and after data could be compared for each. However, such an approach would still only be providing evidence about the impact of the intervention in a single organization. In addition, because the MS are intended to be flexible and tailored to suit the needs of different organizations, the findings would not necessarily be comparable between or generalizable across different organizations. There are a number of other issues with this type of research including ethical issues: it would mean that some parts of the organization would be receiving a potentially beneficial treatment whilst others were not and it may be difficult to find an organization that is sufficiently homogenous so that its different parts would be directly comparable.

Because robust evaluations and research studies of organizational interventions (such as the case control design) are likely to continue to prove difficult to achieve for reasons such as organizations being unable or unwilling to commit to long-term evaluation studies and not having the resources or capabilities to conduct their own robust studies, it might be useful to consider how we can make the best use of existing evidence sources. It would be much harder to meet the requirements of rigorous evaluation at the macro societal level therefore it is unlikely that we will be able to say what the impact of the approach has been on levels of work-related stress in wider society. It is difficult to address the issue of causality at the societal level because it is not possible to evaluate the counterfactual, or what would have happened anyway. However, the Psychosocial Working Conditions Survey focuses specifically upon the stressor areas covered in the MS and the baseline from before the launch of the approach is very valuable because it provides trend data. Although so far the survey has found some significant changes relating to psychosocial working conditions since the launch of the MS, evidence suggests that tangible changes to working conditions following implementation of organizational interventions may require a longer gestation period than originally anticipated so it will be interesting to continue monitoring the results from the survey in the long term if possible. To end on a positive note, and based on current activity that continues across UK industry, what we have described should perhaps be considered as "work in progress".

References

Adams, J., & White, M. (2005). When the population approach puts the health of individuals at risk. *International Journal of Epidemiology, 34*, 40–43.

Anderson, J., Huppert, F., & Rose, G. (1993). Normality, deviance and psychiatric morbidity in the community: A population-based approach to General Health Questionnaire data in the Health and Lifestyle Survey. *Psychological Medicine, 23*, 475–485.

Bartram, D., Yadegarfar, G., & Baldwin, G. (2009). Psychosocial working conditions and work-related stress among UK veterinary surgeons. *Occupational Medicine. 59*, 334–341.

Black, C. (2008). *Review of the health of Britain's working age population: Working for a healthier tomorrow*. London: TSO.

Broughton, A., Tyers, C., Wilson, S., & O'Regan, S. (2009). *Managing stress and sickness absence: Progress of the Sector Implementation Plan – Phase 2* (Research Report RR694). Health & Safety Executive.

Constable, S., Coats, D., Bevan, S., & Mahdon, M. (2009). *Good jobs*. (Research Report 713). Norwich: HMSO.

Cousins, R., Mackay, C. J., Clarke, S.D., Kelly, C., Kelly, P., & McCaig, R. H. (2004). 'Management Standards' and work-related stress in the UK: Practical development. *Work & Stress, 18*(2), 113–136.

Cox, T. (1993). *Stress research and stress management: Putting theory to work* (Research Report 61). Health and Safety Executive.

Cox, T., Griffiths, A., Barlow, C., Randall, R., Thomson, T., & Rial-González, E. (2000). *Organizational interventions for work stress: A risk management approach*. (Research Report 286/2000). Sudbury: HSE Books.

Cox, T., Karanika, M., Mellor, N., Lomas, L., Houdmont, J., & Griffiths, A. (2007). *Implementation of the Management Standards for Work-related Stress: Process Evaluation SIP1 Technical Report T/6267*. University of Nottingham: Institute of Work, Health and Organizations.

Cox, T., Karanika-Murray, M., Griffiths, A., Wong, Y. Y. V., & Hardy, C. (2009). *Developing the Management Standards approach within the context of common health problems in the workplace: A Delphi study*. Health and Safety Executive, Contract Research Report 687: HSE Books.

Crawford, J., George, P., Graveling, R., Cowie, H., & Dixon, K. (2010). *Good work, good health*. (Research Report 603-00944). Edinburgh: Institute of Occupational Medicine.

Edwards, J., Webster, S., van Laar, D., & Easton, S. (2008) Psychometric analysis of the UK Health and Safety Executive's Management Standards work-related stress indicator tool. *Work & Stress, 22*, 96–107.

Egan, M., Bambra, C., Petticrew, M., & Whitehead, M. (2009). Reviewing evidence on complex social interventions: Appraising implementation in systematic reviews of the health effects of organizational-level workplace interventions. *Journal of Epidemiology and Community Health, 63*(1), 4–11.

European Commission. (1996). *Guidance on risk assessment at work*. Luxembourg: Office for Official Publications of the European Communities.

Goldberg, D. P. (1972). *The detection of psychiatric illness by questionnaire*. London: Oxford University Press.

Griffiths, A. (1999). Organizational interventions: Facing the limits of the natural science paradigm. *Scandinavian Journal of Work, Environment & Health, 25* (6, Special Issue), 589–596.

Head, J., Ferrie, J., & Brunner, E. et al. (in press) *Supporting the development of management standards for work-related stress: impact on health and sickness absence*. Health & Safety Executive.

Health and Safety Executive (1995). Stress at Work: A Guide for Employers. HSG116. Sudbury: HSE Books.

Health and Safety Executive (2001). Tackling work-related stress: A managers' guide to improving and maintaining employee health and well-being HSG218. Sudbury: HSE Books.

Health & Safety Executive. (2009). *Health and safety statistics 2008/2009*. www.hse.gov.uk/statistics/overall/hssh0809.pdf. Last accessed on 18/07/2010.

Jordan, J., Gurr, E., Tinline, G., Giga, G., Faragher, B., & Cooper, C. (2003). *Beacons of excellence in stress prevention*. Sudbury: HSE Books.

Kerr, R., McHugh, M., & McCrory, M. (2009). HSE Management Standards and stress related work outcomes. *Occupational Medicine*, *59*, 574–579.

LaMontagne, A. D., Keegel, T., Louie, A. M., Ostry, A., & Landsbergis, P. A. (2007). A systematic review of the job stress intervention evaluation literature: 1990–2005. *International Journal of Occupational & Environmental Health*, *13*(3), 268–280.

Leka, S., Cox, T., & Zwetsloot, G. (2008). The European Framework for Psychosocial Risk Management (PRIMA – EF). Nottingham: I–WHO Publications.

Leka, S., Jain, A., Iavicoli, S., Vartia, M., & Ertel, M. (2011). The role of policy for the management of psychosocial risks at the workplace in the European Union. *Safety Science*, *49*(4), 558–564.

Mackay, C. J., Cousins, R., Kelly, P. J., Lee, S., & McCaig, R. H. (2004). 'Management Standards' and work-related stress in the UK: Policy background and science. *Work & Stress*, *18*, 91–112.

McLaughlin, J., & Jordan, G. (1999). Logic models: A tool for telling your program's performance story. *Evaluation and Program Planning*, *22*, 65–72.

Mellor, N., Smith, P., Mackay, C., & Palferman, D. (under review). The Management Standards for stress in large organizations. *International Journal of Workplace Health Management*. Manuscript submitted for publication.

Nielsen, K., Randall, R., Holten, A.-L., & Rial-González, E. (2010). Conducting organizational-level occupational health interventions: What works? Work & Stress, 24(3), 234–259. Doi: 10.1080/02678373.2010.515393

Packham, C., & Webster, S. (2009). Psychosocial working conditions in Britain in 2009. Health and Safety Executive. See: www.hse.gov.uk/statistics/pdf/pwc2009.pdf

Rogers, P. (2000). Causal models in program theory evaluation. In P. Rogers, T. Hacsi, A. Petrosino, & T. Huebner (Eds.), *Program theory in evaluation: Challenges and opportunities, new directions in program evaluation* (pp. 47–55). San Francisco, CA: Jossey-Bass Publishers.

Rose, G. (1985). Sick individuals and sick populations. *International Journal of Epidemiology*, *14*, 32–38.

Rose, G. (1989). The mental health of populations. In P. Williams, G. Wilkinson, & K. Rawnsley (Eds.), *The scope of epidemiological psychiatry: Essays in honour of Michael Shepherd* (pp. 77–85). London: Routledge.

Rose, G. (1992). *The strategy of preventive medicine*. Oxford: Oxford University Press.

Rydstedt, L., Ferrie, J., & Head, J. (2006). Is there support for curvilinear relationships between psychosocial work characteristics and mental well-being? Cross-sectional and long-term data from the Whitehall II study. *Work & Stress 20*, 6–20.

Stansfeld, S. A., Woodley-Jones, D., Rasul, F., Head, J., Clarke, S., & Mackay, C. (2008) Work-related distress in the 1990s: A real increase in ill-health? *Journal of Public Mental Health*, *7*(1), 26–31.

Tyers, C., Broughton, A., Denvir, A., Wilson, S., & O'Regan, S. (2009). *Organizational responses to the HSE Management Standards for work related stress: Progress of the sector implementation plan – Phase 1*. (Research Report 693). Health and Safety Executive.

15 Moving policy and practice forward

Beyond prescriptions for job characteristics

Kevin Daniels, Maria Karanika-Murray, Nadine Mellor, and Marc van Veldhoven

The purposes of this chapter are two-fold. The first purpose is to review critically yet constructively policies and guidance for organizational practice based on prescriptions for job redesign. The job or workplace as the units of analysis typify many national and supra-national monitoring systems for work-related stress and, in guidance and policy, is probably at its most sophisticated in systems such as Great Britain's Health and Safety Executive's (HSE) Management Standards for Work-Related Stress (MS) (www.hse.gov.uk/stress/standards; the development of the Management Standards is presented in detail by Mackay et al., this volume). The second purpose is to suggest areas that policy makers and practitioners may explore to ensure an evolution in guidance, policies and practices that better reflects current knowledge on job design, stress and well-being.

We start by describing and evaluating the MS from the employer and practitioners' perspective. We then look at what the MS can be expected to achieve. We do this by reviewing the historical case of a similar approach adopted in the Netherlands prior to the implementation of the MS in Great Britain. In the final part of the chapter, we explore areas in which policy, guidance and practice might be extended to reflect better the current "state-of-the-science" in respect to job design, work-related stress and well-being.

The Health and Safety Executive's Management Standards for Work-Related Stress

The World Health Organization (WHO), the International Labour Organization (ILO), and the European Union (EU) have all issued guidance emphasising the need to assess the causes of stress and take preventive action to eliminate these causes at source (ETUC, 2004; ILO, 2001; Leka and Cox, 2008; Leka, Griffiths and Cox, 2003). Embedded in this guidance is the notion that adverse job characteristics are risk factors for poor well-being and ill-health (MacKay et al., 2004). Examples include high job demands, low job control, low support from co-workers and lack of role clarity (Cousins et al., 2004). These adverse job characteristics are also known as psychosocial hazards, because they are deemed to be risk factors for stress-related problems. Such job characteristics are often treated as if they are objective features of the work environment (MacKay et al., 2004). Treating such job characteristics as objective aspects of the work environment enables adverse job characteristics to be treated like many other health and safety risks, in which workers' exposure to the hazards are determined and then primary prevention strategies developed to limit exposure to the hazards (Cox et al., 2000).

This approach to managing stress has realized one of the most sophisticated approaches to policy: the Management Standards for Work-Related Stress (HSE 2007; MacKay et al., 2004; also Mackay et al., this volume). Implementing such an approach at national level has rarely been done and raises process issues that will be considered below. The MS are concerned with regulating six psychosocial hazards: job demands, job control, support, relationships at work, role clarity and management of change. The MS provide a validated questionnaire (Edwards et al., 2008) to enable employers to assess psychosocial hazards in their workforce (an indicator tool; www.hse.gov.uk/stress/standards/pdfs/indicatortool.pdf), a process to take action when hazards are identified, and target states to be achieved in the workforce as a whole as indexed by the indicator tool (Cousins et al., 2004; MacKay et al., 2004). Compliance with the MS is not necessarily tied with the use of the HSE indicator tool; other assessment tools are allowable within this approach. The MS are not legally enforceable. The regulatory framework is limited to the assessment of the risks posed by workplace factors and the mitigation of their possible effects. The MS are based on a stepwise participative problem-solving approach to translating risk assessment into organizational action.

In Great Britain, the MS approach was launched at the end of 2004. A variety of interventions were adopted to disseminate the MS approach and encourage organizations to tackle work-related stress. Over a three-year period, work-related stress was a national priority programme and several pilot and implementation programmes were put in place. HSE's specific aim regarding the introduction of the MS was to encourage organizations to adopt good management practices for tackling work related stress; targeting a reduction in the incidence of stress cases and stress related sickness absence (Cousins et al., 2004).

The MS have many positive features, including: the development of the Standards from a review of the existing scientific evidence (Rick et al., 2002); the care, attention to detail, and involvement of multiple stakeholder group inherent in operationalizing the MS (Cousins et al., 2004); the relative ease with which the MS allow organizations to comply with UK and EU Health and Safety Legislation in the area of psychological health and well-being; and that the MS are diagnosis-based, solution-orientated and participatory (Kompier, 2004).

There are three levels at which the MS can be evaluated:

i   by policy progress and national statistics relevant to the MS;
ii by examining practitioners' experiences with putting the MS into practice; and
iii by examining the assumptions underpinning the MS.

MS policy progress

At national level, surveys of self-reported work-related illness indicate a 7 per cent drop in the estimated incidence of stress cases since the introduction of the MS in 2004 (HSE, 2009a). Looking over time, both the prevalence and incidence rate of self-reported work-related stress, depression or anxiety in the workforce has remained broadly level over the period 2001/02 to 2008/09, with the exception of 2001/02 which was found to have a level that was statistically significantly higher than that of 2008/09 (HSE, 2009a). Days lost estimates for stress, depression or anxiety have followed a similar pattern to the prevalence and incidence rates: days lost per worker have been broadly level over the period 2001/02 to 2008/09 with the exception of 2006/07 where the rate was statistically significantly higher than in 2008/09 (HSE, 2009a).

Another trend indicator of stress levels can be seen from the results over a six-year period (2004–2009) of the psychosocial working conditions survey in Great Britain (Packham and Webster, 2009). The first survey was completed shortly before the MS were launched, hence providing a baseline for working conditions. Results show positive changes on *management of change* and *manager support* but no significant change over the six-year period for *demands, control, peer support, relationships* or *role clarity*.

A survey in 2007 of 500 human resources and occupational health professionals from public sector organizations who attended HSE training workshops on how to implement the MS approach revealed that organizations had made changes to their policies and procedures, had a renewed focus on stress and absence issues and "trainees" were more confident in dealing with these issues and in conducting stress risk assessments (Broughton et al., 2009).

Some organizations succeeded in their effort of reducing stress as shown in case studies, reporting lower levels of absence or fewer stress-related cases (HSE, 2009b) whilst other workplaces may have been less successful given the many barriers to overcome. In sum, the figures perhaps indicate an increase in some activities for tackling stress at work, but considerable policy support and guidance was needed to help organizations implement the MS.

The view from practice

In this section, we outline the practitioners' experiences and summarize the main criticisms of the MS. Process issues pertaining to the MS approach were investigated during the early piloting of the approach in "volunteer" organizations (Cox et al., 2007; Tyers et al., 2009). Later on, to evaluate the effectiveness of the approach outside these pilot organizations, additional data were collected from nine case studies and a survey of 500 organizations (Broughton et al., 2009). This latter work converged with earlier pilot work in identifying persistent barriers to implementation. These barriers included a lack of resources, information and training; lack of commitment to implementing changes; lack of senior management buy-in; and lack of openness around stress. Enablers included the existence of underpinning stress and absence policies; management involvement and commitment to implementing change; a preventative approach; good data collection; and a supportive environment (including external support). A further study summarized the experts' views (including UK and EU policy-makers, researchers, and practitioners) of the MS, their perceived usefulness and practicality (Cox et al., 2009).

It is interesting to note that Bourbonnais, Jauvin, Dussault and Vézina (this volume), and other authors, corroborate the barriers and enablers to intervention implementation found in the research on this MS. Drawing from their work on intervention implementation in correctional establishments, Bourbonnais et al. also emphasize the importance of support from senior management; the existence and allocation of relevant financial and human resources; openness, trust, commitment, and communication; and a sound methodology including prior data collection. This and similar work on a range of sectors and establishments (e.g. Mikkelsen, Saksvik and Landsbergis, 2000; Murta, Sanderson and Oldenburg, 2007; Nytrø et al., 2000; Semmer, 2006) show that researchers' and practitioners' recommendations on intervention implementation do seem to converge on the role of a range of specific process and context issues. We summarize some of the barriers to the implementation of the approach in reference to this work under four broad headings: Sensitivity to the organizational context, resources, implementation, and competence.

First, the narrow focus of the Indicator Tool (the 35-item questionnaire developed by the HSE to assess the six primary stressors as part of the Management Standards for Work Related Stress, see Chapter 14) on six psychosocial hazards highlights the issue of sensitivity of assessment and intervention decisions to the organizational context. For example, experts have identified that a different, more focused and simpler approach that does not require a questionnaire may be necessary for small organizations, where there is stronger communication, support, participation and feedback, and where it is easier to carry out focus groups (Cox et al., 2009). One size does not fit all in risk assessment and management. Rather, risk management should be able to accommodate different organizational needs, circumstances and resources. As mentioned, organizations may be confused about the best way to accommodate the MS approach within existing initiatives; they may not follow the MS systematically, or adapt the conventional steps to their specific needs (e.g. Cox et al., 2007, 2009; Tyers et al., 2009). The degree to which the process is used as a preventative or a reactive tool also varies across organizations (Broughton et al., 2009). A flexible and accommodating process (perhaps a "modular" approach; Tyers et al., 2009) would be more effective and better able to promote ownership and organizational learning.

Second, some organizations experience problems following through and implementing interventions (Cox et al., 2009; Tyers et al., 2009). This common barrier to practice is also identified in European case studies (e.g. European Agency for Safety & Health at Work, 2009). A number of factors can make implementing the MS complex and taxing, including a lack of resources, the size or structure of the organization, inability to maintain the momentum and difficulties communicating the process (Cox et al., 2007, 2009; Tyers et al., 2009).

Third, practitioners have indicated that it would be desirable to develop a flexible and comprehensive toolbox to support the MS process (Cox et al., 2007). Such a toolbox should be sensitive to the organizational context and needs, and also include a range of supplementary assessment tools, guidance, the business case and a set of decision-making rules to support the skills and competencies for implementing the MS. It might be possible to develop knowledge and to help organizations learn by highlighting good practice examples and experiences (e.g, European Agency for Safety & Health at Work, 2009).

Fourth, the issue of resources, roles, competencies and capacity-building in organizations has been frequently raised in research evaluating the MS (Broughton et al., 2009, Cox et al., 2007, 2009; Tyers et al., 2009). While senior management commitment is one key ingredient for successful risk management (Cox et al., 2007; Tyers et al., 2009), those carrying out the actual work may often lack the necessary seniority or competencies to implement the MS or communicate the MS to senior management (Cox et al., 2009; Tyers et al., 2009). Thus, developing management commitment and practitioner competencies becomes a priority for capacity-building in organizations (Cox et al., 2009), which can be supplemented with workable integrated policies in organizations (Tyers et al., 2009). Indeed, in the early years of the MS, external support was made available to organizations (Broughton et al., 2009) and many organizations used external consultants (Cox et al., 2007). Continuing such support through affordable specialist intervention services and external expertise might be important and well received (Cox et al., 2009). Another way to build capacity would be to devolve more power to OH services, broadening their remit (from risk assessment to job design, education and higher-level decision-making) and equipping them with necessary skills, competencies and knowledge. This can include management skills, but also negotiation and analytical skills (Broughton and Tyers, 2008).

To conclude, whilst there is consensus that the MS provide a useful framework (Tyers et al., 2009) and embody some good principles and a good management approach (Cox et al., 2009), practitioners concur that "the approach works well in principle but less so in practice" (Cox et al., 2009, p. 27). As mentioned, the MS approach has had a positive impact on raising awareness of stress, readiness to identify and deal with the issue, understanding how stress can affect employees, and management training (Tyers et al., 2009). From the practitioners' perspective, there is still work to be done in terms of the quality of implementation of the MS and the resources, skills and competencies required to support implementation (Cox et al., 2007, 2009).

The view from theory

The MS were developed from the approach to research on work-related stress that was dominant in the immediate run-up to the development of the MS (Rick et al., 2002). This approach, which had been arguably dominant in organizational research since the 1960s, still dominates the literature. It is based on assessing job characteristics as if they were entities independent of workers and assesses their relationships with various indicators of well-being or stress-related problems. Sometimes these relationships are assessed with statistical controls for workers' personal characteristics such as personality, and sometimes by examining statistical interactions with workers' personal characteristics. Daniels (2011) has listed four major assumptions underpinning the scientific basis of the MS and has questioned the tenability of these assumptions.

The first of these assumptions holds that job characteristics are objective properties of jobs and relatively stable. Without this assumption, it would not be possible to assess workers' exposure to adverse job characteristics at one point in time and then develop procedures to minimize exposure to these adverse job characteristics at a future point in time and that will also apply to other workers doing the same job. However, evidence suggests that job characteristics are neither objective nor stable. One argument against the objectivity of job characteristics concerns the wide-spread use of self-reports of job characteristics (Daniels, 2006). Indeed the MS indicator tool is a self-report instrument. Alternative methods have also been criticized as measures of shared perceptions or the product of institutionalized ways of thinking (Daniels, 2006).

However, in a strict form, objectivity relates to job characteristics being the properties of jobs, not properties of the people performing those jobs: Two people performing the same job under the same conditions at the same time should be subject to the same job characteristics. However, there is growing evidence that people shape their jobs, individually or collectively, over the longer term (Berg, Wrzesniewski and Dutton, 2010; Leana, Appelbaum and Shevchuk, 2009) and enact specific job characteristics for specific purposes over the shorter term (e.g. Daniels et al., 2009). That people shape and enact their work environment indicates that job characteristics are not independent of the people performing those jobs. That people shape their jobs also questions the stability of job characteristics. Moreover, there is growing evidence from diary studies that indicates that job characteristics such as control, demands and support vary from week-to-week, day-to-day and even within the working day (e.g. Daniels et al., 2009).

The second assumption is that job characteristics have probabilistic relationships with well-being and health. Whilst acknowledging the large literature supporting this assumption, Daniels (2011) raises Rick and Briner's (2000) concern that the probabilities may not be high enough or of enough specificity to be able to develop organizational practices that will reliably improve health and well-being. This is tied up with the third assumption that organizations can take action in redesigning jobs and organizational processes that will yield improvements in well-being. Whilst the evidence does indicate that some interventions can lead to improvements, not all interventions do (e.g. for recent reviews of intervention studies see: Bambra et al., 2007; Egan et al., 2007; LaMontagne et al., 2007; Richardson andRothstein, 2008). At present, we do not know how many interventions fail to show beneficial effects because of deficiencies in the processes of implementing changes (see preceding section) or because of deficiencies in the conceptual basis of the interventions.

The final assumption is that differences between individuals are largely irrelevant to the prevention of stress-related ill-health at work. A meta-analysis indicates that individual differences have stronger predictive relationships with stress-related symptoms than self-reports of job characteristics (Fergusson, Daniels and Jones, 2006). Other evidence suggests that differences between individuals are important because how individuals appraise and cope with adverse events at work are critical processes in protecting well-being (Daniels, Harris and Briner, 2004).

In summary then, Daniels (2011) indicated that the MS and other legislative approaches' dependence on the dominant approach to stress in the organizational sciences has meant that two important principles have been largely overlooked in policy. These are that people are active in shaping their work environment and also that people are active in interpreting and responding to what happens to them at work. Evidence suggests acknowledging this agency and interpretation is important for interventions as well as everyday working activities (see e.g. Nielsen, Randall and Albertsen, 2007; Randall, Nielsen and Tvedt, 2009).

We have now reviewed the MS approach in terms of policy progress, practice and theory. All in all, we can conclude that the MS have generated a lot of debate and activity in the UK. Given the amount of debate, one may wonder what the MS might be expected to achieve in the long run, and how they may evolve. In order to be able to discuss this issue later in this chapter, we now present a historical account of a management standards approach in the Netherlands that has been in place for 20 years.

The Dutch experience

In 1980, absenteeism peaked in the Netherlands at an astonishing 10 per cent. The Dutch Labour Conditions Law (dating from 1934) was in need of revision. There was a widespread belief in possibilities to make work and workplaces in the Netherlands better, and that important decreases in absenteeism and work incapacity were possible. Stress was then still somewhat of a taboo topic in Dutch society.

During the 1980s there was a growing conviction that attention for psychosocial working conditions was necessary, and that standards in this area were necessary. A new Working Conditions Act (WCA) was launched in 1983, initiating a shift from Government regulation to self-regulation by employers and employees, supported by professionals (anticipating the EU framework for working conditions of 1989, which advocated a similar shift). Rather than the terms "psychosocial hazards" or "psychosocial risks", at the time the term "conditions for well-being at work" was preferred.

After much political debate and a whole series of exploratory studies, it was decided to anchor Dutch standards for "well-being at work" in objective terms, e.g. in terms independent from the individual workers' experience and evaluation of work. This decision was strongly influenced by a dominant research stream in the Netherlands at the time, that was based on socio-technology (Kuipers and Amelsvoort, 1992) and German work regulation theory (Hacker, 2003).

Explicit mention of Dutch standards for well-being at work was included in the Dutch WCA in 1990. The law stated that an employer should:

- Adapt workplaces, work methods and work tools to employees in ways that are ergonomically sound.
- Divide and allocate tasks to take into account the personal characteristics of employees in terms of age, gender, physical and mental health status, experience, professionalism and knowledge of the Dutch language.
- Design work in such a way that it contributes to the development of skills and competencies in employees.
- Design work in such a way that employees can perform tasks according to their own insight, have social contact with colleagues during working hours, and are informed about goals, results and demands in relation to their job performance.
- Avoid repetitive tasks with short cycles, as well as avoid machine-paced work for employees.

The introduction of these Dutch standards was carefully planned. Implementation followed only after developing and testing an instrument that could operationalize the facets mentioned in the law. This was the WEBA-method (Projektgroep WEBA, 1989). The WEBA is an expert rating instrument (several days of training are required to be able to apply it) that assesses seven factors for any job:

1. Completeness – the presence of sufficiently supportive and preparatory processes for task execution.
2. Organizing tasks – opportunities to solve problems with supervisors and/or colleagues as well as opportunities for participation in decision-making.
3. Monotonous, short-cycled work.
4. Complexity – sufficient use of skills and creativity.
5. Timing control – no machine pacing.
6. Contact possibilities during work.
7. Information about work, workplace and organization.

So, the Netherlands had its specific form of standards for psychosocial working conditions and job stress in place about 20 years ago. In the beginning, uptake of the standards was rather limited, mostly involving organizations with a strong motive to offer their employees the best possible working conditions. The uptake process depended very much on "visionary" managers willing to experiment in this area.

The WCA also required organizations to make use of professional assistance for improving working conditions and occupational health. For example, the WCA prescribed services related to risk assessment and risk evaluation, absenteeism counselling and periodic health examination. This provided legal grounds for occupational health/working conditions experts and services. There was a strong growth of this sector in the period 1990–2005. In 1993, again after a considerable time of preparation and study, the WCA was amended to require every occupational health/working conditions service to employ at least one work and organizational expert (alongside an occupational physician, an occupational hygienist and an occupational safety expert). This work and organizational expert had a series of tasks which were explicitly mentioned in the law (Peereboom, Ludding, and Van der Woude, 1997): job design; work organization; absenteeism interventions; and design and maintenance of a systematic approach to managing working conditions. A professional register was formed to recognize experts qualifying for this role. Registration was only possible after finishing a two-year post-academic teaching program that included courses, assignments and practice.

By the mid-1990s, absenteeism was on the decline, although still high at about 7 per cent. At that time, the long-term work incapacity percentage was peaking at 12 per cent of the national work force. The country was diagnosed to be suffering from "the Dutch disease" of sickness absence and work disability (Geurts, Kompier, and Gründemann, 2000). Politicians and public debate called for further action, and rather than further improving working conditions, the political climate and collective opinion now changed towards shaping the financial contingencies for absenteeism and work incapacity. There was a call for revision of the social security system in relation to work-related health. In the second half of the 1990s, the laws on working conditions and social security in respect of work incapacity were integrated. There was a gradual shift of costs for absenteeism/work incapacity from society (tax payer) to the "polluter" (employers). These changes took place gradually over a 10 year period since (Popma and Kwantes, 2006). The net result was that employers started to take the management of working conditions and absenteeism/work incapacity more seriously, as to do so made sense from a cost containment perspective. This further stimulated the growth of occupational health care practice in the Netherlands.

Against the background of changes in social security and accountability, the Dutch standards approach started evolving in practice. There were severe resource implications of applying the original WEBA-method for assessing well-being at work: It took a full day for a job to be rated in a particular setting. Employers found this to be quite expensive. In addition, the method was limited in terms of which jobs could be rated: the original method appeared to be relevant mostly for less complex jobs. For more complex jobs, the "standards" were so low as to be uninformative. In response to these limitations and the increasing demand from organizations and professionals for a practical tool, alternative survey-based methods were developed that included an expansion of factors evaluated. Examples of these are the NOVA-WEBA (Kraan et al., 2000) and the VBBA (Van Veldhoven and Meijman, 1994), but many more were developed at the time. The expansion of the range of factors considered was most notable in the direction of (subjective) work stressors such as job demands, role stressors, relationships at work and organizational change. From this point on, in terms of content, rather similar factors were addressed in Dutch instruments as in the current MS approach in Great Britain. Aggression, violence and sexual intimidation were explicitly mentioned in the WCA revision of 1998, however.

Survey-based approaches were mainly applied to medium and larger organizations. In smaller organizations, even more "lightweight" methods were needed (to have a good cost-benefit ratio). Here, the dominant approach adopted was the inclusion of psychosocial factors in overall working conditions and safety checklists for the workplace. These methods are based on expert checklist ratings that are completed during a brief visit to the workplace, often combined with one or more brief interviews with workplace representatives.

The national initiative against psychosocial job risks and stress at work reached its peak when the Dutch Government initiated the so-called "covenants on health and safety at work" (1999–2005), which were Government sponsored systems of measuring and intervening at the level of branches of industry (efforts were targeted at high-risk branches of industry, and nine branches were targeted in the case of job pressure/job stress, see Taris, van der Wal and Kompier, 2010). At this stage, many branches of industry further adapted the available general instruments for their own purposes and contexts. Most covenants contained five elements:

1 A state of the art baseline study.
2 Signing of the covenant, that included a set goal to achieve, e.g. a specific percentage reduction in the number of employees exposed to certain risks or outcomes.
3 Initiation of a committee to develop intervention plans.
4 Implementation of interventions.
5 Evaluation of interventions.

One of the conclusions reached by Taris, van der Wal and Kompier (2010) about the covenants targeted at job stress is that their effectiveness was limited, and depended on whether there were concurrent efforts at the organizational level. Nevertheless, in terms of amount of effort invested and specificity of risks checked, this was probably the most active period in the Netherlands.

Around 2005 the major revisions in social security systems for absenteeism/work incapacity (which had started around the mid-1990s) were completed. Also, a heightened level of societal attention and action towards managing working conditions was securely in place. The results were strong: absenteeism had fallen to around 5 per cent by 2005 (and has decreased somewhat further since), and work incapacity was down to 9 per cent (and has also decreased further since). The "Dutch disease" had been partially cured.

Against the background of lowered costs for absenteeism and work incapacity, the obligation to hire expert advice on working conditions/occupational health was gradually faded out from the WCA. Since then, the number of experts employed in occupational health/working conditions services has nearly halved. Attention for working conditions is still deemed relevant by all parties involved, but a lower level of priority has become common sense. In line with this general feeling of successful societal risk management, in 2007 the term "well-being at work" was finally deleted from the WCA. In effect, this is also an end to the original Dutch standards approach in the WCA.

The new term adopted in the WCA is that of "psychosocial workload". This encompasses a series of topics that the employer should monitor and manage in order to protect his/her employees. These specific topics are bullying, sexual intimidation, aggression, violence, work pressure and job strain. The shift away from job content-related factors and towards factors relating to social workplace risks is easy to recognize. This reflects the results of monitoring systems at the national level that indicate such psychosocial hazards are currently important for the Dutch workforce (Smulders, 2006). Instruments dating from the 1990s have been updated or new instruments have been created to match this shift. Psychosocial and/or job design factors that are not mentioned explicitly anymore in the WCA are still part of the general obligation of an employer to take good care of employee health and safety, and monitor and assess risks accordingly. Among these are factors like job control, social support at work, role conflict and organizational change which are acknowledged as key areas in the MS in Great Britain. In effect, only the MS key areas on job demands and relationships are currently acknowledged explicitly in the Dutch WCA. This is consistent with earlier experiences in the Netherlands (like the covenants) that detailed specification of factors and standards are best left to specific local parties involved, thus signifying further deregulation of working conditions law.

What have the Dutch standards achieved?

Probably the most important effect has been that the Dutch standards in the WCA have encouraged psychosocial job characteristics to be taken more seriously by organizations and by professionals. This has been an important step in "normalising" the issue of job stress in Dutch society. In the 1980s the topic was still taboo in many organizations, whereas it is now more or less a normal topic for management policy and practice at all levels.

A second effect has been the development of applied instruments, sharing common ground in what to measure (whether by expert rating, survey or quick checklist). The NOVA-WEBA and VBBA have been used as the basis for many tailor-made instruments in particular branches of industry and types and sizes of organizations. The adoption of these basic instruments has established common terminology among researchers/professionals and as a result more consistent communication towards employers and employees. Also, organizations have been willing to share data to develop benchmarks for the purpose of adequate risk assessment, indirectly stimulating research on data that has accumulated. For survey results to be interpretable at the job or workplace level, the need arose to have reference data for the Dutch workforce, but also for each branch of industry or type of job. The quality of the benchmark became an issue that went beyond academic debate: Managers got involved because results might have implications for their department and/or organization. The Dutch standards have decisively stimulated the development and adoption in practice of measures for psychosocial working conditions and their effects; although the factors included and the kinds of measurement applied (expert rating, survey or checklist) have evolved considerably from the original. In terms of measurement, the standards have been flexibly distributed across the workforce.

Third, the Dutch standards have contributed, directly or indirectly, to the reduction of absenteeism and work incapacity, as well as to better quality of jobs in the Netherlands. As this is completely interwoven with other concurrent changes in society, especially in social security legislation, it is impossible to quantify exactly the contribution of the standards in this context.

Fourth, the Dutch standards have stimulated the insight that once a certain degree of consensus and a basic *modus operandi* has been achieved, methods, and standards are less important than interventions and results. In the beginning, much of the effort went into discussions about terminology, validity and reliability of instruments. While such debates are important for advancing policy and research, they are less relevant for practice. Once a certain point was reached, more liberty was given in the WCA and in Dutch Government-promoted instruments to monitor sector-specific or job-specific risks, and to formulate sector- or job-specific standards with a legal basis. At the level of national policy, a shift has taken place from close specification on the work conditions to assess and change, and how to assess those work conditions to a more flexible framework focused on monitoring and addressing the important factors for each organizational context. The MS key area of job control serves as a case in point. Whereas this was a major factor in the original well-being at work standards, explicitly recognized in the WCA text and in related instruments (1989/1990), this is now (2007 onwards) a factor which is only monitored where necessary without explicit legal basis.

This being said, a number of basic psychosocial risks remain, that do require a strong legal framework that requires action if risks are high. This is exactly what the change in terminology and sub-factor specification of 2007 is trying to achieve. Psychosocial workload risks now specified relate to job pressure and strain on the one hand and to workplace bullying, aggression, violence and sexual intimidation on the other. Even so, instruments to monitor these issues are not prescribed. Any organization in the Netherlands has options available for this purpose, either from accepted general instruments or instruments specifically constructed by organizations or bodies representing specific branches of industry.

Moving policy and practice forward

On the basis of our review of policy progress, practice and theory, as well as the historical account of the Dutch experience with a similar approach, how can the MS in Great Britain be judged and what is the best prognosis for their development? The Dutch experience indicates that short-term improvements in absence or illness rates need not be the best metrics for evaluating the MS in its early years. Instead, reasonable attainments might be: raised awareness; sharing of terminology and knowledge amongst practitioners; and improved management practices. The Dutch and the UK experiences both indicate the need to provide guidance and support in implementing solutions when it is needed. And last but not least, the Dutch experience indicates the need to evolve policy that is adaptable to specific contexts, to changing societal and work conditions and to specific needs for intervention. To this, we would add that policy needs to adapt to empirical, conceptual and methodological developments in the research base, so that policy advances beyond the static, job design perspective that characterized research in the run up to the MS (Daniels, 2011). How then might policy and policy-led practice move forward? Based on our analysis above, we believe there are a number of areas in which the MS might be usefully developed.

Implications for risk assessment

One theme that emerges from assessing the MS from all perspectives is the need to develop guidance on developing risk assessments that are tailored to specific contexts. Whilst the Dutch experience indicates benefits in having a common general terminology, a common terminology does not imply uniformity in assessment tools, uniformity in the psychosocial hazards most relevant in any given contexts or the absence of nuanced differences between contexts. Indeed, in the Dutch case, there has been a shift in the kinds of psychosocial hazards targeted by policy for specific branches of industry, organizations and/or jobs. Evidence from policy research and practitioners is that guidance should include material on tailoring risk assessments to specific contexts, generating risk assessments that provide some indication of links to harm, and developing skills and competences in overcoming organizational barriers to conducting risk assessments. Developments in theory indicate risk assessments need to take into account how workers interpret and shape their work environment.

Putting these together, we can draw specific implications for risk assessments. First, alterations to the Dutch standards indicate that assessing exposure to psychosocial hazards is still necessary, even though the nature of what are considered to be the most "toxic" hazards may change. However, tailoring assessments to specific contexts might be more appropriate, so that risk assessments are made in terms that workers understand and can see are particularly relevant to them (Cox et al., 2000). Another relevant issue may be the development of benchmarks which are not at the national level but which are specific for a branch of industry, large organization and/or job. Most risk assessments, including the MS Indicator Tool, ask workers to assess their overall average levels of exposure. However, given evidence for variability in psychosocial hazards, it may seem sensible to extend assessments of exposure not just to assess average levels, but also to indicate variability in exposure.

Assessing exposure is important, but may not always be enough in itself. Linking assessment to harm need not necessarily entail educating practitioners in complex statistics. Rather, direct indicators of perceived links to harm could be used instead. Perhaps two important candidates here are asking workers to indicate whether they perceive their current levels of exposure to psychosocial hazards has adverse impacts on their ability to pursue their personal goals at work and their affective experiences of work. This is because of evidence that perceptions of psychosocial hazards' impact on goals and affective experiences may influence affective well-being, which in turn may be a lead indicator of other aspects of well-being (Daniels, 2011). Given that workers may be able to self-regulate exposure to some hazards or their impact, risk assessments might also include assessments of coping potential (Biron et al., 2006).

One avenue for research is to develop a questionnaire and other kinds of methods that can easily be implemented in a wide-range of contexts that capture factors such as psychosocial hazards' average levels of exposure, variability, impact on goals, impact on affect and perceived coping potential and yet are still flexible enough to be sensitive to local contexts. Finally, in addition to work on overcoming barriers to action following risk assessment (Biron, Gatrell, and Cooper, 2010), there may need to be further work on overcoming barriers to psychosocial risk assessment. Developing tools that can help determine organizations' capability in carrying out risk assessments and implementing solutions might be important during the initial phases of risk assessment to help practitioners plan for subsequent developments.

Implications for risk management

Although the MS may have evolved from theories of job design, the MS do not prescribe job redesign as the only means of reducing exposure. Extending the scope of guidance to cover a wider range of potential interventions may seem sensible, and in keeping with the MS. The importance of individual differences, interpretations and agency indicates that multilevel interventions may be useful (Tetrick andQuick, this volume). However, acknowledging interpretation and agency in the management of psychosocial hazards leads to a wide array of potentially relevant interventions, many of which could integrate interventions into existing human resource management (HRM) and operational practices. Some potential candidates are discussed here.

Performance and development reviews (often known as performance appraisals) are at the interface of HRM and operations. Perhaps because the popular view is that these can be stressful occasions, potential benefits may not have been explored. However, the normative HRM literature indicates that performance and development reviews should be opportunities for workers and their managers to discuss, negotiate and shape performance goals, personal goal and career progress, the content and process of work, and training and other forms of support for skill development (Armstrong, 2009). Given that workers may shape their own work to regulate psychosocial hazards and their impact, and the importance of personal goals to well-being (see e.g. Daniels, 2011), there may exist the opportunity for integrating the development review aspects of performance management with the management of psychosocial hazards. Of course, development reviews require managers to engage with the process and provide support and advice as follow-ups to actions agreed at developmental reviews, underscoring the importance of line managers and management training (Donaldson-Fielder, Yarker, and Lewis, this volume).

Development reviews often form a vital function in career management and training. Career development, skill acquisition and learning may relate to personal goals associated with perceived competence and personal growth, and so be linked to better well-being (cf. Ryff andKeyes, 1995). Therefore, aspects of HRM linked to career development, training and knowledge sharing through team working practices and communities of practice may all have a role to play in minimising psychological harm and enhancing well-being. Also, such practices are often considered to be linked more closely to HRM and operational practices than to health and safety management.

Together with line management training and job redesign then, there is the possibility of multiple and integrated HRM practices being used to enhance well-being that go beyond simply regulating psychosocial hazards through job redesign. The notion of high performance work practices (HPWP) refers to an integrated and mutually reinforcing set of HRM practices that subsumes the practices discussed in this section, and are associated with enhanced organizational performance (Combs et al., 2006), safety (Zacharatos, Barling, and Iverson, 2005) and well-being at work (Butts et al., 2009). However, other work indicates HPWP could be detrimental to well-being (Kroon, van de Voorde, and van Veldhoven, 2009). Therefore, if HPWP offer a potential route forward that encapsulates more than just reductions in psychosocial hazards, then research is needed to find out when and for whom HPWP contribute to enhanced well-being.

One potential candidate for conditions that encourage the beneficial effects of HPWP on well-being could be the extent to which workers and managers have strong and collaborative relationships in the implementation of HPWP (cf. Rizov and Croucher, 2008). Of course, there are many other factors to consider in the implementation of interventions, so developing guidance on implementing interventions might be just as important as developing a tool box with a range of interventions.

Conclusions

Gaps between policy and effective organizational practice in managing psychosocial workplace factors have been previously identified and explained, at least partly, by the perception that psychosocial issues are too complex to deal with and difficulties in developing consensus around issues (Iavicoli et al. 2004; Leka et al., 2011). However, experience from the Netherlands and the UK suggests that policy can be developed that can influence organizational practice.

Our examination of the MS and the Dutch standards suggests that job redesign to minimize psychosocial hazards is just one organizational practice amongst potentially many that can enhance well-being at work. HRM practices that focus on individual development and personal goals, such as training, career management and developmental appraisals might be useful supplements to job redesign or provide a way of supporting workers to redesign their own jobs. However, developing a wider array of interventions is dependent upon risk assessments that go beyond exposure to a narrow range of psychosocial hazards, but are capable of detecting the nuances of specific contexts, as well as a range of other important factors related to variability, impact and coping potential.

The suggestions made in this chapter for advancing policy and practice in relation to work-related stress are predicated both on a critical yet appreciative review of the MS as well as on developments in research and theory. Clearly, such suggestions need research both on the content of interventions and the processes of implementing interventions. In turn, we expect such research to introduce new insights for policy. New insights should not surprise us: As we hope our review has shown, one key to successful policy is its evolution.

Acknowledgements

Kevin Daniels' contribution to this chapter has been partially supported by Engineering and Physical Sciences Research Council grants EP/D04863X/1 and EP/F02942X/1. Maria Karanika-Murray's contribution has been partially supported by the Economic and Social Research Council's First Grant Scheme grant number RES-061-25-0344.

Note

Correspondence concerning this chapter should be addressed to: Professor Kevin Daniels, Norwich Business School, University of East Anglia, Norwich Research, Park, Norwich, NR4 7TJ, United Kingdom. Tel:+44 1603 4561. Email: k.daniels@uea.ac.uk.

References

Armstrong, M. (2009). *A handbook of human resource management practice* (11th Ed.). London: Kogan Page.

Bambra, C., Egan, M., Thomas, S., Petticrew, M., & Whitehead, M. (2007). The psychosocial and health effects of workplace reorganization 2. A systematic review of task structuring interventions. *Journal of Epidemiology and Community Health, 61*, 1028–1037.

Berg, J. M., Wrzesniewski, A., & Dutton, J. E. (2010). Perceiving and responding to challenges in job crafting at different ranks: When proactivity requires adaptability. *Journal of Organizational Behavior, 31*, 158–186.

Biron, C., Gatrell, C., & Cooper, C. L. (2010). Autopsy of a failure: Evaluating process and contextual issues in an organizational-level work stress intervention. *International Journal of Stress Management, 17*, 135–158.

Biron, C., Ivers, H., Brun, J. P., & Cooper, C. L. (2006). Risk assessment of occupational stress: Extensions of the Clarke and Cooper approach. *Health, Risk & Society, 8*, 417–429.

Broughton, A., & Tyers, C. (2008, June). *Managing stress and absence: Evaluating the impact of Health & Safety Executive initiatives*. Paper presented at the Institute of Work Psychology international conference Work, Well-being and Performance: New Perspectives for the Modern Workplace, Sheffield, UK.

Broughton, A., Tyers, C., Denvir, A., Wilson, S., & O'Regan, S. (2009). *Managing stress and sickness absence. Progress of the Sector Implementation Plan – Phase 2. Research Report RR694*. HSE books.

Butts, M., Vandenberg, R. J., DeJoy, D. M., Schaffer, B. S., & Wilson, M. G. (2009). Individual reactions to high involvement work processes: Investigating the role of empowerment and perceived organizational support. *Journal of Occupational Health Psychology*, *14*, 122–136.

Combs, J., Liu, Y., Hall, A., & Ketchen, D. (2006). How much do high performance work practices matter? A meta-analysis of their effects on organizational performance. *Personnel Psychology*, *59*, 501–528.

Cousins, R., MacKay, C., Clarke, S., Kelly, C., Kelly, P., & McCaig, R. (2004). 'Management Standards' and work-related stress in the UK: Practical development. *Work & Stress*, *18*(2), 113–136.

Cox, T., Griffiths, A., Barlow, C. A., Randall, R., Thomson, L., & Rial-González, E. (2000). *Organisational interventions for work stress: a risk management approach*. Sudbury: HSE Books.

Cox, T., Karanika-Murray, M., Griffiths, A., Wong, Y. Y. V., & Hardy, C. (2009). *Developing the management standards approach within the context of common health problems in the workplace: A Delphi study* (Research Report RR687). Sudbury: HSE Books.

Cox, T., Karanika, M., Mellor, N., Lomas, L., Houdmont, J., & Griffiths, A. (2007). *Implementation of the Management Standards for work-related stress: Process evaluation*. Report to the Health & Safety Executive. Nottingham: Institute of Work, Health & Organizations, University of Nottingham.

Daniels, K. (2006). Rethinking job characteristics in work stress research. *Human Relations*, *59*, 267–290.

Daniels, K. (2011). Stress and well-being are still issues and something still needs to be done: Or why agency and interpretation are important for policy and practice. In G. P. Hodgkinson and J. K. Ford (Eds.), *International review of industrial and organizational psychology* (Vol. 25). Chichester: Wiley.

Daniels, K., Boocock, G., Glover, J., Hartley, R., & Holland, J. (2009). An experience sampling study of learning, affect, and the Demands Control Support model. *Journal of Applied Psychology*, *94*, 1003–1017.

Daniels, K., Harris, C., & Briner, R.B. (2004). Linking work conditions to unpleasant affect: cognition, categorisation and goals. *Journal of Occupational and Organizational Psychology*, *77*, 343–364.

Edwards, J. A., Webster, S., Van Laar, D., & Easton, S. (2008). Psychometric analysis of the UK Health and Safety Executive's Management Standards work-related stress Indicator Tool. *Work & Stress*, *22*, 96–107.

Egan, M., Bambra, C., Petticrew, M., Whitehead, M., Thomas, S., & Thompson, H. (2007). The psychosocial and health effects of workplace reorganization 1: A systematic review of interventions that aim to increase employee participation or control. *Journal of Epidemiology and Community Health*, *61*, 945–954.

ETUC. (2004). *Framework agreement on work-related stress*. Brussels: European Trade Union Confederation.

European Agency for Safety & Health at Work. (2009). *Assessment, elimination and substantial reduction of occupational risks*. Luxembourg: Office for Official Publications of the European Communities.

European Agency for Safety & Health at Work. (2010). *Mainstreaming OSH into business management*. Luxembourg: Office for Official Publications of the European Communities. http://osha.europa.eu/en/publications/reports

Fergusson, E., Daniels, K., & Jones, D. (2006). The relative contributions of work conditions and psychological differences to health measures: A meta-analysis with structural equations modelling. *Journal of Psychosomatic Research*, *60*, 45–52.

Geurts, S. A. E., Kompier, M. A. J., & Gründemann, R. G. W. (2000). Curing the Dutch disease? Sickness absence and disability in the Netherlands. *International Social Security Review*, *53*, 79–103.

Hacker, W. (2003). Action regulation theory: a practical tool for the design of modern work processes? *European Journal of Work and Organizational psychology*, *12*, 105–130.

Health and Safety Executive (HSE). (2007). *Managing the causes of work-related stress: A step by step approach using the Management Standards*. HSG 218.

Health and Safety Executive (HSE). (2009a). www.hse.gov.uk/statistics/overall/hssh0809.pdf. Accessed on 25 May 2010.

Health and Safety Executive (HSE), (2009b). Business Solution Case studies. www.hse.gov.uk/stress/casestudies.htm. Accessed on 25 May 2010.

Iavicoli, S., Deitinger, P., Grandi, C., Lupoli, M., Pera, A., & Rondinone, B. (2004). Fact-finding survey on the perception of work-related stress in EU candidate countries. In: S. Iavicoli, P. Deitinger, C. Grandi, M. Lupoli, A. Pera, M. Petyx (Eds.), *Stress at work in enlarging Europe* (pp. 81–97). Rome: ISPESL.

ILO. (2001). *Guidelines on occupational safety and health management systems*. Geneva: International Labor Office.

Kompier, M. (2004). Commentary: Does the 'Management Standards' approach meet the standard? *Work & Stress*, *18*(2), 137–139.

Kraan, K. O., Dhondt, S., Houtman, I. L. D., Nelemans, R. & de Vroome, E. M. M. (2000). *Handleiding NOVA-WEBA: Hernieuwde versie [Manual NOVA-WEBA: new version]*. Hoofddorp: TNO Arbeid.

Kroon, B., van de Voorde, K., & van Veldhoven, M. (2009). Cross-level effects of high-performance work practices on burnout. *Personnel Review*, *38*, 509–525.

Kuipers, H., & van Amelsvoort, P. (1992). *Slagvaardig organiseren. Inleiding in de sociotechniek als integral ontwerpleer [Decisive organizing: an introduction to sociotechnology as an integrated approach to design]*. Deventer: Kluwer.

LaMontagne, A. D., Keegel, T., Louie, A. M., Ostry, A., & Landsbergis, P. A. (2007). A systematic review of the job-stress intervention evaluation literature 1990–2005. *International Journal of Occupational and Environmental Medicine*, *13*, 268–280.

Leana, C., Appelbaum, E., & Shevchuk, I. (2009). Work process and quality of care in early childhood education: The role of job crafting. *Academy of Management Journal* *52*, 1169–1192.

Leka, S., & Cox, T. (2008). *PRIMA-EF: Guidance on the European framework for psychosocial risk management. A resource for employers and worker representatives*. Geneva: World Health Organization.

Leka, S., Griffiths, A., & Cox, T. (2003). *Work organization and stress: Systematic problem approaches for employers, managers and trade union representatives*. Geneva: World Health Organization.

Leka, S., Jain, A., Iavicoli, S., Vartia, M., & Ertel, M. (2011). The role of policy for the management of psychosocial risks at the workplace in the European Union. *Safety Science*, *49*(4), 558–564.

MacKay, C., Cousins, R., Kelly, P., Lee, S., & McCaig, R. (2004). 'Management Standards' and work-related stress in the UK: Policy background and science. *Work & Stress*, *18*(2), 91–112.

Mellor, N., & Hollingdale, K. (2006). *Sector Implementation Plan 1. Stakeholder Feedback Process Evaluation Study*, Buxton: Health and Safety Laboratory Research Report.

Mikkelsen, A., Saksvik, P. Ø., & Landsbergis, P. (2000). The impact of a participatory organizational intervention on job stress in community health care institutions. *Work & Stress, 14*, 156–170.

Murta, S. G., Sanderson, K., & Oldenburg, B. (2007). Process evaluation in occupational stress management programs: A systematic review. *American Journal of Health Promotion, 21*, 248–254.

Nielsen, K., Randall, R., & Albertsen, K. (2007). Participants' appraisals of process issues and the effects of stress management interventions. *Journal of Organizational Behavior, 28*, 793–810.

Nytrø, K., Saksvik, P. Ø., Mikkelsen, A., Quinlan, M., & Bohle, P. (2000). An appraisal of key factors in the implementation of occupational stress interventions. *Work & Stress, 13*, 213–225.

Packham, C., & Webster, S. (2009). *Psychosocial Working Conditions in Britain in 2009*. Statistics Branch. Health and Safety Executive.

Peereboom, K. J., Ludding, J. J. M., & Van der Woude, M. A. (1997). *De arbeids- en organisatiedeskundige in de arbodienstverlening [The work and organizational expert in working conditions/occupational health services]*. Den Haag: SDU.

Popma, J., & Kwantes, J. H. (2006). Essentials of labour law and social security legislation. In: P. G. W. Smulders (Ed.), *Worklife in the Netherlands* (pp. 25–42). Hoofddorp: TNO Work & Employment.

Projektgroep WEBA. (1989). *Functieverbetering en organisatie van de arbeid [Job redesign and work organization]*. Den Haag: Ministerie SZW.

Randall, R., Nielsen, K., & Tvedt, S. D. (2009). The development of five scales to measure employees' appraisals of organizational-level stress management interventions. *Work & Stress, 23*, 1–23.

Richardson, K. M., & Rothstein, H. R. (2008). Effects of occupational stress management intervention programs: A meta-analysis. *Journal of Occupational Health Psychology, 13*, 69–93.

Rick, J., & Briner, R. B. (2000). Psychosocial risk assessment: Problems and prospects. *Occupational Medicine, 50*, 310–314.

Rick, J., Thomson, L., Briner, R., O'Regan, S., & Daniels, K. (2002). *Review of existing scientific knowledge to underpin standards of good practice for key work-related stressors – phase 1*. Sudbury: HSE Books.

Rizov, M., & Croucher, R. (2008). Human resource management and performance in European firms. *Cambridge Journal of Economics, 33*, 253–272.

Ryff, C. D., & Keyes, C. L. M. (1995). The structure of psychological well-being revisited. *Journal of Personality and Social Psychology, 69*, 719–727.

Semmer, N. K. (2006). Job stress interventions and the organization of work. *Scandinavian Journal of Work and Environmental Health, 32*, 515–527.

Smulders, P. G. W. (2006). *Worklife in the Netherlands*. Hoofddorp: TNO Work & Employment.

Taris, T. W., van der Wal, I., & Kompier, M. A. J. (2010). Large-scale job stress interventions. In J. Houdmont, & S. Leka (Eds.), *Contemporary occupational health psychology: Global perspectives on research and practice, Volume 1* (pp. 77–97). Chichester: Wiley-Blackwell.

Tyers, C., Broughton, A., Denvir, A., Wilson, S., & O'Regan, S. (2009). *Organisational responses to the HSE management standards for work-related stress. Progress of the Sector Implementation Plan – Phase 1*. Research Report 693. HSE Books.

Van Veldhoven, M., & Meijman, T. F. (1994). *Het meten van psychosociale arbeidsbelasting met een vragenlijst: de vragenlijst beleving en beoordeling van de arbeid (VBBA). [The measurement of psychosocial job demands with a questionnaire: the questionnaire on the experience and evaluation of work (QEEW)]*. Amsterdam: Dutch Institute for Working Conditions.

Zacharatos, A., Barling, J., & Iverson, R. D. (2005). High-performance work systems and occupational safety. *Journal of Applied Psychology, 90*, 77–93.

16 Evidence-based practice – its contribution to learning in managing workplace health risks

Andrew Weyman

Introduction

The notion that policy decisions, whether in government or work organizations, should be rational and based on sound scientific evidence, rather than rooted in ideology and rhetoric has strong intuitive appeal. Indeed, the term *evidence-based practice* has become so embedded in regulatory and public health policy speak in Britain over the last decade that few question its volition or its implications. Clearly, the concept is far from new, in so far as it reflects the basic tenets of post enlightenment science, the rationalist models of economists over investment and consumption behaviour and is a fundamental principle within medicine in the trialling of drugs and treatments. However, what is new is the perspective that these principles can, and should, be applied more broadly to policy development and delivery, in government, its agencies and quangos and though embedding these principles amongst employers. Arguably, it is here that we can detect an ideological perspective, in so far as the mantra of evidence-based practice appears, in many respects, to be a uniquely British phenomenon. This is not to suggest that it is not encountered elsewhere, merely that the profile it enjoys in the UK owes much to the policy delivery model propagated under the New Labour administration (1997–2010) (Solesbury, 2001).

In the UK, the *Modernising Government* White Paper[1] (1999) played a pivotal role in popularising the concept and embedding its principles amongst those responsible for delivering policy agendas. In public health, occupational health and workplace safety, as in other regulatory and policy delivery domains, a key focus over the last decade has been on finding ways to socially engineer the world, a focus on discovering *"what works?"*, what motivates the public, employers and employees to behave in the ways that they do? And what kinds of intervention might induce them to behave in ways that *we* would *prefer*? (see Davis et al., 2000).

What the White Paper called for was the ability to adopt an informed, more strategic approach, in which the vision was to use scientific evidence, essentially, in four ways:

- to identify what needed to change;
- to appraise alternative options for intervention;

- to monitor the trajectory of change following intervention; and
- to evaluate the impact.

This new perspective, in particular the acute focus on engendering behavioural change in target groups (the general public, employees or employers) brought with it new challenges to policy makers, in so far as where technician expertise was present this tended to be in the natural sciences, medicine and engineering, rather than the social and behavioural sciences (see Bullock, Mountford, and Stanley, 2001). Key to all of this was the ability to gather and interpret socio-demographic evidence and to draw upon published social science research evidence to inform decision making over options for intervention and the need to gather robust post-hoc evidence of impact.

Strident attempts were made to take the perspective beyond strategic decision making within each Government Department, to make it an embedded principle within the policy delivery process, in central government, its agencies, local government and, in particular, the National Health Service (NHS). We have also witnessed attempts to embed evidence-based principles in others, notably professional bodies, third sector organizations and employers, vestiges of the approach, for example being apparent within Dame Carol Black's recent review of occupational health (Black, 2008).

At times, intense debate surrounds the net result of the translation of the politico-ideological ambition and the variable practice that resulted across government Departments and Agencies over the last decade, including claims that: an acute focus on realising service delivery targets risked diverting attention from the primary intent; costs of gathering evaluation evidence could be disproportionate relative to delivery budgets and, more cynically, that the perspective is little more than a further manifestation of the audit society (see Power, 2004). Intentionally stepping aside from these debates, and issues surrounding the durability of the Labour Government's (1997–2010) legacy as a model for public service delivery, few would argue against the premise that good decision making is informed decision making, based upon robust evidence. But, what do we mean by evidence and how might the principles of evidence-based practice benefit employers in managing risks to employee health and well being?

Calls for a prevention focus

A core feature of the Modernising Government White Paper (1999, see www.nationalschool.gov.uk/policyhub/docs/modgov.pdf) was the intensified emphasis on preventative activity. This change of emphasis was most acutely apparent in the public health domain, where the traditional curative model was cast as partial, failing to reap the health (and socio-economic) benefits of addressing the causes of ill health, prominent amongst which were lifestyle characteristics, such as those associated with diet and smoking behaviour. In essence, the focus here was on risk reduction; mitigating the potential for harm and consequent onset of ill health. A comparable paradigm shift was called for in the Black review, with respect to the practice of health professionals' and employers' approach to occupational health (Black, 2008). The degree of change in perspective called for in both the public and occupational health domains should not be under-estimated. It heralded a major reorientation, beyond the familiar reactive territory of treating sick individuals, to the adoption a more strategic approach, based on discovering which groups of individuals are potentially exposed to harm and finding ways to mitigate this, either by reducing it at source or by motivating some change in behaviour of those at risk, or those who place others at risk.

A risk (assessment and control) based approach has represented the cornerstone of UK, and EU, health and safety regulation for over three decades, to the extent that a prevention orientation has become an embedded characteristic of mainstream employer practice for addressing technological hazards, dangerous environments, exposure to dangerous substances, and physical-workload elements of musculoskeletal risks (HSE, see: www.hse.gov.uk/risk). However, whether attributable to the traditional health (treatment) paradigm, or the tendency for intervention focus to exhibit a *lifestyle-drift* at the expense of upstream determinants of (ill) health (Hunter et al., 2009), it seems that occupational health professionals and employer orientations remain rooted in a reactive biomedical treatment model, focused on the individual as the unit of intervention and change (Black, 2008). Proactive, prevention orientated interventions tend to be limited to the comparatively narrow domains of attempting to influence employee health and lifestyle choices (Black, 2008; Dugdill et al., 2008).

The premise here is not to suggest that the traditional treatment orientation of occupational health is outmoded, or that lifestyle interventions are inappropriate, merely that if this represents the boundaries of employer perspectives then the best that can be achieved is partial solutions, at the expense of underplaying causal influences attributable to the organization of work and systems of support (LaMontagne et al., 2007; Jordan et al., 2003). Moreover, for a prevention orientation to have maximal impact, in common with the public health domain, there is a need for decision makers to be data-rich in terms of how they identify priorities and solutions and their value, i.e. demonstrating impact and contribution to the well-being of employees and the employing organization.

While a risk-based approach enjoys a significant degree of embeddedness within the workplace safety domain, its effectiveness remains under-evidenced in the majority of organizations, in terms of its impact on working practices, at least beyond the inevitably partial picture afforded by auditing procedures and outcome data (accidents, ill health and near misses). However, there is evidence of good practice (see, for example, Nielsen et al., 2010) Here we encounter something of a disjunctive between a management approach based upon prevention and a feedback process that is typically based on sparse, highly chance related data, reflecting failure, rather than success in managing risk.

Managing risk in the workplace is about intervention, attempting change in the natural or prevailing order of things, whether this be in traditional domains, such as machinery guarding and manual handing (reduction) systems, or in newer, less familiar areas, such as mental health and work related stress. Proactive intervention is about managing the potential for undesired outcomes, and thereby mitigating the risks to employee health, safety, and well-being.

To summarize:

- There is growing recognition of the need for a prevention orientated, risk-based, approach to employee ill-health prevention/management and promotion of well-being.
- A risk based approach requires a paradigm shift in occupational health practice, with a primary emphasis on prevention, and maintaining well-being, rather than treatment.
- A prevention focus for work-related health offers alignment with the EU risk assessment/control model and contemporary employer hazard management practice.
- Proactive, prevention orientated intervention has potential to extend beyond issues of employee health and lifestyle choices.
- Managing risk at work is about intervention to change the prevailing order of things, in traditional domains such as machinery guarding and manual handing (reduction) systems and newer areas, such as mental health and work related stress.

Data rich or data poor – the limitations of outcome data

The relative scarcity and chance related peakiness of outcome data is problematic for employers in so far as it makes it difficult to identify trends, particularly over quarterly and annual accounting periods, and *risks* engendering a reactive rather than strategic approach, where, in the absence of supplementary evidence, the corporate emphasis is prone to lurching from one issue to the next as ticking-bombs detonate (see Hale and Glendon, 1987).

The limitations of outcome data, attributable to its relative scarcity, chance relatedness and the potential for reporting biases have long been recognized within the safety domain (Heinrich, Peterson, and Roos, 1980), but are brought into acute focus in the occupational health context. The manifestation of negative health outcomes is complicated by the fact that many dose-response relationships are poorly understood, not only by employers and but also regulators due to gaps in the underpinning science; symptomatology may be incompletely mapped and may also be prone to social amplification/attenuation effects, as well as the fact that reliable data on individual exposure is often difficult to establish (Burton et al., 2008; HSE, 2005a; Weyman, 1997). Beyond this generic susceptibility, life events and interactions between work and non-work elements, combined with the significant delay between exposure and the onset of symptoms for many conditions, all serve to cloud the picture.

It is also apparent that systems for the classification of accident and ill health commonly used by employers tend to be of limited utility in terms of their capacity for interrogation, to identify causes and underlying trends (Anderson and Weyman, 1998). Typically, the data collected is limited to manifest injury, or health condition, moreover, where casual information is collected this tends to reflect immediate, rather than deeper causal influences. This situation is almost certainly reinforced in Britain by statutory reporting obligations on employers under the *Reporting of Diseases and Dangerous Occurrences Regulation* (RIDDOR, 1995) and the outcome-orientated taxonomic parameters of that system, which itself is of limited utility to the Health and Safety Commission/Executive (HSC/E) in setting regulatory priorities, for fundamentally equivalent reasons (HSE personal communication).

Despite intuitive appeal as a measure of success in risk management, and the inevitable focus that results in the aftermath of failure, put succinctly, accident and particularly health outcome data is of limited utility to employers as feedback on risk management performance. Worse than this, these trailing indicators possess the potential to subvert a strategic approach, where single or penny numbers of contiguous common cases divert attention from other important agendas, or where good fortune may give rise to complacency, in the absence of negative events (Pidgeon, 1997). Clearly, to suggest that employers (and regulators) should disregard accident and ill health outcome data is untenable, however, if organizations are to gain timely, statistically reliable feedback and learning at an appropriate level of granularity on their hazard management performance, there is a strong case for suggesting that they would benefit from additional, supplementary, measures of their risk management performance, e.g. climate surveys; behavioural (including antecedent) audits (see below).

The recognition of the disjuncture between risk (potential) and outcomes (manifest consequences) has stimulated significant interest in the scope for developing tools and techniques that operate as lead indicators. The traditional workspace safety domain reflects greater maturity in this respect, e.g. organizational focused psychometric measures of health and safety culture/climate (Cox and Flin, 1998; HSE, 1999; HSE, 2005b; HSL, 2010; Nieve and Sorra, 2003), and more accomplished incarnations of the behavioural safety/behavioural auditing traditions (HSE, 2000a). However, more recent developments such as the HSE's *Management standards for work related stress* (Cousins et al., 2004; Kerr, McHugh, and McCrory, 2009) reflect equivalent prevention based principles in the pursuance of an informed, managed approach to mitigating harm and promoting well-being.

The potential, though seemingly rarely realized (Cox and Flin, 1998; HSE 2000b; Weyman et al., 2006), strength of such measures and tools when used by employers is that they offer feedback and learning on key issues such as compliance with good practice, managerial priorities and behavioural norms. A characteristic they share is being based on attitudinal and/or behavioural data gathered from employees. Such approaches have gained significant ground within the safety arena, only recently have we witnessed a broadening of the perspective to psychosocial health domains (Cousins et al., 2004; Edwards, Van Laar, and Easton, 2009; Van Larr, Edwards, and Easton, 2008), with there remaining considerable scope for further development. A caveat, however, relates to the widely recognized limitations to attitudinal evidence, in terms of the scope for respondent attribution bias and weaknesses in predicting behaviour. This raises questions over the utility of such data in absolute terms, but does not undermine its role as a barometer of change, where repeated sampling can reasonably be assumed to embody common error, i.e. measurement of relative change is reliable, even if absolute values should be treated with caution.

To summarize:

- The scarcity and chance relatedness of outcome data makes it difficult to identify trends, particularly over quarterly and annual accounting periods.
- The limitations of outcome data (trail indicators) are brought into acute focus in the occupational health context due to issues of delay between exposure and manifestations of ill health, variability in individual susceptibility.
- Employer systems for the classifying of accidents and ill health tend to be limited in terms of their capacity for interrogation, to identify root causes and underlying trends.
- Accident and ill health outcome data alone offers a limited and partial picture of risk management performance.
- Recognition of the disjuncture between risk (potential) and outcomes (manifest consequences) has stimulated significant interest developing tools and techniques that operate as lead indicators, focused on risk assessment, management and reduction.

Intervention design – social engineering in the workplace

Learning organizations are "... skilled at creating, acquiring, and transferring knowledge, and at modifying [their] behaviour to reflect new knowledge and insights" (Garvin 1993, p. 80). While establishing systems for intelligence gathering though the development of a comprehensive set of trail and lead indicators, offers organizations the capacity to benchmark and monitor risk management performance, key issues surround what the organization does with this intelligence. Beyond the traditional risk control domains of exclusion, substitution and engineering control, intervention options relate to the design of work, leadership style, formal and informal systems of social support climate and culture, and embody the need to engage with a potentially complex array of psycho-social variables.

While physical and environmental challenges to employee health and well-being can be addressed through extension of the established engineering-control model of intervention that underpins the safety-risk management tradition (see Anderson, 2010) and employee lifestyle choices can be addressed with public health models, it is apparent that many organizations find psychosocial hazards more challenging to address, not least because the subject matter transparently lies beyond the comfort zone of many safety professionals, but also because the science relating to "solutions" is less well-mapped. This situation is complicated by the fact that, historically, social science research on health has been disproportionately focused on public health contexts, rather than the world of work. While these domains are not exclusive, and there is significant scope for read across, there are a number of key differences with respect to volition over exposure to harm and options for mitigating this.

Two key contrasts that can be drawn between public and occupational health settings relate to: (i) the scope for controlling variables resulting in exposure, (ii) the volitional or imposed nature of exposure to harm. In public health contexts, beyond prohibition, e.g. banning smoking in public places (HM Government Health Act 2006), advice to the public via obligations on suppliers to label hazardous products (Chemicals Hazard Information and Packaging for Supply Regulations 2008; EC No 1272/2008), or voluntary codes, such as supermarkets labelling unhealthy foods (FSA, 2007), the focus is on the individual as the unit of change. Moreover, the dominant, and default, perspective has been one of education and awareness raising, implicitly based upon saturation and knowledge deficit assumptions, broadly underpinned by value expectancy science (see Ajzen, 1985; Becker, 1974; Fishbein and Ajzen, 1975), combined with emergent interest in re-castings of early behavioural-decision theory insights to nudge individuals in the desired direction (Thaler and Sunstein, 2008).

A by-product of the restricted scope for intervention in the public health domain, in particular the modest scope for control over exposure, is that this has had the effect of restricting mainstream perspectives to cognitive models of behaviour change, in particular the varied incarnations of value expectancy approaches, and consequent restriction of focus to the individual as the unit of change.

While there is scope for work organizations to draw upon public health models and practice, and the extensive review evidence that has been commissioned by government Departments and Agencies over the last decade (see, for example GSR 2008a, 2008b; NICE, undated), its applicability to the workplace is for the most part restricted to domains where exposure to harm is volitional, or embodies volitional elements, e.g. behavioural aspects of exposure to harmful substances and health/lifestyle choices. More fundamentally, the scope for applying such approaches to the headline causes of absence, work-stress, and musculoskeletal disorders is limited and while relatively easy to operationalize, and *conveniently* aligned with traditional individual-treatment orientations of occupational health practitioners, a disproportionate emphasis on lifestyle health issues risks diverting attention away from addressing deeper, more fundamental, influences on employee health that relate to the design of work and associated socio-cultural arrangements. In terms of impact, the overwhelming evidence from the public health domain is that metrics of sustained behavioural change arising from educational initiatives are modest (Vanwesenbeeck et al., 1999; Weyman and Kelly, 2000), particularly where there are strong structural and cultural drivers that motivate continued exposure. A modest rate of change does not necessarily render such initiatives redundant, as rates of behaviour change/health improvement as low as 1 per cent or 2 per cent can yield significant net gains at a national level. However, equivalent rates in a single work organization will deliver barely discernible gains, in the short-medium term, and likely render the business case for investment illusive. In particular, the scope for such approaches to have a significant impact on the headline causes of absence (HSE, 2000c), work-stress and musculoskeletal disorders, is limited.

The consideration of intervention options requires engagement with the foreseeable outcomes arising from alternative change theories and pathways. *Theory of change* relates to the process of options appraisal and the consideration of, if not modelling of, alternative futures and their outcomes. This requires:

- A critical perspective on the underpinning scientific evidence and consideration of this with reference to the nature of the behaviour(s) to be addressed and the context of change.
- Consideration of the context of intervention and pathways to influence.
- Identifying physical, socio-technical and cultural barriers to change.
- Defining the causal pathway(s) from the current position to the achievement of defined objectives, by specifying what must happen for goals to be achieved, e.g. line managers' conduct annual staff health reviews, employees participate in risk assessment.
- Articulation of underlying assumptions which can be tested and measured.

(After HM Treasury Guidance, 2010)

However, engagement with the theory of change elements should not be restricted to the consideration of evidence and pathways to influence, it is also necessary to engage with how much influence is required to achieve the desired metric(s) of change. Specially, where there is a desire to realize a defined rate of change, against which impact might be assessed, and perhaps the business case demonstrated, it is necessary to consider published evidence or comparable activity.

For example, if we were to set objectives of a 5 per cent reduction in the consumption of high fat foods in the works canteen/refectory, or a 10 per cent increase in employee involvement in risk assessment, we need to consider what the underpinning social science evidence tells us about likely metrics of impact on behaviours for a given intervention design. For an organization of 1000 employees, 5 and 10 per cent respectively equate to engendering sustainable change in 50 and 100 individuals. Where the evidence of a given intervention type is that it engenders sustainable change in, say, 4 per cent of the target group, combining the outcome target with known metrics of change permits a proximate calculation of the total number of employees who must be exposed to the intervention for it to stand a viable chance of realising its objectives. In our example organization, this equates to engagement with 650 individuals in the case of fatty foods and, an unattainable in this instance, 1300 for involvement in risk assessment.

Interventions can fail due to conceptual flaws in the intervention logic and/or simply because the scale of activity was insufficient to generate the metrics of change required to achieve headline objectives (Biron, Gatrell, and Cooper, 2010; Nazaruk, Weyman, and Hellier, 2009). Even intrinsically sound interventions are at risk of being abandoned where they fail to deliver headline objectives (in experimental terms, a type 1 error). This situation serves to muddy the waters for practitioners regarding *"what works"*.

To summarize:

- Establishing systems for intelligence gathering though trail and lead indicators, offers the capacity to benchmark and monitor risk management performance.
- Many organizations find psychosocial hazards challenging, due to limited in-house expertise and because the science of "solutions" is less well-mapped.
- Key contrasts between public and occupational health settings relate to the scope for controlling exposure and the volitional or imposed nature of exposure to harm.
- Health behaviour change models relate to volitional exposure to harm and are of limited utility for addressing headline workplace health issues.
- There is a risk that even well designed interventions can be considered to have failed if they are weakly/unconvincingly evidenced.

The application evidence deficit

Beyond simple awareness raising education-based interventions, more embracing and sophisticated approaches require the configuration of some form of delivery mechanism. Inevitably, this involves discovering ways to impact on, not just the behaviour of target groups, but the behaviour of those personnel responsible for rolling out the intervention. While policy makers are seemingly awash with review evidence over *what works* in terms of ways to engender behaviour change in target groups, the social science evidence base makes a very modest contribution to the related science of delivery and *making it happen*. Routinely, intervention architects are data poor, in terms of actively monitoring and managing intervention performance.

Thus, in addition to grappling with the science underpinning the theory of change, intervention designers have to engage with the intervention logic, specifically key elements relating to pathways to change, staff roles and responsibilities and evidence of change. Moreover, there is a need to recognize that effective intervention delivery is typically iterative and cyclical, rather than linear, informed by process evaluation evidence that tests assumptions and delivery performance, i.e. an evidence-based approach is more than the capacity to demonstrate impact, it is an integral part of the change management process and contributions to organizational learning (see Figure 16.1).

An evidence-based approach is about possessing the capacity to plan and revise delivery pathways, while gathering feedback evidence that actively tests assumptions over effectiveness, e.g. through piloting to test initial assumptions and the use of intermediate performance measures, combined with a preparedness to refocus interventions in the light of new insights. Without this there is a danger that interventions may either fail to germinate, drift off course, or simply absorb resources to no useful purpose. Potential points of failure in the roll-out/delivery process are multiple, not least because almost all large scale interventions bring with them the need for some degree of organizational change, e.g. new

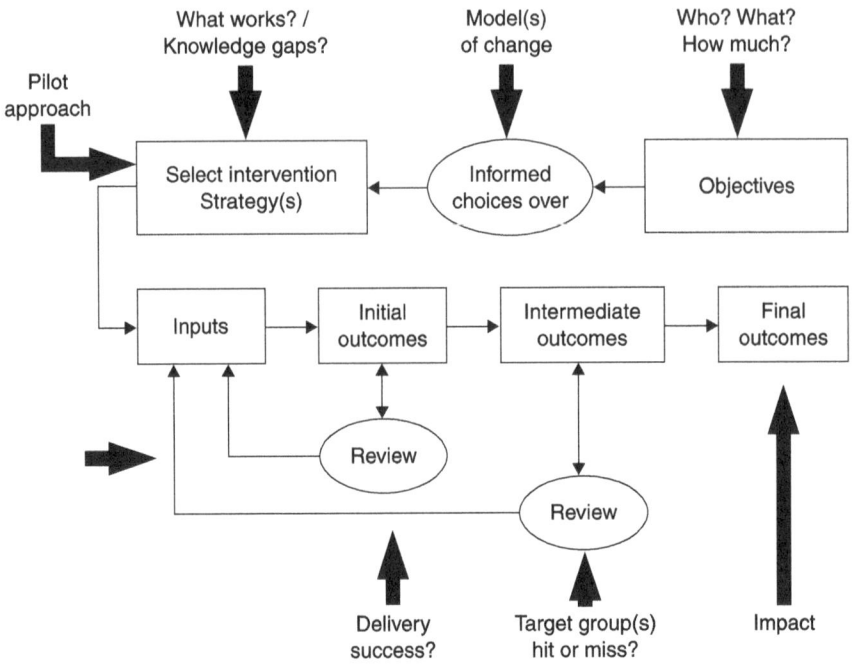

Figure 16.1 The intervention evidence cycle.

procedures, new responsibilities, changes in relationships with others, cessation of established (possibly highly culturally valued) working practices.

The relative paucity of published evidence and guidance on successful intervention roll-out in both public and occupational health contexts means that practitioners are left to find their own way and learn their own lessons. Here, the need to performance manage interventions comes into acute focus. It is critical in establishing what works, to establish how and in what ways interventions are working (or not working). Common experience is that most organizations are weak in this respect, their approach to employee health (and workplace safety) intervention is, perhaps somewhat uncharitably, characterizable as one of *hit and hope*, leaving their champions unsighted and vulnerable in instances where their good intentions are left languishing, lost somewhere in the long grass. The range of reasons why workplace health interventions can fail is extensive, and not limited to this domain, but include aspects surrounding cultural readiness for change, possibly extending to active employee resistance, fundamental flaws in the understandings of systems/the organization of work, social relationships and incomplete models of exposure to harm (Whysall, Haslam, and Haslam, 2006; Nazaruk, Weyman, and Hellier, 2009).

Evidence-based practice 343

The capacity to identify and respond to a comprehensive array of lead indicators that test assumptions, monitor progress and identify barriers to change mitigates the risk of intervention failure. Options include gathering an array of behavioural and attitudinal evidence, much of which may be qualitative, e.g. interviews, focus groups or workshops, or more formal approaches such as surveys or behavioural audits, involving delivery staff and/or representatives of the target group(s).

Attempting to conduct an OHS intervention in the absence of a well configured plan for monitoring delivery performance and progress towards defined objectives not only risks interventions stalling, drifting off course, or even having counter productive results, it leaves OHS practitioner agendas vulnerable to detractors. Where possible, outcome objectives should be defined in terms of behavioural anchors, or end-points, that can be mapped onto an agreed pathway to delivering risk reduction measures. For example, borrowing from the HSE's Management standard for workplace stress, the delivery pathway might relate to the degree of saturation of the employee baseline stress audit questionnaire; the proportion of line managers who have engaged with staff over the causes of job-stress; the proportion of departments that have implemented job-design changes; their impact on staff job-stress ratings, etc. Similarly, for lifestyle health initiatives, a baseline survey of staff food consumption habits; monitoring the availability of healthy options in on-site food outlets; monitoring the take-up of healthy food options, etc.

The evidence gathered does not need to be of the same status and resistant to the same degree of scrutiny as that gathered in public policy delivery contexts, there is no need to demonstrate the counter-factual here. However, it does need to be reliable and sufficiently robust to demonstrate the value of the intervention and to be useful in informing strategic decision making over future OHS policy and priorities.

To summarize:

- All interventions require some form of delivery mechanism. This involves discovering ways to impact on the behaviour of target groups *and* the behaviour of those responsible for rolling out the intervention.
- Intervention designers need to engage with the theory of change, the intervention logic, delivery staff roles and responsibilities and evidence of impact. Effective intervention delivery is typically iterative and cyclical, rather than a linear process.
- An evidence-based approach is about possessing the capacity to plan and revise delivery pathways, while gathering feedback evidence that actively tests assumptions over effectiveness, leading to refocusing where necessary.
- The capacity to identify and respond to a comprehensive array of lead indicators that test assumptions, monitor progress and identify barriers to change mitigates the risk of intervention failure.
- Where possible, outcome objectives should be defined in terms of behavioural anchors/end-points that map onto an agreed pathway for delivering risk control measures.

Risk management issue or risk management culture?

A characteristic of contemporary pro-active (prevention orientated) workplace health intervention activity is the tendency to pursue relatively narrow topic-specific, or themed agendas, rather than an embracing, holistic, systems approach on a broad front. As insights from the workplace safely culture framework demonstrate, where intervention success rests upon securing significant re-orientations in risk management practice that is in conflict with the established management culture, style and approach this can pose a significant barrier to the realization of objectives (Pidgeon, 1997).

A number of variables likely contribute to the enduring popularity of themed approaches:

- they possess strong intuitive appeal as *the* answer to presenting issues, such as an "outbreak" or long-term pattern of ill health;
- they provide a focus for the typically limited resources available, avoiding the risk of dilution were intervention on a broader front attempted;
- the OHS and HR functions that routinely constitute the point of initiation routinely have a modest influence on engendering fundamental change in management style and the organization of work to achieve cultural change on a broad front;
- likely most saliently, risk management priorities and practice tend to follow the regulatory emphasis.

In Britain, the last decade has witnessed increased regulatory emphasis on a topic based/themed approach, most acutely within the health domain, with priorities set on the basis of prevalence rates (HSE, 2000b). However, a themed approach is in some tension with the established EU perspective on OHS, focused on precursors and embedding broad-based risk-management principles amongst employers (The Management of Health and Safety at Work Regulations 1999; Framework Directive 89/391/EEC).

From both a regulatory and employer perspective, we can observe that pathways to success in themed approaches rest upon the realization of a common set of management practices. Irrespective of the topic, we routinely encounter evaluation evidence that cites the need for a mature health and safety culture, high levels of senior management commitment, the key role played by line managers, the benefits of engagement with, and involvement of, employees over sources of harm, etc. (Gadd and Collins, 2002; Kelly and Mellor, 2005; Lunt et al., 2008). The principal limitation of a piecemeal approach is that if the requirements for success are in conflict with the established climate the chances of success will be severely blunted. This should not be taken to infer that a themed approach is inherently doomed to failure, simply that if success necessitates fundamental changes in *the way we do things around here* (ACSNI, 1993), in the absence of a fertile social environment, there is a high probability that a series of themed interventions will run into a common set of cultural barriers.

Only recently have health researchers begun to engage with established culture/climate elements and to build upon the holistic evidence-based safety risk management perspective, broadening this to the management of non-traditional domains such as mental health, work stress and psychosocial elements (MacKay et al., 2004; Kohler and Munz, 2006; Van Larr, Edwards, and Easton, 2007). Critically, developments in this vein do not mean an end to themed approaches, but highlight the need to address more fundamental precursor variables relating to employer and OHS professionals' approaches to managing health risks, in order that themed approaches might realize their potential.

However, a significant limitation of the health culture/climate perspective is that it remains underdeveloped, in terms of the related science of application and intervention (see Nielsen et al., 2010). Specially, it seems that it is not so much a shortage of risk management performance profiling/assessment tools that constitutes a barrier to the realization of the potential of such approaches, but the scarcity of guidance on their application and the contribution of the evidence produced to organizational learning, strategic decision making and intervention. While performance measures possess the potential to provide intelligence on who is at risk and in what way – and offer direction for more detailed investigation, through for example, staff engagement activity, they are essentially barometers, with the capacity to highlight and monitor *hot topics* or *hot spots* that would benefit from intervention and change. In this respect they are merely an initial stage in the change process. Change rests with finding ways to technically and/or socially engineer solutions, through intervention. Even fundamental issues such as interpreting demographic differences across an organization by department, job-role or function do not appear to come naturally to the majority of risk managers, whose skill and expertise typically lies in non-social science domains (Weyman, 2007). In common with much of the health behaviour research, a failing of the risk management/climate perspective is that the primary academic emphasis, to date, has been on discovering and defining core constructs and their statistical properties. While this is appropriate there is a relative dearth of guidance regarding the science of application, specially, how the evidence they produce can be used to inform strategic decision making and set agendas for intervention and improvement in workplace risk management practice.

Here we can observe parallels with evidence-based approaches in the public health domain. While policy makers are seemingly awash with review evidence over *what works* in terms of ways to engender behaviour change in target groups, the social science evidence base makes a very modest contribution to the related science of delivery and *making it happen*.

To summarize:

- Behavioural change rests with finding ways to technically and/or socially engineer solutions, through intervention.
- Irrespective of the topic, criteria for intervention success routinely include a mature health and safety culture, high levels of senior and line management commitment, and active systems for employee engagement/participation.

- The principal limitation of a themed approach is that if the requirements for success are in conflict with the established climate the chances of success will be severely blunted.
- A themed approach is in some tension with the established EU perspective on OHS, focused on precursors and embedding broad-based risk-management principles amongst employers.
- Some health researchers have begun to build upon the holistic evidence-based safety risk management perspective, to address mental health, work stress, psychosocial elements and well-being.
- A limitation of the health culture/climate – risk management – perspective relates to the relative paucity of evidence on the science of application.

Conclusions

An evidence-based prevention orientated approach has strong intuitive appeal, in terms of the potential operational and business benefits of a more informed strategic approach to workplace health and well-being risk management. However, the challenge for twenty-first century workplace regulators and OHS professionals and their employers in realizing these benefits is significant and multidimensional. On the one hand they need the capacity to adopt a critical perspective in navigating their way though academic and practitioner accounts of *"what works"*, in terms of motivating behavioural change in target groups, while at the same time drawing upon a complementary, and currently scantily populated, evidence base on *making it happen*. Beyond this there is need for further engagement with social science tools and techniques to gather robust evidence to performance manage the intervention delivery process and demonstrate impact.

Many workplace interventions fail simply because their protagonists are unsuccessful in gaining sufficient support at senior management level to adequately resource and culturally permission their interventions, and this is undoubtedly a source of great frustration for many. If these professionals are to maximize their success in securing access to funds to pursue headline prevention agendas, such as works stress, and musculoskeletal disorders, they need to have the capacity to equip themselves with strong evidence, in order to convince the senior decision making function of the efficacy of their endeavour. This is a major challenge for a community of professionals with predominantly engineering, natural science, or health treatment backgrounds, particularly as they are being *asked* to engage with new (population based) perspectives and issues, such as mental health, psychosocial elements and issues of well-being, areas where the science of cause and effect is incomplete, and where *answers* are not well mapped (Lunt et al., 2008). Failure to achieve this risks retreat to narrow low-cost, typically low-impact topics, possibly to the extent of risking trivialization of their contribution and increased marginalization.

OHS policy makers and professionals not only need to possess or have access to social science skills to recognize and devise worthwhile interventions but also the ability to demonstrate their value to others who can permission them, provide access to funds and facilitate the necessary cultural profile within the organization. Intuitively, equipping OHS professionals with the necessary skills to review and interpret social science evidence is a potentially valuable area for continuous professional development, mirroring the skills based training that brigades of public policy professionals have undertaken over the last decade. Assuming the presence of appropriate good quality training, this is resolvable. However, the content needs to be supplemented with guidance on operationalizing interventions, techniques for gathering performance management evidence, demonstrating its value to others, topics on which the academic and practitioner literatures remain largely silent. It is almost certainly the case that good practice exists in this area, however it seems that the secrets of success reside in the psyche of OHS practitioners rather than in published works with the result that this knowledge remains substantially untapped (Nazaruk, Weyman, and Hellier, 2009).

Plugging the gap, through the provision of good quality guidance on the gathering and use of social research evidence to set agendas for, and delivery of, change has significant potential to strengthen the case for intervention advanced by OHS and related practitioners and to raise the profile of their contribution to organizational performance.

A holistic approach to occupational health at work must necessarily focus, not only on the development of a culture that promotes lifestyle health and well-being, but on the creation of organizational systems that minimize or mitigate challenges to employee health and promote well-being on a broad front (Semmer, 2006). In practice, comprehensive holistic management systems, that combine primary, secondary, and tertiary intervention strategies, that address both the causes and consequences of health at work offer the significant promise in terms of gains in employee health and benefits to employers (Jordan et al., 2003; LaMontagne et al., 2007; Taris, van der Wal, and Kompier, 2010), but remain the most rarely encountered. From the small number of pioneering studies to date it seems that there is significant scope to build upon and expand the established safety-risk model of hazard assessment and control, using this as a platform to broaden employer perspectives and practice in dealing with the psychosocial and mental health agendas.

Note

1 Mirroring European Commission practice, in the UK a White Paper sets out the Governent's policy intent to act on a defined issue or area.

References

Advisory Committee on the Safety of Nuclear Installations (1993). *Organising for Safety, Health and Safety Commission.* ACSNI Human Factors Study Group, 3rd Report.

Ajzen, I. (1985). The theory of planned behaviour. *Organizational Behavior and Human Decision Processes, 50*(2), 179–211.

Anderson, M. (2010). *Thinking about behavioral safety.* Available online at: www.hse.gov.uk/humanfactors/resources/articles/behavioural-safety.htm

Anderson, M., & Weyman, A. K. (1998) *Development of an accident classification taxonomy for UK mines*, Health & Safety Executive Report No: IR/L/EBS/98/15.

Becker, M. H. (1974). The Health Belief Model and personal health behavior. *Health Education Monographs*, 2(4), 324–473.

Biron, C., Gatrell, C. & Cooper, C. L. (2010). Autopsy of a failure: Evaluating process and contextual issues in an organizational-level work stress intervention. *International Journal of Stress Management*, 17(2); 135–158.

Black, Dame C. (2008). *Working for a healthier tomorrow – Review of the health of Britain's working age population.* HMSO, London.

Bullock, H., Mountford, J., & Stanley, R. (2001). *Better Policy Making.* Centre for Management and Policy Studies, London.

Burton, A. K., Kendall, N. A .S., Pearce, B. G., Birrell, L. N., & Bainbridge, L. C. (2008). *Management of upper limb disorders and the biopsychosocial model.* Health and Safety Executive, Research Report 596.

Chemicals Hazard Information and Packaging for Supply (CHIP) Regulations 2008.

Cousins, R., MacKay, C. J., Clarke, S. D., Kelly, C. J., Kelly, P., & McCaig, R. H. (2004). Management Standards and work-related stress in the UK: Practical development. *Work & Stress*, 18(2), 113–136.

Cox, S., & Flin, R. (1998). Safety culture: Philosopher's stone or man of straw? *Work and Stress*, 12(3), 189–201.

Dugdill, L., Brettle, A., Hulme, C., McCluskey, S., & Long, A. F. (2008). *Workplace Physical Activity Interventions: a Systematic Review.* White Rose Research. See: eprints.whiterose.ac.uk/3578/

European Community (2008). *EC Regulation 272/2008 of the European Parliament and of the Council on classification, labelling and packaging of substances and mixtures, amending and repealing Directives 67/548/EEC and 1999/45/EC, and amending Regulation (EC) No. 1907/2006.*

European Community (1989) *Framework Directive 89/391/EEC.*

Edwards, J., Van Laar, D. L., & Easton, S. (2009). The Work-Related Quality of Life (WRQoL) scale for Higher Education employees. *Quality in Higher Education*, 15(3), 207–219.

Fishbein, M., & Ajzen, I. (1975). *Belief, attitude, intention, and behavior: an introduction to theory and research.* Reading, Mass.: Addison-Wesley Pub.

Food Standards Agency (2007). *Traffic light food labels.* Available online at: www.fph.org.uk/uploads/ps_food_labelling.pdf

Gadd, S., & Collins, A. M. (2002). *Safety culture: A review of the literature.* Health and Safety Laboratory Report HSL/2002/25.

Garvin, D. A. (1993). Building a learning organization. *Harvard Business Review*, 71(4), 78–91.

Government Social Research (2008a). *GSR Knowledge Review – Behaviour Change.* See: www.gsr.gov.uk

Government Social Research (2008b). *GSR Behaviour Change Knowledge Review Reference Report: An overview of behaviour change models and their uses.* Darnton, A. See www.civilservice.gov.uk/Assets/Behaviour_change_reference_report_tcm6-697.pdf

HM Government Health Act 2006 *Statutory Instrument (C.28).* See: www.statutelaw.gov.uk/content.aspx?activeTextDocId=2573453

HM Government (1999). *The Management of Health and Safety at Work Regulations* 1999.

HM Treasury Guidance, 2010 *Magenta book – Guidance Notes on Policy Evaluation.* See: www.nationalschool.gov.uk/policyhub/magenta_book

Hale, A. R., & Glendon, A. I. (1987). Individual behaviour in the control of danger. Elsevier, New York.

Heinrich, H. W., Peterson, D., & Roos, W. (1980). *Industrial Accidents Prevention* McGraw-Hill, New York.

Health and Safety Executive (1999). *Summary guide to safety climate tools*. Offshore technology report 1999/063. See: www.hse.gov.uk/research/otopdf/1999/oto99063.pdf

Health and Safety Executive (2000a). *Behaviour Modification to improve safety*. Offshore technology report 2000/003. See: www.hse.gov.uk/research/otopdf/2000/oto00003.pdf

Health and Safety Executive (2000b). *Revitalising Health and Safety Strategy Statement* June. See: www.hse.gov.uk/revitalising/strategy.pdf

Health and Safety Executive (2000c). *Safety culture maturity model*. Offshore technology report 2000/049. See: www.hse.gov.uk/research/otopdf/2000/oto00049.pdf

Health and Safety Executive (2005a). *Influencing Change – Health and Safety Executive Business Plan 2008/09*. See: www.hse.gov.uk/aboutus/strategiesandplans/bplan0809

Health and Safety Executive (2005b). *A review of safety culture and safety climate literature for the development of the safety culture inspection toolkit*. Research Report 637. HSE Books.

Health and Safety Laboratory (2010). *Safety Climate Tool (SCT)*. See: www.hsl.gov.uk/news/hsl-launches-new-safety-climate-tool-.aspx

Hunter, D. J., Popay, J., Tannahill, C., Whitehead, M., & Elson, T. (2009) Marmot Review Working Committee 3 Cross-cutting sub-group report *Learning Lessons from the Past: Shaping a Different Future*. Available online at: www.dur.ac.uk/resources/public.health/news/FinalSynthesisedReporttoMarmotReview-WC3subNov09.pdf

Jordan, J., Gurr, E., Tinline, G., Giga, S. I., Faragher, B., & Cooper, C. L. (2003). *Beacons of Excellence in Stress Prevention* (Research Report 133). London, UK: HSE Books.

Kelly, C. J., & Mellor, N. (2005) *Evaluation of the Management Standards Pilot Study*. Health and Safety Laboratory WPS/04/05.

Kerr, R., McHugh, M., & McCrory, M. (2009) HSE Management Standards and stress-related work outcomes. *Occupational Medicine* 59, 574–579 Published online 7 October 2009 doi:10.1093/occmed/kqp146.

Kohler, J. M., & Munz, D. C. (2006). Combining individual and organizational stress interventions. *Consulting Psychology Journal: Practice & Research*, 58(1), 1–12.

LaMontagne, A. D. Keegel, T. Louie, A. M. Ostry, A., & Landsbergis, P. A. (2007). A systematic review of the job-stress intervention evaluation literature, 1990–2005. *International Journal of Occupational and Environmental Health*, 13, 268–280.

Lunt, J., Bates, S., Bennett, V., & Hopkinson, J. (2008). *Behaviour change and worker engagement practices within the construction sector*. HSE Books. See: www.hse.gov.uk/research/rrpdf/rr660.pdf

MacKay, C. J., Cousins, R., Kelly, P., Lee, S., & McCaig, R. H. (2004). Management Standards' and work-related stress in the UK: policy background and science. *Work & Stress*, 18(2) 91–112.

Modernising Government (1999). *HM Government White Paper*. Available online at: www.hse.gov.uk/pubns/web42.pdf

National Institute for Clinical Excellence (undated). *Resilience, coping and salutogenic approaches to maintaining and generating health a review*. Emily Harrop, Samia Addis, Eva Elliott, Gareth Williams. See: www.cardiff.ac.uk/socsi/cishe/pages/Publications/Behaviour_Change-Review_on_Resilence_coping_and_salutogenic_approaches_to_health.pdf

Nazaruk, M., Weyman, A., & Hellier, E. (2009). Beyond behaviours: factors behind successful implementation of safety interventions. Paper presented at the *Behavioral Safety Now Conference*, 7–8 October 2009, Orlando, USA.

Nielsen, K., Randall, R., Holten, A.-L., & Rial-González, E. (2010). Conducting organizational-level occupational health interventions: What works? *Work & Stress*, 24(3), 234–259. Doi: 10.1080/02678373.2010.515393

Nieve, V. F. & Sorra, J. (2003). Safety culture assessment: a tool for improving patient safety in healthcare organizations. *Quality and Safety in Health Care*, 12, 17–23.

Pidgeon, N.F. (1997). The limits to safety culture, politics, learning and man-made disasters. *Journal of Contingencies and Crisis Management*, 5(1), 1–14.

Power, M. (2004). *The Risk Management of everything: rethinking the politics of uncertainty*. DEMOS. See: www.demos.co.uk/files/riskmanagementofeverything. pdf?1240939425

Reporting of Injuries, Diseases and Dangerous Occurrences Regulations (RIDDOR) (1995). See: www.hse.gov.uk/riddor

Semmer, N. K. (2006). Job stress interventions and the organization of work. *Scandinavian Journal of Work, Environment and Health*, 32(6), 515–527.

Solesbury, W. (2001). *Evidence Based Policy: Whence it Came and Where it's Going*. ESRC UK Centre for Evidence Based Policy and Practice: Working Paper 1. See: http://kcl.ac.uk/content/1/c6/03/45/84/wp1.pdf

Taris, T. W., van der Wal, I., & Kompier, M (2010). Large-scale job stress interventions. In J. Houdmont and S. Leka (Eds.), *Contemporary occupational health psychology: Global perspectives on research and practice*. Chichester, UK: John Wiley & Sons (pp. 77–97).

Thaler, R. H., & Sunstein, C. R. (2008). *Nudge: Improving decisions about health, wealth, and happiness*. Yale University Press.

Van Larr D., Edwards, J. A., & Easton, S. (2007). The Work-Related Quality of Life scale for healthcare workers. *Journal of Advanced Nursing*, 60(3) 325–333.

Vanwesenbeeck, I., van Zessen, G., Ingham, R., Jaramazovic, E., & Stevens, D. (1999). Factors and processes in heterosexual competence and risk and integrated review of evidence. *Psychology and Health*, 14(1), 25–26.

Weyman, A. K. (1997). *Psycho-social determinants of work related upper limb disorders – A review of literature*. HSE Health Directorate Report No: IR//EWP/97/07.

Weyman, A. K. (2007). The cost of accidents: Investing in a positive safety culture. *Proceedings of the Midland Institute of Mining Engineers*, Safety Seminar, Sheffield, April.

Weyman, A. K., & Kelly, C. J. (2000). *Risk perception and risk communication in the workplace. A review of literature*. HSE Contract Research Series Report. CRR198/2000. HSE Books, HMSO.

Weyman, A. K., Pidgeon, N. F., Jeffcott, M., & Walls, J. (2006). *Organizational dynamics and safety culture in UK train operating companies*. Research Report 421. HSE Books.

Whysall, Z., Haslam, C., & Haslam, R. (2006). Implementing health and safety interventions in the workplace: An exploratory study. *International Journal of Industrial Ergonomics*, 36(9), 809–818.

Part 4
Conclusions

17 Concluding comments

Distilling the elements of successful organizational intervention implementation

Maria Karanika-Murray, Caroline Biron, and Cary L. Cooper

Effective and efficient organizational interventions for stress and well-being are important not only for developing a healthy and productive workforce but also because business organizations are increasingly viewed as vehicles for achieving improvements in population health (Black, 2008; Fleming, 2007; Tetrick and Quick, 2002). An engaged workforce in well-managed organizations can contribute to enhanced productivity, quality products, and a high-performing resilient workforce, which in turn can contribute to a well-functioning society (Black, 2008). However, developing, implementing and evaluating organizational interventions with a view to reduce employee stress and build healthy organizations can be a very complicated matter.

As highlighted by the chapters in this volume, complexity here is encountered at a number of levels (see also Karanika-Murray, 2010). First, work-related stress and well-being are almost invariably multivariate, since they relate to many facets of their context (Everitt, 1978), the nature of the work-health relationship is often dynamic and discontinuous (Rick et al., 2002), and stress and well-being are nested within multiple individual, organizational and social systems. Furthermore, some of the relationships between work and health are known with little certainty. For example, a cursory examination of the literature shows that our best models often explain up to 40 per cent of the variance in job satisfaction whilst 5 per cent of sickness absence explained is considered good. When certainties are so low, there is a need to build conservative margins into our stress and well-being management and interventions. In addition, organizations are complex adaptive systems where changes in one element can lead to changes in another. No intervention, no matter how discrete or local, is stationary (Hayes, 2002), nor can it be isolated from the broader organization and its functioning (Johns, 2006). Organizations can be complex and adaptive regardless of whether centralised control exists, because they consist of individuals. As activation theory (Gardner, 1986; Gardner and Cummings, 1988) postulates, every individual organism has a characteristic level of activation or homeostasis, and deviations from this homeostatic level determine its efficiency or performance and also motivation to maintain it. This adds to the point made by Biron in this volume in regards to realistic evaluation applied to organizational interventions: "the real engine for change in social programs is the process of differently resourced subjects making choices about the opportunities provided" (Pawson and Tilley, 1997, p. 46). As highlighted by Baril-Gingars et al. and Tvedt and Saksvik in this volume, any organizational-level intervention, whether it is focused on health promotion or on work redesign, any change to "the way things work", can have a range of potential "side-effects" which are rarely confined to the target groups of the intervention. Furthermore, often, resources are finite and shared between competing interventions, such that the agents of change have to be inventive and to draw from different sources in order to support an intervention. Finally, an additional layer of complexity is added by the fact that the initiators, the implementers and the recipients of organizational interventions are often different groups of stakeholders (Schrader-Frechette, 1998). Thus, those who make the decisions about resource allocation are not those who are affected by these decisions or who are in need of the changes in their work methods or stress and well-being levels, making intervention implementation more difficult.

The sources of complexity listed above add to the long list of methodological difficulties that have been highlighted in relation to designing, implementing and evaluating organizational interventions (e.g., Biron, this volume; Bourbonnais et al., this volume; Briner and Reynolds, 1999; Cox et al., 2007; Nytrø et al., 2000; Cooper, Dewe, and O'Driscoll, 2001; Dewe and Kompier, 2008; Griffiths, 1999; Kompier, Cooper, and Geurts, 2000; Kompier and Kristensen, 2001; LaMontagne, Noblet, and Landsbergis, this volume; Reynolds and Shapiro, 1991). It is therefore of little surprise that although we know that work is good for health and well-being (Waddell and Burton 2006), we know much less on how to develop, implement, and evaluate changes to create healthy workplaces. When it comes to delivering successful and sustainable interventions, we cannot tell with certainty what works, for whom or under what circumstances (Hill et al., 2007). This is not necessarily because the interventions do not work, but because the available tools that we have to understand what works and when are not adequate for purpose.

It is therefore important to find ways to harness some of this complexity and to develop a language and toolbox of methods specific to organizational interventions for stress and well-being. Some may argue that we do not know enough about the relationships between work and health in order to develop practical actions (e.g. Briner and Reynolds, 1999). But, as this book demonstrates, a substantive body of work on interventions has developed in recent years, and we believe that it is possible to make this tacit and dispersed knowledge explicit. Taking into account the process and contextual issues that can impact on the design, implementation, and evaluation of organizational interventions for stress and well-being is one step towards reducing this complexity. Consideration of process and context is, among others, about consideration of the characteristics of the people involved and of the boundaries within which the intervention takes place. Any action requires an implementation plan, and taking into account process and contextual issues is about carefully considering resources, boundaries, and contingencies that can impact on the success or failure of our interventions. By collating the experiences and reflections of some of the leading scholars and practitioners in the area, this book can help to achieve a step-change in organizational interventions for stress and well-being that take into account the range of process and contextual issues that can impact on their successful implementation.

Distilling the elements of successful implementation of organizational interventions

The contributors to this book have dispensed needed knowledge on how the implementation process and the wider organizational context can shape the success or failure of organizational interventions for stress and well-being. Here, based on these chapters and on the broader literature on organizational interventions (e.g. Biron, this volume; Biron, Gatrell, and Cooper, 2010; Briner and Reynolds, 1999; Cox et al., 2007; Cooper, Dewe, and O'Driscoll, 2001; Dewe and Kompier, 2008; Griffiths, 1999; Kompier, Cooper, and Geurts, 2000; Kompier and Kristensen, 2001; Mellor, Karanika-Murray, and Waite, this volume; Mikkelsen, Saksvik, and Ladsbergis, 2000; Nytrø et al., 2000; Reynolds and Shapiro, 1991; Semmer, 2006; Kompier et al., 1998) we attempt to distil some of the elements of successful implementation of organizational interventions. We have identified eight, which are discussed below.

ELEMENT 1

Use of multimodal comprehensive interventions that combine individual and organizational-level actions. The importance of combining individual and organizational-level actions, which provide a primary, a secondary and a tertiary focus, cut across workplace hierarchies, and that take into account all elements of the system, has been stressed here (e.g. Mellor et al., Tetrick, Campbell, and Gilmore, this volume) and in other major works (e.g. LaMontagne et al., 2007; Kohler and Munz, 2006).

ELEMENT 2

Use of multi-disciplinary approaches and integration of health promotion with risk management. Multi-disciplinary approaches can help to cast light on both macro (e.g. restructuring, organizational change) and micro (e.g. ergonomics, health-related behaviours, individual differences) influences on stress and well-being, to engage all necessary skills and resources for developing multimodal interventions, and to ensure the commitment of major stakeholders (human resources, senior management, occupational health, health and safety, well-being, etc.) (e.g. Biron this volume; Baril-Gingras et al., this volume; Mellor et al., this volume).

ELEMENT 3

Use of participative approaches. Participation is an essential component for establishing commitment and ownership of problems and solutions because it involves all stakeholders in intervention design and implementation. It is important that the stakeholders involved have clearly defined roles, a coordinating group is established, there is wide participation in designing and implementing an intervention, and support is secured at both local (e.g. Lewis, Yarker, and Donaldson-Feilder, this volume) and organizational levels (especially senior management commitment) (e.g. Bourbonnais et al., this volume).

ELEMENT 4

The importance of a carefully developed implementation framework cannot be stressed enough. Any solution or action plan is only as good as its implementation. The pragmatics of intervention implementation clearly shows that ongoing monitoring of the implementation process, having specific, achievable, and measurable goals; clear allocation of roles and responsibilities; and clear communication plans are all essential elements of well-implemented interventions.

ELEMENT 5

A strong evidence base coupled with an assessment of the needs of the organization and its employees. Any solution needs to be tailored to the specific organizational context, based on an understanding of the relationships between work and stress or well-being and between individuals and their organizations. It also requires an accurate diagnosis of the symptoms, problems and their root causes; assessment of stakeholders' attitudes and readiness for change; addressing any differences in perceptions and motives that can act as barriers to implementation; and monitoring the target groups' uptake and awareness of the intervention. Although off-the shelf measures and solutions can be useful in some situations, in most cases these do not help to understand how and why the solution has worked or not, and therefore a tailored fit for purpose approach is most often appropriate (e.g., Biron, this volume; Klein Hasselink et al., this volume; Randall and Nielsen, this volume).

ELEMENT 6

Integration of solutions into normal business practice. Use of add-on off-the-shelf packages more often than not only provides temporary solutions. Addressing the causes of stress and well-being requires solutions that target underlying and often more permanent aspects of work design or organizational systems, policies, and procedures. Often, integration of occupational safety and health into business management requires a certain level of organizational maturity (European Agency for Safety & Health at Work, 2010). This is based on the ability to detect obstacles early, to initiate and implement corrective action and review, to redefine the way things work, and to learn from failure.

ELEMENT 7

External forces are often instrumental for challenging and spurring change at the more local organizational level. As the Nordic and Dutch cases have shown (see chapters by Tvedt and Saksvik, and by Daniels, Karanika-Murray, Mellor, and van Veldhoven; see also chapters by Tetrick, Quick, and Gilmore; Mackay et al.; and Weyman), regulation and enforcement can often act as major stimulants for developing organizational responsibility, ownership, and a proactive approach to stress and well-being at work. Combined with a strong business case for investing in employee well-being, governmental strategies (supported by legislation) and senior management initiatives (providing decision-making and resources) have acted as important ways to establish well-being as a priority in many countries, and to address sources of resistance to change. External pressure at this level is important for initiating but not necessarily sustaining changes in attitudes, agendas, and priorities. This is because prolonged external pressure may adversely affect organizations' motives for investing in employee health and well-being.

ELEMENT 8

Clear conceptualizations and definitions are important for a succinct language of organizational interventions. Interventions aiming to improve the working environment and working conditions arise from either the stress or the well-being fields, each of which is equipped with a set of specific tools, methods and practices. As the title of the book suggests, and as mentioned by Norbert Semmer in the Foreword, this volume is about creating healthy, fulfilling, and meaningful workplaces, and how to bring about these organizational changes. Instead of focusing on the existing debates regarding the differences between approaches emerging from the stress vs the more positive approaches (i.e. well-being), we aimed to provide knowledge on the ways in which process and context can inform research and practice. Without clear definitions and operationalization of process and contextual issues, research on this topic will remain consensus weak, as evidenced by Egan et al. (2009).

Developing a language of organizational interventions

As the chapters in this book have shown, the pragmatics of intervention implementation dictate that it is often difficult to combine methodological rigour with practical relevance. In the case of developing and implementing organizational interventions for stress and well-being, practice that is relevant and useful for the target populations will remain a priority over methodological rigour. This is until, of course, better ways to bridge the practitioner-researcher divide are developed; see Anderson, Herriott, and Hodgkinson, 2001; Anderson, 2007). The chapters in this book offer a way forward for appreciating process and contextual issues and for developing a needed "language" of organizational intervention implementation, supplemented by strong theoretical models, consistent terminology, specific methods, and purposefully-developed tools.

One of the shortcomings of organizational intervention theory and the reason for its failure to address the important *how*, *why* and *when* questions, is that so far, it has offered factor models, focusing on content and criterion outcomes, as opposed to the process and context of an intervention. Mohr (1982) makes a useful distinction between factor, or variance, and process models: variance models aim to explain the variance in the outcomes and provide sufficient conditions for the outcome, whereas process models describe how and when things happen and the necessary conditions for an outcome. It has been acknowledged that variance models have dominated research in the social sciences to the extent that true process theories are rare (Gilbert, 1995).

Organizational intervention theory has, in this respect, provided factor or variance models which describe the effects of particular interventions or the factors affecting particular outcomes (for example, see review by Hill et al., 2007). Unlike organizational change theory (see chapters by Tetrick et al., and Tvedt and Saksvik), organizational intervention theory has so far failed to explain the process of change and the boundary conditions for the impact of the intervention (content) on the desired outcome (criterion). Process theories are of great value for developing knowledge because they acknowledge the importance of context and time (Karanika-Murray, 2010). The good news is that organizational intervention theory is developing along these lines, a case in point being Randall and Nielsen's (this volume) model of intervention fit in complex organizational environments. Furthermore, emerging understanding of how change at one level can affect outcomes at another (e.g. senior management support and individuals' readiness for change, or participation and organizational development/learning) is an important contribution towards developing multilevel process theory of organizational interventions (see Biron, this volume). Of course, we should note that it is important to also understand the relationship between context and process. As Griffin (2007) notes, the context can operate in a number of ways, including shaping opportunities and constraints, shaping the task characteristics of operational uncertainty, providing contingencies for effective decision-making, encouraging specific processes, and providing enhancing and constraining processes. Further conceptual and empirical work is definitely needed to develop organizational intervention theory.

In the last few years there have also been notable changes in research methods and the development of needed measures that are essential for understanding the process and context for successful intervention implementation. Work on the diagnosis and measurement of stakeholders' attitudes and readiness for change, perceptions and motives, uptake and awareness of the intervention (see Tvedt and Saksvik, this volume) is paramount for understanding the particulars and developing process models of intervention implementation. This work can be used as a starting point for future integration of knowledge. We need to know how different stakeholders experience change, and what factors influence their ownership of the change process and of the intervention programme. Future work could help to develop comprehensive evidence-based diagnostic and implementation tools that provide tailored solutions and that incorporate the stakeholders' needs and resources.

Conclusions

Work-related stress, health and well-being is a pragmatic and functional field. It has evolved from and is shaped by real life concerns, is called on to solve real life problems, and its methods are determined by practical constraints and opportunities (for a historical account see Barling and Griffiths, 2010). This book's contributors have shown that:

1. we have so far developed a good understanding of the process and context issues that can determine the success or failure of organizational interventions for stress and well-being,
2. a language and new theory of organizational interventions is starting to emerge,
3. researchers and practitioners' experiences converge into a set of elements that characterize the successful implementation of organizational interventions.

This book is a small step towards understanding process issues in organizational interventions for stress and well-being. We look forward to new, insightful research and practical solutions on how to take into account process and contextual issues for developing successful and sustainable interventions. We hope that this volume will provide both practitioners and academics with actionable knowledge for developing practical and effective interventions.

References

Anderson, N., Herriott, P., & Hodgkinson, G. P. (2001). The practitioner–researcher divide in Industrial, Work and Organizational (IWO) psychology: Where are we now, and where do we go from here? *Journal of Occupational & Organizational Psychology), 74,* 391–411.

Anderson, N. (2007). The practitioner-researcher divide revisited: Strategic-level bridges and the roles of IWO psychologists. *Journal of Occupational & Organizational Psychology, 80*(2), 175–183.

Barling, J., & Griffiths, A., 2010. A history of occupational health psychology. In: Quick, J. C. & Tetrick, L. E., ed., *Handbook of Occupational Health Psychology* 2nd ed. American Psychological Association, Washington, D.C.

Biron, C., Gatrell, C., & Cooper, C. L. (2010). Autopsy of a failure: Evaluating process and contextual issues in an organizational-level work stress intervention. *International Journal of Stress Management, 17*(2), 135–158.

Black, C. (2008). *Working for a healthier tomorrow*. London: TSO.

Briner, R. B., & Reynolds, S. (1999). The costs, benefits, and limitations of organizational level stress interventions. *Journal of Organizational Behavior, 20,* 647–664.

Cooper, C. L., Dewe, P. J., and O'Driscoll, M. P. (2001). *Organizational Stress: A Review and Critique of Theory, Research and Applications*. Thousand Oaks, California: Sage Publications.

Cox, T., Karanika, M., Griffiths, A., & Houdmont, J. (2007). Evaluating organizational-level work stress interventions: Beyond traditional methods. *Work & Stress, 21,* 348–362.

Dewe, P., & Kompier, M. (2008). *Foresight Mental Capital and Wellbeing Project. Wellbeing and work: Future challenges.* The Government Office for Science, London.

Egan, M., Bambra, C., Petticrew, M., & Whitehead, M. (2009). Reviewing evidence on complex social interventions: appraising implementation in systematic reviews of the health effects of organisational-level workplace interventions. *Journal of Epidemiology and Community Health, 63*(1), 4–11. doi: 10.1136/jech.2007.071233

Egan, M., Bambra, C., Thomas, S., Petticrew, M., Whitehead, M., & Thomson, H. (2007). The psychosocial and health effects of workplace reorganization 1. A systematic review of organisational-level interventions that aim to increase employee control. *Journal of Epidemiology & Community Health, 61,* 945–954.

European Agency for Safety & Health at Work (2010). *Mainstreaming occupational safety and health into business management.* Luxembourg: Office for Official Publications of the European Communities.

Everitt, B. S. (1978). *Graphical Techniques for Multivariate Data.* New York: North-Holland.

Fleming, P. (2007). Workplaces as settings for public health development. In: A. Schriven & S. Garman (Eds.), *Public health: Social context and action* (pp. 166–179). London: Open University Press.

Gardner, D. G. (1986). Activation theory and task design: An empirical test of several new predictions. *Journal of Applied Psychology, 71,* 3, 411–418.

Gardner, D. G., & Cummings, L. L. (1988). Activation theory and job design: Review and reconceptualization. In B. M. Staw & L. L. Cummings (Eds.), *Research in Organizational Behavior, Vol. 10.* Greenwich, CT: JAI Press.

Gilbert, G. N. (1995). Emergence in social simulation. In G. N. Gilbert & R. Conte (Eds.), *Artificial societies: The computer simulation of social life* (pp. 144–156). London: UCL Press.

Griffin, M. (2007). Specifying organizational contexts: Systematic links between contexts and processes in organizational behavior. *Journal of Organizational Behavior, 28*(7), 859–863.

Griffiths, A. (1999). Organizational interventions: Facing the limits of the natural science paradigm. *Scandinavian Journal of Work, Environment & Health, 25*(6), 589–596.

Hayes, J. (2002). *The theory and practice of change management.* Basingstoke: Palgrave.

Hill, D., Lucy, D., Tyers, C., & James, L. (2007). *What works at work? Review of the evidence assessing interventions to prevent and manage common health problems.* Institute for Employment Studies, UK.

Johns, G. (2006). The essential impact of context on organizational behavior. *Academy of Management Review, 31,* 386–408.

Karanika-Murray, M. (2010). Work and health: Curvilinearity matters. In J. Houdmont & S. Leka (Eds.). *Contemporary occupational health psychology: Global perspectives on research, education, and practice,* Vol. 1. Chichester, England: Wiley-Blackwell.

Kohler, J. M., & Munz, D. C. (2006). Combining individual and organizational stress interventions. *Consulting Psychology Journal: Practice & Research, 58*(1), 1–12.

Kompier, M. A. J., Cooper, C. L., & Geurts, S. A. E. (2000). A multiple case study approach to work stress prevention in Europe. European Journal of Work & Organisational Psychology, 9, 371–400.

Kompier, M. A. J., Geurts, S. A. E., Grundemann, R. W. E., Vink, P. & Smulders, P. W. G. (1998). Cases in stress prevention: The success of a participative and stepwise approach. *Stress Medicine, 14,* 155–168.

Kompier, M., & Kristensen, T. S. (2001). Organizational work stress interventions in a theoretical, methodological and practical context. In J. Dunham (Ed.), *Stress in the workplace: Past, present and future*. London: Whurr.

LaMontagne, A., Keegel, T., Louie, A. M., Ostry, A., & Landsbergis, P. (2007). A systematic review of the job-stress intervention evaluation literature, 1990–2005. *International Journal of Occupational and Environmental Health, 13*, 268–280.

Mikkelsen, A., Saksvik, P. Ø., & Landsbergis, P. (2000). The impact of a participatory organizational intervention on job stress in community health care institutions. *Work & Stress, 14*(2), 156–170.

Mohr, L. B. (1982). *Explaining organizational behavior*. San Francisco: Jossey-Bass.

Nytrø, K., Saksvik, P. Ø., Mikkelsen, A., Bohle, P., & Quinlan, M. (2000). An appraisal of key factors in the implementation of occupational stress interventions. *Work & Stress, 14*, 213–225.

Pawson, R., & Tilley, N. (1997). *Realistic evaluation*. London: Sage Publications.

Reynolds, S., & Shapiro, D. (1991). Stress reduction in transition: Conceptual problems in the design, implementation, and evaluation of worksite stress management interventions. *Human Relations, 44*, 717–733.

Rick, J., Thomson, L., Briner, R., O'Regan, S., and Daniels, K. (2002). *Review of existing supporting scientific knowledge to underpin standards of good practice for key work-related stressors – Phase 1*. Brighton: The Institute for Employment Studies.

Schrader-Frechette, K. S. (1998). *Risk and rationality: Philosophical foundations for populist reforms*. Berkeley: University of California Press.

Semmer, N. K. (2006). Job stress interventions and the organization of work. *Scandinavian Journal of Work, Environment, & Health, 32*, 515–527.

Tetrick, L. E., & Quick, J. C. (2002). Prevention at work: Public health in occupational settings. In: J. C. Quick & L. E. Tetrick (Eds.), *Handbook of occupational health psychology* (pp. 3–18). Washington, DC: American Psychological Association.

Waddell, G., & Burton, A. K. (2006). *Is work good for your health and well being?* London: The Stationery Office.

Index

absenteeism 7, 43, 44, 60, 165, 170, 292, 315–16; in the Netherlands 319, 321, 322, 324; sickness 1, 103, 188, 287, 298, 299, 303, 306, 308, 309, 314, 351
absenteeism counselling 320
abusive supervision 221
action plans 86–7, 95, 146, 153, 156, 172–4, 176, 196, 198, 202, 230, 285–6, 290, 307, 309, 354
activation theory 351
alcohol 39
ambulance service 239, 241–3
Andrews' principle 177
anti-discrimination obligations 25
anxiety 22, 46, 188, 221, 239, 250, 286, 287, 292, 298, 304, 315
applied psychology 2
Arbitration Conciliation Advisory Service (ACAS) 297, 302, 307

back pain 9, 49
behavioural science 4, 334
beyondblue initiative 25
"black box" 6, 9, 163, 164
bullying 79, 82, 83, 220, 222, 323, 324
burnout 82, 188, 189, 203, 207, 218, 221, 249

cardiovascular disease 9, 189
career development 47, 326–7
causality 12, 80, 168, 263, 310; generative view of 168; successionist view of 168
change agents 65–6; horizontal 63; vertical 63
Change Impact Factor (CIF) 107, 110, 113
change management 10, 59, 60–1, 66, 68, 69, 84, 113, 315, 341
Clinton, Bill 72
cognitive adaption theory 109
cognitive appraisal 65, 121

communication 26–8, 31, 33, 51–2, 64, 65, 69, 77, 79–80, 91–4, 123, 150, 170, 197, 199, 204–7, 209, 217, 218, 220, 222, 230–2, 250, 271, 273–7, 278, 292, 294, 297, 301, 316, 323, 354
compensation 24, 33, 72, 79, 95, 141
conflict management 106, 225, 230
constructivism 167
contingency planning 31, 180
contingency theory 166
Copenhagen Burnout Inventory 193
Copenhagen Psychosocial Questionnaire (COPSOQ) 29
coping style 40, 45–6, 49, 84, 121
coping strategies 125
coronary artery disease 39
correctional officers 187–210
creativity 45, 60, 320
crisis management 202, 247–8
critical incident stress debriefing (CISD) 247–50
critical incident stress management (CISM) 247–9

deficit approach 44
Delphi approach 302–3
demand-control-support (DCS) model 82, 188, 192
depression 9, 22, 24, 46, 52, 189, 221, 250–1, 287, 292, 298, 304, 315
deregulation 93, 141, 323
diabetes 39
diagnosis (organizational) 61–4, 122, 139, 145–6, 148–50, 175, 314, 354, 356
diary studies 318
diet 46, 50, 334
disability 47, 179, 321
downsizing 22, 103, 122, 124
"Dutch disease" 321, 322

effort-reward imbalance model (ERI) 188, 192–3, 196, 197, 203
egalitarian culture 105
emergency services 12, 238–51; gender and race in 244–5; industrial relations in 243; male domination of 242–3
emotional first aid 247
emotional intelligence 45
empiricism 167
employee assistance programmes 46
employee buy-in 177
employment law 25
empowerment 10, 28, 29, 39, 45, 51, 113, 114, 209, 220, 225
entropy 93, 94
environment–intervention fit 120, 121, 123, 124–5, 127–30
epidemiology 136
ergonomic training 143
ergonomics 135, 136–7, 140, 353
eudemonic tradition 44
European Agency for Safety & Health at Work (EU-OSHA) 39, 46
European Commission 286
European Union (EU) 47, 313, 314, 316, 319, 335, 336, 344, 346
evidence based practice 333
extended health erosion path 82
extended motivational path 82
extraversion 126

family 93, 238, 245–6, 251; fictive 245–6
Fire Brigades Union (FBU) 243
fire service 238–43
fitness for purpose 5
flexible working 26, 47
Future Inquiry 29

General Health Questionnaire 239, 295
group cohesion 42
group norms 84

harassment 83, 193, 203–4, 207, 208, 240
health 10, 170–1, 219, 263–4, 336, 351; definition of 60; mental 1, 24–5, 39, 43, 46, 52, 135, 143, 145, 149, 151, 155, 156–7, 170, 187, 189, 191–2, 197, 203, 207, 222, 244–5, 263, 286, 320, 335, 345, 346; physical 1, 9, 39, 46, 52, 188–9, 203, 207–8, 263, 320; positive 60
"Health Circles" 29
health awareness 47
health care services 47

health interventions *see* interventions
health promotion 1, 9, 21, 25, 39–40, 46–8, 50, 52, 62, 93, 142, 352, 353; worksite 3, 28, 46–7, 165
health and safety 33, 47, 48, 53, 77–80, 85, 135, 150, 152, 174, 189, 223, 224, 229, 230, 233, 285, 287, 289, 298, 307, 314, 322, 323, 327, 333, 335, 337, 344, 345, 353; psychological 78–80
Health and Safety Commission 285, 337
Health & Safety Executive (HSE) 12, 33, 40, 103, 178, 216, 223, 233, 251, 287–9, 291–9, 302–8, 313–16, 335, 337, 343
Healthy Change Process Index (HCPI) 106, 108, 110, 114
Healthy Conducive Production Model 93, 94
high blood pressure 39
high performance work practices (HPWP) 327
human resources 25, 31, 45, 47, 48, 53, 65, 85, 144, 173, 190, 224, 225, 226, 229, 230, 233, 265–7, 298, 304, 307, 315, 316, 326–7, 344, 353
hypothalamic-pituitary-adrenal system 46

ill-health 26, 28, 40–3, 46, 59, 67–8, 70, 239, 287, 294, 313, 319, 336; management of 7, 47, 336
immune functioning 46
Implementation Logical Model (ILM) 285–6
implementation theory 6
indicator tool 224–6, 289, 291, 305, 314, 316, 318
individual differences 40, 121, 123, 125, 127, 241–2, 260, 353
Individual Job Impact Index 107
industrial relations 136, 243, 249, 298
injuries 46, 49, 52, 67–9, 77, 79, 80, 95, 103, 137, 140, 190, 245, 287, 336
intention to leave 10, 63, 66, 88, 221
International Labour Organization (ILO) 39, 46, 47, 313
interpersonal relationships 40, 71, 175, 198, 201, 202, 208, 209
interpretive perspectives 3–4
intervention appraisal 123, 124, 127, 129–31, 279
intervention evaluation 21, 130, 135, 154, 171, 176, 178, 189, 258, 276–9, 300–1, 304, 307, 310, 322
intervention fit 11, 120–31

intervention logic model (ILM) 292–3, 296, 308
"intervention paradox" 131
intervention research 2–4, 9, 11–12, 21, 77, 108–10, 114, 120–5, 127, 128, 130–1, 135–8, 140, 144–5, 152, 164, 166, 180, 189, 191, 207, 258–9, 262–3, 277–9, 357
intervention theories 61, 336
interventions (health/stress) 5–6, 8–9, 12, 21, 77–9, 84–5, 95, 104, 109, 144, 163, 169, 177, 238, 261, 290, 308, 316, 333–4, 351–6; absenteeism 321; complexity of 10–11, 351–3; effectiveness of 2, 5, 120–2, 135, 164, 191, 251, 259, 262–3, 352; in emergency services 12, 238–51; individual-level 1, 22, 40–1, 67, 77, 114, 155, 241, 260, 335; multi-disciplinary 11, 136, 233, 297, 303, 353; multi-level 42–3, 47, 170, 353–4, 356; multimodal 353; national 12, 232–3, 308, 309, 324; in the Netherlands 12, 261–2, 264, 265–7, 313, 319–25, 354; organizational 1–2, 8–9, 22, 23–32, 40, 67–71, 77, 120, 122, 126, 152, 163–80, 208, 221, 230–1, 241, 259–79, 285, 301, 303, 307–9, 328, 345, 354–7; and organizational change 102–14, 260; participative approaches to 10, 11, 102, 104, 111, 114, 145, 151, 170, 178, 187, 192, 195, 197–9, 203, 204, 208–11, 264, 276, 314, 353–4; planning of 122, 124, 127–8, 189; primary 40–1, 77–8, 83, 95, 154, 170, 194, 240–1, 246, 261; risk evaluation of 109; secondary 41, 77, 83, 95, 154, 170, 240–1, 246, 261; small group 258, 263–79; study of 85–95, 124, 135–47, 154, 157n4, 171–9; success/failure of 120–3, 129–31, 353–7; tertiary 41, 77, 83, 95, 154, 170, 246–7, 261; threats to 30

job content 266, 271, 289, 309, 323
Job Content Questionnaire (JCQ) 192
job control 22, 27, 43, 95, 125, 222, 125, 266–7, 313, 318, 314, 315, 323, 324
Job-Demand Control (JD-C) model 40, 42–3, 81–2
Job-Demand Resources (JD-R) model 40, 42–3, 81
job demands 40, 42–3, 51, 81, 107, 125, 126, 266–7, 270, 295, 313–14, 315, 318, 321, 323

job design 13, 124, 127, 288, 289, 309, 313, 317, 318, 321, 323, 325, 326, 327, 343
job resources 81
job role 122, 126, 345
job satisfaction 40, 44, 88, 92, 125, 147, 170, 178, 190, 243, 267, 351
job security 114, 156, 193

Leader Behaviour Description Questionnaire (LBDQ) 224
leadership 26, 29, 45, 50, 63, 66, 70–1, 204, 217–22; change-specific 66; group 42; laissez-faire 220–1; negative 218, 220–1; transactional 29, 50–1, 220; transformational 50–1, 66, 70, 217, 218, 220
leadership styles 26, 50–1, 71, 147, 217–20
lean production 32
lifestyle 47, 335, 336, 347; unhealthy 46
Lincoln Industries 59, 71–2, 73
line managers 12, 42, 146, 147, 174–5, 177, 216–34, 301, 303, 306, 308, 326, 327, 340, 343, 344

macho culture 83, 242
macroergonomic theory 68–9
macro-level challenges 22, 29
management behaviours 217–34
management buy-in 24, 79, 216, 233, 290, 315
Management Competencies for Preventing and Reducing Stress (MCPARS) programme 223–6
management development 27, 226, 229, 232
Management Standards approach 12, 285, 287–96, 298–301, 304, 307–10, 313–19, 321, 324–8
management style 27, 43, 217, 344
management support 2, 23–4, 33, 63, 68, 79, 85, 143, 217–19, 232, 305, 315, 316, 356
mental capital 10, 39
mental illness 24, 39
"mental stress" claims 24, 33
micro-level challenges 22, 29
middle management 23–4, 30, 104
Multifactor Leadership Questionnaire (MLQ) 224
musculoskeletal problems 39, 69, 135–6, 149, 298, 335

needs assessment 25, 29, 170
New Labour 333

New Public Management (NPM) 32–3, 105
Nordic participative tradition 10, 102, 104, 114, 354
Nottingham Health Profile (NHP) 193
NOVA WEBA survey 265–6, 270, 276, 321, 323

obesity 39, 142
occupational health 2, 5, 21, 48, 53, 70, 155, 171, 173, 223, 229, 245, 302, 307, 315, 320, 333, 335–6, 347, 357
occupational health climate 39, 48–52
occupational health and safety (OHS) 11, 28, 31, 77, 80, 83–5, 135, 136–45, 147, 149–51, 153–7, 343–7; culture of 80
occupational identity 244–6
occupational psychology 260, 297
occupational safety and health (OSH) 47–8
open systems theory 105
operationalization 40, 91, 104, 108, 139, 148, 295, 314, 320, 339, 347, 355
optimism 45, 108, 109, 113, 220
organizational behaviour 44; informal 104; positive 60, 69
organizational change 10, 23–4, 30, 42, 59, 61–7, 71, 73, 84, 102–14, 148, 153–5, 166, 175, 191, 195, 198, 208–9, 260, 309, 353, 356; and management support 23–5, 288; and participatory processes 27–30; resistance to 23
organizational climate 40, 45, 78, 84, 201
organizational commitment 23, 24, 26, 28, 32, 44, 125
organizational culture 40, 42, 84, 144, 156, 229, 238, 242–3
organizational development 10, 24, 32, 59, 61, 67–8, 70–2, 95, 216, 356
organizational health 23, 25, 29, 31–2, 60–1, 71, 175
organizational learning 28
organizational justice 44, 71
organizational performance 1
organizational psychology 4, 21, 113, 298

"paradigm wars" 3–4, 163, 164–6, 179
participation 63–4
participatory processes 27–30
peer support 245, 249, 304–5, 315
performance management 72, 326, 347
person–environment fit 122, 240, 245
person–intervention fit 120, 121, 123, 125–9, 131
person–job fit 126

person–role fit 245
personal development 7, 81, 93, 251, 327
physical exercise 46, 50
policing 83, 168–9, 238, 239, 241–2
positive psychology 44
positive affect 45
positive emotions 45
positivism 3–4, 167
postmodernism 4
post-positivism 167
posttraumatic stress disorder (PTSD) 82, 239, 247
presenteeism 1, 7, 170
PRIMA-EF 33–4
privatization 22
"process" 5–6
process evaluation 5
productivity 7, 24, 32, 40, 44, 49, 67, 78, 143, 218, 286
project fatigue 144
Psychiatric Symptom Index (PSI) 193
psychological capital 44
psychological debriefing 247
psychological disorders 82
psychological distress 7, 10, 83, 95, 178, 188, 190, 193, 196, 203, 239
psychological first aid 248
psychological flexibility 45, 114, 125
psychological health 10, 60, 77–80, 82–3, 95, 177, 188
psychological resource theory 44
psychological risk 39
psychological uncertainty 65–6
psychosocial health 11, 25, 26, 43, 46–7
psychosocial hazard/risk 1, 7, 10, 21–5, 27, 29, 32, 33–4, 39–40, 46, 52, 77, 81–2, 141, 164, 216, 217–18, 221–3, 226, 230, 233, 287–8, 313, 319, 322, 324, 325, 327
psychosocial safety climate (PSC) 10, 77–96
psychosocial work environment 1, 22, 46–7, 53, 61, 77–8, 135, 152, 154, 178, 187, 189–90, 192, 197, 198, 203, 207, 210, 310, 319, 323, 345, 346
psychosocial working conditions survey 304, 315
psychosocial workload 323
psychosomatic complaints 9
public health 2, 21, 40, 59–60, 72, 333–5

qualitative studies 3–4, 11, 67, 135, 136–40, 187, 191–2, 195, 198, 225

quantitative studies 3–4, 67, 187, 192, 196
Quebec Health Survey (QHS), 192–3, 196

randomized controlled trials (RCTs) 3, 165, 168, 249–50
readiness to change 62–3, 108–9; creating 65
realistic evaluation (RE) theory 11, 163–4, 166–71, 180
recruitment 5, 47, 145, 242
rescue personality 241
resilience 41, 44, 60, 93, 108, 114, 241, 251
restructuring 22, 30, 122, 124, 353
Revitalising Health and Safety (RHS) 286
risk assessment 25, 95–6, 154, 171–5, 205, 217, 288–90, 296, 300, 303, 307, 308, 314, 315, 316, 325–7, 335
risk behaviours 26, 27
risk factors 7–8, 25–6, 68, 86, 145, 152, 276, 288, 290, 313 see also psychosocial hazard/risk
risk management 25, 78, 85–6, 95, 122, 176, 178, 217, 287, 289, 296, 316, 336–7, 344
risk reduction approach 7
role clarity 103, 313–15

safety climate 10, 49–51, 77–81, 83, 88–91, 93, 96n1
safety culture 96n1, 337, 344, 345
San Antonio Air Logistics Center 72–3
Sector Implementation Plans (SIPs) 286, 296–9, 302–3, 306–7
self-efficacy 45, 63, 70, 84, 108, 125, 126, 129, 147, 220
self-esteem 46, 251
self-realization 44
Self-Reported Working Conditions survey 304
senior management 23, 48
shift work 22
Short-Form Health Survey 193
sickness absence 10, 39, 88, 95, 287, 298, 299, 304, 306, 308, 309, 314, 321, 351
sleep problems 193, 203, 239, 243
smoking 26, 39, 46, 47, 50, 72, 334
social exchange theory 65, 82
social experiments 259–60
social level controllers 95–6
social sciences 4, 167, 334, 346–7, 356
social support 1, 9, 41, 81, 83, 125, 131, 152, 188, 190, 192–3, 197, 203, 207, 222, 245–6, 295, 323, 338

socio-demographic factors 40, 143, 334
socio-technology 319
staff turnover 24, 92, 113, 141, 144, 170, 188, 197, 206, 209, 218
strain 7, 9, 24, 44, 188, 203, 207, 217, 250–1
stress 8–9, 26, 32, 40, 44–6, 51, 67, 82, 121, 171–2, 218, 229, 232, 298, 304, 317, 319; approaches to 7, 13, 21, 299–300; cybernetic theories of 124; and management 217; occupational 9, 12, 86, 136, 143, 151, 163–4, 171, 174, 188, 217, 222, 238–40, 242, 259, 261; posttraumatic 238, 240; prevalence of 224, 239; prevention of 1–2, 33, 40–1, 48, 67, 77–9, 85, 87, 91, 94, 95, 96, 154, 156, 189, 226, 233, 242, 250–1, 261, 287, 289, 297, 303, 308–9, 313, 319, 346, 351; terminology of 6–7; transactional theories of 121, 124, 131; traumatic 239–40; work-related 8, 12, 21–2, 24–5, 33, 44, 46, 77, 78, 82, 86, 95, 121–3, 144, 145, 147, 154, 155, 164, 188, 193, 199, 207, 217, 224, 245, 258, 261, 266, 270–1, 275–9, 278, 285–9, 292–7, 300–3, 306, 308, 310, 313–15, 317, 320, 322–3, 328, 335, 343, 345, 346, 351–5
stress agents 77
stress audits 42, 243, 343
stress climate 96n1, 337, 344, 345
stress crossover 240
stress interventions see interventions
stress management 26, 28, 136, 138, 147, 152, 155–6, 171, 177–8, 223–8, 232, 238, 241, 242, 247, 251, 296; preventive 60, 72, 314
stress models 8, 40
stress moderators 40
stress outcomes 78, 79, 84, 88, 91
stress prevention model 85, 95
stress research 3, 9, 190–210, 222–6, 258–9, 261–79
stress risk assessment 171–3, 315
stress theories 7, 82, 121
stressors 1, 7–9, 40–1, 44, 49, 60, 81, 172, 200, 217, 238–42, 250, 261, 263, 264, 287; extreme 246; psychosomatic 22

task design 40
team approaches 29, 33, 42, 68, 79, 82–4, 112, 125, 148, 171–4, 190, 202, 206–7, 219, 222, 225, 228–31, 245–6, 297–8, 327
terminology 6–7, 355

Theory of Planned Behaviour 64
total quality management 32, 48
trade unions 29, 53, 92, 94, 96, 137, 141, 150, 155, 190, 194, 196, 204, 205, 206, 243, 298
Transformational Leadership Questionnaire (TIQ) 224
Transtheoretical Model of Change (TTM) 68–70
trauma risk management (TRiM) 248–9

United States Air Force 59, 72–3

VBBA tool 321, 323
victimization 168–9
violence 193, 197, 239, 321, 323, 324; psychological 202
volunteering effect 176–8

weight loss 46
WEBA-method 320
well-being 23, 25, 26, 40, 44–5, 59–60, 67, 71, 81, 122, 126, 129, 144, 147, 163, 164, 170, 190, 217–18, 222, 229, 233, 263, 286, 289, 303, 309, 313, 315, 318, 319–20, 327, 346, 351, 353–7; definition of 59–60; fluctuations in 41; promotion of 1–2, 5, 7, 9, 13, 23, 32, 41–2, 47–8, 67–8, 70, 218, 231, 233, 337, 347; psychological 44, 77, 82; subjective affective 44
wellness, culture of 71–2
wellness programmes 72–3
Wider Implementation Plan (WIP) 286, 296, 300
women and work 28, 143, 196, 242, 245
work design/redesign 9, 216, 222, 264, 352, 354
work environment 9, 10, 22, 40–2, 44, 45, 47, 49, 51, 53, 60, 61, 73, 84, 87, 104, 105, 107, 109, 113, 121, 125, 128, 138, 152, 154, 187, 191, 192, 196, 210, 217, 260, 264, 268, 276, 314, 318, 319, 325
work psychology 4, 121, 166
work regulation theory 319
work-related musculoskeletal disorders (WRMD) 69
work roles 40
work scheduling 22
Working Conditions Act (WCA) (the Netherlands) 319–20, 323, 324
Working Environment Act (Norway) 104, 106
workplace coaching 69–70
workplace illness 8, 32–3, 46, 67, 137, 152, 155, 239, 315
World Health Organization (WHO) 39, 46, 313